F/ ESC60180R

RS.08 (Issue 2)

Corus UK Limited
Swinden Technology Centre

Library and Information Services
Moorgate, Rotherham   S60 3AR

NOTES TO BORROWERS

1. This publication should be returned or the loan renewed within 21 days, i.e. by the last date shown below.

2. As this Library is primarily a reference tool for the Technology Centre, borrowed publications should be readily accessible on demand.

3. It is a condition of this loan that, in case of damage or loss, this publication shall be replaced at the cost of the borrower

Name	Address	Date Due	Initial Before Returning
Y. Spasov		16/09/06	

Process
Fluid
Mechanics

PRENTICE-HALL INTERNATIONAL SERIES
IN THE PHYSICAL AND CHEMICAL ENGINEERING SCIENCES

NEAL R. AMUNDSON, EDITOR, *University of Minnesota*

ADVISORY EDITORS

ANDREAS ACRIVOS, *Stanford University*
JOHN DAHLER, *University of Minnesota*
THOMAS J. HANRATTY, *University of Illinois*
JOHN M. PRAUSNITZ, *University of California*
L.E. SCRIVEN, *University of Minnesota*

AMUNDSON *Mathematical Methods in Chemical Engineering:
Matrices and Their Application*
AMUNDSON AND ARIS *Mathematical Methods in Chemical Engineering:
Vol. II, First Order Partial Differential Equations and Applications*
ARIS *Elementary Chemical Reactor Analysis*
ARIS *Introduction to the Analysis of Chemical Reactors*
ARIS *Vectors, Tensors, and the Basic Equations of Fluid Mechanics*
BALZHISER, SAMUELS, AND ELIASSEN *Chemical Engineering Thermodynamics*
BRIAN *Staged Cascades in Chemical Processing*
CROWE ET AL. *Chemical Plant Simulation*
DENN *Process Fluid Mechanics*
DENN *Stability of Reaction and Transport Processes*
DOUGLASS *Process Dynamics and Control: Vol. I, Analysis of Dynamics Systems*
DOUGLASS *Process Dynamics and Control: Vol. II, Control System Synthesis*
FOGLER *The Elements of Chemical Kinetics and Reactor Calculations:
A Self-Paced Approach*
FREDRICKSON *Principles and Applications of Rheology*
FRIEDLY *Dynamic Behavior of Processes*
HAPPEL AND BRENNER *Low Reynolds Number Hydrodynamics
with Special Applications to Particulate Media*
HIMMELBLAU *Basic Principles and Calculations in Chemical Engineering, 3rd edition*
HOLLAND *Fundamentals and Modeling of Separation Processes:
Absorption, Distillation, Evaporation, and Extraction*
HOLLAND *Multicomponent Distillation*

Process Fluid Mechanics

MORTON M. DENN

Allan P. Colburn Professor
Department of Chemical Engineering
University of Delaware

Keywords
1. Fluid flow
2. Fluid mechanics
3. Viscosity

30·K-86

PRENTICE-HALL, INC.

Englewood Cliffs, New Jersey 07632

Library of Congress Cataloging in Publication Data

DENN, MORTON M   1939—
  Process fluid mechanics.

  (Prentice-Hall international series in the physical and
chemical engineering sciences)
  Includes bibliographies and indexes.
  1. Fluid mechanics. I. Title.
TA357.D47 1980   620.1'06   79-16319
ISBN   0-13-723163-6

Printed in the United States of America
10 9   8   7   6   5   4

Editorial production and interior design by Theodore Pastrick
Interior artwork by Judy Katzer
Manufacturing buyer: Gordon Osbourne

PRENTICE-HALL INTERNATIONAL, INC., *London*
PRENTICE-HALL OF AUSTRALIA PTY. LIMITED, *Sydney*
PRENTICE-HALL OF CANADA, LTD., *Toronto*
PRENTICE-HALL OF INDIA PRIVATE LIMITED, *New Delhi*
PRENTICE-HALL OF JAPAN, INC., *Tokyo*
PRENTICE-HALL OF SOUTHEAST ASIA PTE. LTD., *Singapore*
WHITEHALL BOOKS LIMITED, *Wellington, New Zealand*

Contents

4. Flow of Particulates

Part III   Macroscopic Problems

5. Macroscopic Balances

6. Applications of Macroscopic Balances

Part V   Approximate Methods

11. Ordering and Approximation   *233*

12. Creeping Flow   *243*

13. The Lubrication Approximation   *265*

Preface

Process Fluid Mechanics is a text for a first course in fluid mechanics for engineering students who are oriented towards process applications. The book was developed for use in an undergraduate chemical engineering course taught in the first semester of the third year at the University of Delaware, and drafts of the first sixteen chapters have been used in that course since 1972. Portions of the text have also been used in the first half of a one-semester course offered to first year chemical engineering graduate students at Delaware in order to provide a common framework for the more advanced material subsequently covered.

The goals of the book are described in Chapter 1, but a few words are appropriate here. Process engineers have two somewhat distinct needs for fluid mechanics. They must be prepared to deal with macroscopic problems, such as pressure drop and power calculations, flow distribution, and the use of correlations. They must also have a reasonable appreciation of detailed flow structure, including boundary layers and turbulence; this is needed in part as a basis for a firm understanding of heat and mass transfer. These two requirements place severe demands on the student in a one-semester course, but the modern engineering curriculum allows no more time for the subject. *Process Fluid Mechanics* is an attempt to provide an orderly treatment of the essentials of both the macro- and micro-problems of fluid mechanics. (It was a surprise to me when I first began to teach the subject to find that it is the macroscopic problems that cause the most difficulty. In retrospect this is reasonable; macroscopic equations are underdetermined, requiring the student to use intuition in making approximations while he or she is first developing the physical understanding that leads to intuition. I have tried, in Chapter 6, to formalize the approach somewhat. In contrast, the solution of microscopic balance problems is more logically structured, though placing a greater demand on the student's quantitative ability.)

Our third year students at Delaware have completed differential equations, a one semester course based on the first half of Russell and Denn: *Introduction to Chemical Engineering Analysis*, and the first semester of a two-semester sequence based on Sandler: *Chemical and Engineering Thermodynamics*. They have had an introduction to dimensional analysis and to mass and energy balances for flowing systems in these courses, but their only introduction to mechanics has been in their general physics courses. They have had only a limited introduction to vector analysis, and vectors are used primarily as a nomenclatural tool in the fluid mechanics course. Class time in our undergraduate fluid mechanics course is spent approximately as follows:

CHAPTER	TOPIC	APPROXIMATE CLASS HRS
2	Physical properties	1–2
3	Pipe flow	3–4
4	Flow of particulates	3
5	Macroscopic balances	3
6	Applications of macroscopic balances	4–5
7	Microscopic balances	4–5
8	One dimensional flows	4
9	Accelerating flow	3
10	Converging flow (primarily Secs. 10.2 and 10.7)	$1\frac{1}{2}$
11	Ordering and Approximation	1
12	Creeping flow	2
15	Boundary layer approximation (including portions of Chap. 14)	4
16	Turbulence	$4\frac{1}{2}$

Reading and problem assignments are used to supplement the classroom time and to provide a more balanced coverage of those topics that require less student-teacher interchange. A different selection of topics and relative emphasis would, of course, be appropriate for students with different backgrounds or for courses of different duration.

Our new graduate students come from a variety of backgrounds. Some have emphasized macroscopic problems in their undergraduate courses and some have emphasized microscopic problems; few have had a balanced treatment. We cover Chapters 5 through 17 in an accelerated seven week introduction, with heavy reliance on out-of-class reading. The remainder of our first graduate course emphasizes continuum mechanics, non-Newtonian fluids, and low Reynolds number flows relevant to the processing of polymers.

I have received a great deal of help over the years in formulating and revising my views regarding the teaching of fluid mechanics. I am particularly indebted to my colleague and mentor Arthur Metzner, who has been one of the leading practitioners of process fluid mechanics for two decades; this book is dedicated to him as tangible recognition of his contribution to my development. My full-time and occasional Delaware colleagues Gianni Astarita, David Boger, Pino Marrucci, Chris Petrie, Fraser Russell, and Roger Tanner have taught me much about fluid mechanics as we have worked together, and the continuing influence of Robert Pigford on chemical engineering education at Delaware was important in the overall orientation of the text and particularly in the development of Chapter 6 on applications of macroscopic balances. George Alves of the DuPont Engineering Department made helpful suggestions regarding applications of macroscopic balances. I have twice received support from the University of Delaware administration in the form of Improvement of Instruction grants. I am especially grateful to the students in my fluid mechanics classes, who have helped me to understand as they struggled with difficult concepts.

Morton M. Denn
Hockessin, Delaware

Part I
Introduction

Process Fluid Mechanics 1

1.1 INTRODUCTION

Fluid mechanics is a branch of applied physics that is concerned with the motion of fluids, liquid or gas, and the forces associated with that motion. Fluid mechanics plays a role in most engineering disciplines, and certain underlying principles are common to all areas of application. The emphasis differs within each area, however, and certain topics in fluid mechanics take precedence over others in each discipline.

This book is an introduction to *process fluid mechanics*, in which the emphasis is on those areas of fluid mechanics which are required for the solution of problems associated with the process industries and which form a fundamental part of the training of chemical, petroleum, and mechanical engineers. Some typical process fluid mechanics problems are described in subsequent sections of this chapter. In a classroom learning situation the contents of the book would comprise a one-semester, three-credit course in process fluid mechanics, with additions or deletions as appropriate to take into account the prior study in the area and in related subjects.

1.2 MACRO-PROBLEMS

Problems of processing interest can be roughly categorized as macro- and micro-problems. (The categorization should not be taken too seriously, but it is helpful for our preliminary discussion.) This division reflects two rather

3

different needs in the application of results from a fluid mechanics analysis. *Macro-problems* are those in which the end result is some overall characteristic of the motion itself, often the energy requirement, with no interest in the detailed structure of the flow field. The following examples illustrate typical macro-problems.

1.2.1 Pipeline Flow

A fluid with known physical properties is to be pumped from point A to point B. Flow rate and pipe sizes are specified. Determine the power required to pump the fluid, and hence the pumping cost. An interesting corollary problem, which we shall solve in Chapter 3, is to determine the optimal pipe size in order to minimize the sum of operating and investment costs. (Larger pipe costs more per unit length but requires less pumping energy.)

1.2.2 Packed Reactor

Many chemical reactions are carried out in packed reactors, as shown schematically in Fig. 1-1. Gas or liquid flows through a solid packing; the solid may catalyze the reaction, it may take part directly in the reaction (as in a coal gasification reactor), or it may simply be an inert packing which is present for other reasons, such as improving the heat removal. The pressure drop from one end of the reactor to the other is required as an important design parameter. The pressure drop not only affects the power requirement, but it also enters directly into reaction kinetic and equilibrium parameters and hence into reactor product conversions and distributions.

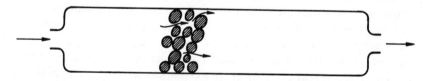

Figure 1-1. Schematic of a packed bed reactor.

1.2.3 Fiber Drawing

A simplified schematic of the process for production of polyester fibers is shown in Fig. 1-2. Molten polymer flows from the last of a series of reactors into an extruder. The extruder is a screw pump, identical in concept to the familiar home meat grinder. As liquid polymer is forced axially along the

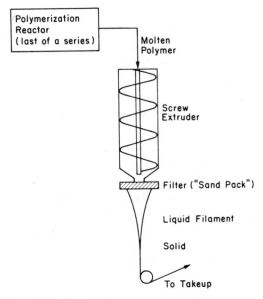

Figure 1-2. Schematic of process for making polyester fiber.

screw it builds up in pressure; this pressure is sufficient to force the fluid through a filter and then to force a liquid filament out through a small hole called a *spinneret*. The filament solidifies some distance from the spinneret. The solid filament is taken up at a faster speed than the liquid filament is extruded, so the filament is drawn down in area. Nearly all of the drawing takes place in the liquid filament prior to solidification.

There are several processing questions of interest here. First, each reactor feeds many extruders, each extruder feeds many spinlines, and each spinline consists of not one but as many as 128 filaments. (Consider the perforated outlet plate of a meat grinder!) Thus, the first set of problems consists of designing flow distribution systems which will ensure equal flow to all branches.

Next, there is a close coupling between the speed with which the extruder screw turns, the torque (and hence the power) required to turn the screw, the liquid properties, the pressure buildup along the screw, and the flow rate through the extruder. This relation is required in order to define two operating variables, the screw speed and the flow rate in a heat exchange system for temperature control.

Finally, the properties of the drawn filament are determined by all the processing variables, particularly the drawing speed and tension. The relation among throughput, liquid properties, drawing speed, and tension is required for design and control purposes.

1.3 MICRO-PROBLEMS

Micro-problems are those in which the detailed structure of the flow field is of importance. This information is often required for the subsequent analysis of other physical processes; in this regard, fluid mechanics provides an essential foundation for the study of heat transfer and mass transfer. The formulation and solution of micro-problems is often different from, and often more complex than the solution of macro-problems. Some typical micro-problems are discussed in the following paragraphs.

1.3.1 Heat Transfer Coefficient

A liquid flowing in a pipe is to be cooled by removing heat through the wall. The detailed flow pattern near the wall is required in order to calculate the heat transfer coefficient (the ratio of heat removal per unit surface area to temperature difference from the inside to the outside of the pipe). The reason that the flow pattern is required can be seen by reference to Fig. 1-3a and b.

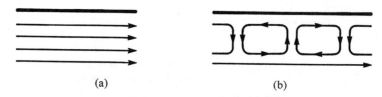

(a)   (b)

Figure 1-3. Possible flow patterns near a wall.

If the flow pattern shown schematically in Fig. 1-3a prevails, fluid which starts at a given distance from the wall never gets any closer to the wall; whereas for the case shown in Fig. 1-3b, fluid moves from the pipe core to the wall and back again. The latter is an effective means of heat transfer, because hot fluid is convected to the colder wall, cooled, and returned to contact the hotter fluid again. In contrast, when the flow lines are straight and no circulation occurs, heat can be removed from the center of the pipe only by conduction to the wall. Conduction is a molecular process and is a much slower way of removing heat than the convective recirculation.

1.3.2 Particle-Fluid Mass Transfer

Particle-fluid mass transfer is similar to the wall heat transfer coefficient problem and is relevant to packed reactors like that in Fig. 1-1 when the solid is a catalyst. For illustrative purposes it is more convenient to consider an isolated spherical particle, as shown in Fig. 1-4. The flow patterns shown in Fig. 1-4a and b can occur, as can other patterns, depending on overall flow conditions which will be discussed in later chapters. Clearly, the rate at which reactants move from the bulk fluid to the catalytic surface depends on the nature of the flow patterns; if the chemical reaction is rapid, the rate at which the catalytic process occurs is controlled by the rate at which reactants move to the surface, and hence by the fluid mechanics around the particle.

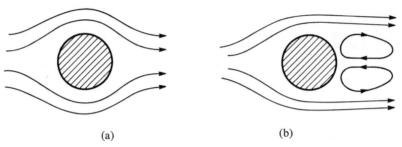

(a)   (b)

Figure 1-4. Possible flow patterns around a solid sphere.

1.4 CONCLUDING REMARKS

The purpose of these introductory remarks has been to define some of the problems that require a knowledge of fluid mechanics for solution and to indicate the broad scope of the subject. The distinction between macro- and micro-problems is a convenient one which we shall retain for a significant portion of the text, although we shall ultimately find that macro-methods are sometimes required for the solution of micro-problems, as well as the converse.

A word about the structure of the remainder of the text is in order here. The introductory portion (Part I) continues with the next chapter, "Physical Properties." Part II, "Dimensional Analysis and Experimentation," and Part III, "Macroscopic Balances," treat the solution of macro-problems and the essential interplay between analysis and experiment. Part IV, "Microscopic Balances," and Part V, "Approximate Methods," treat the solution of micro-problems and are more analytical than the preceding sections. Many of the phenomena introduced experimentally in Part II are obtained from basic principles in Parts IV and V. Part VI deals with more advanced topics and builds on all previous sections.

BIBLIOGRAPHICAL NOTES

Readers of this book will often find it useful to refer to *The Chemical Engineers' Handbook*, published by McGraw-Hill and often referred to as *Perry's Handbook*. The most recent edition available at the time of this writing is the fifth. Other books covering a substantial portion of the material in this text and at approximately the same level, but with different points of view, are:

BENNETT, C. O., and MEYERS, J. E., *Momentum, Heat and Mass Transfer*, 2nd ed., McGraw-Hill Book Company, New York, 1974.

BIRD, R. B., STEWART, W. E., and LIGHTFOOT, E. N., *Transport Phenomena*, John Wiley & Sons, Inc., New York, 1960.

DeNEVERS, N., *Fluid Mechanics*, Addison-Wesley Publishing Company, Inc., Reading, Mass., 1970.

KAY, J. M., and NEDDERMAN, R. M., *Fluid Mechanics and Heat Transfer*, 3rd ed., Cambridge University Press, New York, 1974.

WHITAKER, S., *Introduction to Fluid Mechanics*, Prentice-Hall, Inc., Englewood Cliffs, N.J., 1968.

Physical Properties   **2**

2.1 INTRODUCTION

A treatment of fluid flow requires an understanding of the physical properties of a fluid which affect the motion, and an appreciation of the units in which these properties are measured. The two most important fluid properties are the *density* and the *viscosity*. Density is a familiar concept and is treated only briefly, although there is one subtlety that requires attention. Viscosity is a concept that is generally understood intuitively, but some detailed consideration is necessary in order to establish a precise and useful definition.

2.2 UNITS

Every physical quantity has characteristic *dimensions:* length (L), mass (M), time (Θ), and combinations thereof. Velocity, for example, has dimensions of length per time, $L\,\Theta^{-1}$. Measurement of a physical quantity is carried out by using a system of *units* to characterize the dimensions; the units are arbitrary and do not affect the results and application of a measurement as long as everyone agrees on the means of converting from one set of units to another. For example, time may be measured in seconds (s) or in hours (h), because the conversion factor is universally agreed to be 1 h = 3600 s. Time cannot be measured in months for scientific purposes because there is no precise conversion factor between months and other time units.

A number of systems of units have been in common use throughout the world during the period of development of modern science, and a familiarity with all systems is essential in order to read the scientific and engineering literature. Most scientific writing through the first half of the twentieth century used the metric cgs (centimeter-gram-second) system of units, while engineering literature from North America and Great Britain often used the imperial or English (pound-foot-second) system. Both systems have now been replaced by SI (système international, or international system) units, the accepted worldwide standard for scientific and engineering work. SI units are required in the publications of the American Institute of Chemical Engineers and the American Society of Mechanical Engineers, and they are used throughout this book.

SI is a metric system that uses the meter (m) to measure length, the kilogram (kg) to measure mass, and the second (s) to measure time. The prefix "milli" to denote one-one thousandth is retained, and small lengths may be measured in millimeters (mm). Derived units for other commonly measured quantities have special names, which are given in Table 2-1. Conversion factors from the common cgs and imperial units are given in Table 2-2. Table 2-3 contains some very approximate equivalents between SI and

TABLE 2-1
DERIVED SI UNITS

Measurement	Derived unit	Basic units
Force	newton (N)	1 kg·m/s^2
Pressure, stress	pascal (Pa)	$1 \text{ N/m}^2 = 1 \text{ kg/m·s}^2$
Work, energy	joule (J)	$1 \text{ N·m} = 1 \text{ kg·m}^2/\text{s}^2$
Power	watt (w)	$1 \text{ J/s} = 1 \text{ kg·m}^2/\text{s}^3$

TABLE 2-2
CONVERSION FACTORS FOR UNITS

SI	cgs	Imperial
meter	10^2 centimeters	3.281 ft
kilogram	10^3 grams	2.2046 lb_m
newton	10^5 dynes	0.2248 lb_f
pascal	10 dynes/cm^2	$1.450 \times 10^{-4} \text{ lb}_f/\text{in.}^2$
joule	0.2388 calorie	9.478×10^{-4} Btu
	10^7 ergs	0.737 ft-lb_f
watt	10^7 ergs/s	1.341×10^{-3} hp
		$0.737 \text{ ft-lb}_f/\text{s}$
kg/m^3	$10^{-3} \text{ grams/cm}^3$	$6.242 \times 10^{-2} \text{ lb}_m/\text{ft}^3$

imperial units which might be helpful to North American readers who are still unaccustomed to visualizing quantities in SI units.

TABLE 2-3
APPROXIMATE EQUIVALENTS BETWEEN SI AND
IMPERIAL UNITS

SI unit	Approximate imperial equivalent
meter	10% more than 3 ft
kilogram	10% more than 2 lb_m
newton	10% less than 0.25 lb_f
pascal	10^{-5} atm
joule	10^{-3} Btu
watt	one-third more than 10^{-3} hp

2.3 CONTINUUM HYPOTHESIS

The density provides a convenient means of illustrating the *continuum hypothesis*, which is one of the fundamental principles utilized in fluid mechanics. Density is defined as mass per unit volume, and it is commonly determined by measuring either the mass of a known volume or the volume displaced by a known mass. We recognize, however, that density can vary with position in a process, and a more precise definition is to consider a series of small volumes $\Delta\mathcal{U}$, with corresponding small masses $\Delta\mathfrak{M}$, and to define the density ρ by the limiting process

$$\rho = \lim_{\Delta\mathcal{U}\to 0} \frac{\Delta\mathfrak{M}}{\Delta\mathcal{U}} \tag{2.1}$$

Alternatively, if the density ρ is a function of position, $\rho(x, y, z)$, then the mass contained in a specified volume is the integral of the density over that volume:

$$\mathfrak{M} = \int \rho(x, y, z)\, d\mathcal{U} \equiv \iiint \rho(x, y, z)\, dx\, dy\, dz \tag{2.2}$$

Now, Eqs. (2.1) and (2.2) have implicitly used the *continuum hypothesis*, for they assume that the density is a concept that has meaning for any volume, no matter how small. Yet we know that when the volume is of molecular dimensions we have very rapid spatial variations between dense and non-dense regions, depending on how close we are to a molecule. Hence, the notion of a density function that can be defined by a limiting process such as that of Eq. (2.1) is questionable. *The continuum hypothesis assumes that mathematical limits for volumes tending to zero are reached over a scale that remains large compared to molecular dimensions.* Thus, variations on a molecular scale can be ignored, since they average out over all length scales of interest. Clearly, no result derived using the continuum hypothesis should be expected to apply to a rarified gas, where the distance between molecules may be large.

2.4 VISCOSITY

A useful definition of *a fluid is a material that cannot sustain a shearing stress in the absence of motion.* This definition includes liquids and gases.

The most important physical property of a fluid from the point of view of the study of fluid mechanics is the *viscosity.* The viscosity is the property that determines the ease with which a fluid will flow. Viscosity is a term and a concept that is commonly used in everyday speech; lubricating oils are rated by viscosity, for example. Honey is far more viscous than water; hence, honey pours more slowly from a pitcher than does water. This intuitive notion of viscosity is not adequate for our purposes, however, and we require a precise definition.

2.4.1 Operational Definition of Viscosity

An *operational definition* is one that defines a quantity in terms of an experiment that can be performed to measure the value of the quantity. Only operationally defined quantities are useful in engineering practice. We shall define the viscosity here in terms of one experiment; we shall leave it for subsequent discussion to demonstrate that the quantity so defined is truly a general property of the fluid, and not a very specific response to a particular experiment.

The experiment is shown schematically in Fig. 2-1. Two very large flat plates of area A are kept parallel and are separated by a fluid layer of uniform thickness H. We require that $H \ll \sqrt{A}$, in which case such an arrangement is

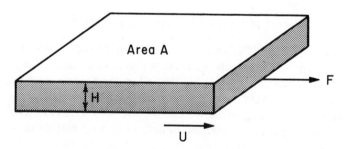

Figure 2-1. Schematic of shearing experiment.

sometimes loosely referred to as "infinite parallel plates." One plate is moved in a straight line relative to the other at a constant speed, U, and the force F required to move the plate at a constant speed is continuously measured. Following a rapid transient at the start of the experiment, F is found in all cases to approach a constant value in time. The following results are observed experimentally when the transient period is past:

1. For constant U and H, the force is directly proportional to the area, A.
2. For constant A, the force is a unique, monotonically increasing function of the *ratio U/H*.

It is convenient to define the *shear stress* and *shear rate* as follows:

$$\text{shear stress:} \quad \tau_s = \frac{F}{A} \quad \left(\frac{\text{force}}{\text{area}}\right) \tag{2.3}$$

$$\text{shear rate:} \quad \Gamma_s = \frac{U}{H} \quad \left(\frac{1}{\text{time}}\right) \tag{2.4}$$

(The symbols γ_s, $\dot{\gamma}$, and $\dot{\varepsilon}$ are also used for the shear rate, and the subscript s is often omitted.) From the two experimental observations we can then write

$$\tau_s = \tau_s(\Gamma_s) \tag{2.5}$$

$$\frac{d\tau_s}{d\Gamma_s} > 0 \tag{2.6}$$

Equation (2.5) states that the shear stress for a given fluid is a unique function of only the shear rate, and Eq. (2.6) states that the function is monotonically increasing.

Figures 2-2, 2-3, and 2-4 show plots of τ_s versus Γ_s measured in our laboratory for three liquids, two aqueous and one organic. Note that the data for three different values of H fall on the same line when Γ_s is used as the ordinate. Note also the temperature sensitivity of the organic oil in Fig. 2-3, where a small effect of a 0.3°C temperature difference is evident. Some systematic deviation is also evident in Fig. 2-4, where temperature control was less careful. The experimental scatter in Fig. 2-4 is typical of results to be expected for a shear stress experiment using most commercial instruments.

The shear stress-shear rate curves in Figs. 2-2 and 2-3 are straight lines passing through the origin. Such a response is often observed and motivates the definition of *viscosity* as the *ratio of shear stress to shear rate*:

$$\text{viscosity:} \quad \eta = \frac{\tau_s}{\Gamma_s} \tag{2.7}$$

(The symbol μ is also used for the viscosity. The fluidity, which is the reciprocal of viscosity, is used on occasion.) Viscosity is measured in pascal-

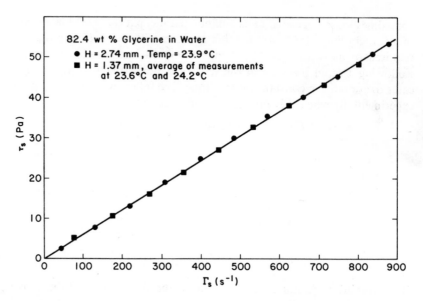

Figure 2-2. Shear stress a function of shear rate, 82.4 wt% glycerine in water.

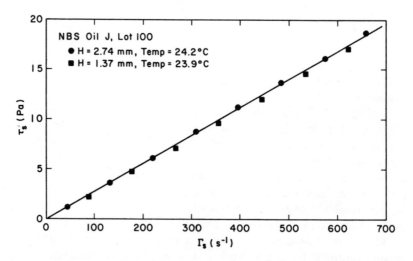

Figure 2-3. Shear stress as a function of shear rate, NBS Oil J, Lot 100.

seconds (Pa·s). Until the adoption of SI units the most common unit for measuring viscosity was the poise (p), after the French scientist Poiseuille, or the centipoise (cp, one-one hundreth of a poise). 1 Pa·s = 10 p = 10^3 cp. Much of the reference literature uses these units. The viscosity of liquid water at room temperature is approximately 1 cp or 10^{-3} Pa·s.

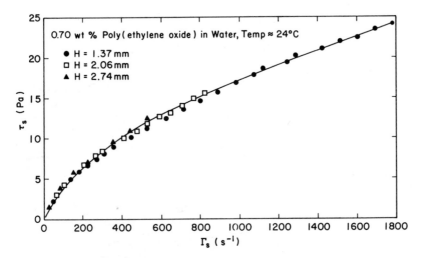

Figure 2-4. Shear stress as a function of shear rate, 0.70 wt % poly(ethylene oxide) in water at approximately 24°C.

2.4.2 Newtonian Fluids

Fluids for which the shear stress-shear rate relation is a straight line passing through the origin are called *Newtonian*, for such a relationship was hypothesized by Newton. The viscosity of a Newtonian fluid is independent of shear rate and depends only on temperature and, to a much lesser extent, on pressure. A viscosity independent of shear rate (usually called a "constant viscosity," with the understanding that the constant depends on temperature and pressure) is a property of only some liquids, as Fig. 2-4 clearly indicates. Typically, low-molecular-weight liquids and all gases have constant viscosities and are Newtonian. High-molecular-weight liquids, including solutions of high-molecular-weight polymers in low-molecular-weight solvents, are often non-Newtonian, in that the viscosity changes with changing shear rate. We will emphasize Newtonian fluids in this first course in fluid mechanics for the pragmatic reason that basic principles are more easily developed for these fluids and, except in the processing of polymers and slurries, we may reasonably expect to handle Newtonian fluids. The study of non-Newtonian fluid mechanics uses Newtonian fluid mechanics as a starting point in all cases.

Figures 2-5 and 2-6 show the viscosity as a function of temperature at atmospheric pressure for some common Newtonian liquids and gases, respectively. The viscosity scale is logarithmic for the liquids and linear for the gases. Note that the viscosity is a decreasing function of temperature for the liquids, and in some cases the temperature dependence is very strong.

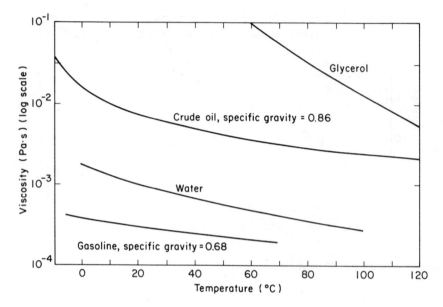

Figure 2-5. Viscosities of common liquids at atmospheric pressure as a function of temperature.

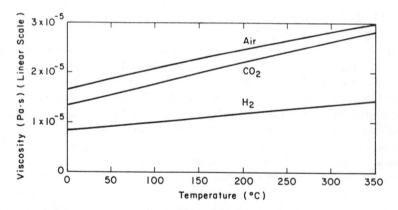

Figure 2-6. Viscosities of common gases at atmospheric pressure as a function of temperature.

The viscosity is an increasing function of temperature for the gases, but the temperature dependence is very weak. The qualitative behavior for gases can be predicted from elementary molecular theories.

The effect of pressure on the viscosity of liquids can generally be ignored for pressures less than 4×10^6 Pa (about 40 atm). The viscosity of gases increases with pressure and can be correlated, like many other physical properties, in terms of reduced variables. Viscosity–pressure correlations may be found in *Perry's Handbook;* roughly, at temperatures more than four times the boiling point in °K, the pressure dependence of gas viscosity is weak for pressures less than 10 times the critical pressure.

2.4.3 Non-Newtonian Fluids

High-molecular-weight liquids, slurries, and suspensions are often non-Newtonian, in that the viscosity is a function of shear rate. The viscosity of the aqueous poly(ethylene oxide) solution from Fig. 2-4 is shown in Fig. 2-7; the individual data points have not been reproduced, and the viscosity

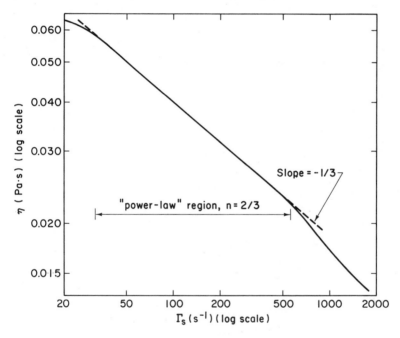

Figure 2-7. Viscosity as a function of shear rate for 0.70 wt % poly(ethylene oxide) in water at approximately 24°C, computed from the line drawn through the data in Fig. 2-4.

function is calculated from the line in Fig. 2-4. Note the log-log scale, which is the way that viscosity data are usually presented. Figure 2-8 shows data on a polymeric system that were obtained over an extremely wide range of shear rates; the approaches to constant values of the viscosity at low and high shear-rate ranges are sometimes referred to as upper and lower Newtonian regions, respectively. The shape of the curve in Fig. 2.8 is typical of data for polymer solutions and melts, but experimental difficulties at low and high shear rates often prevent measurement of the upper and lower limiting values.

When the viscosity is a function of shear rate, we can rewrite Eq. (2.7)

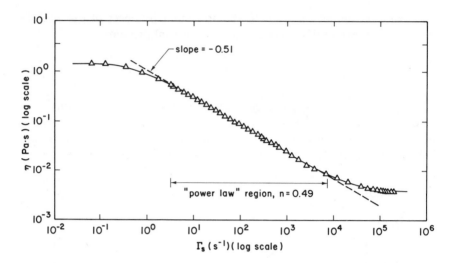

Figure 2-8. Viscosity as a function of shear rate, 0.4 wt % polyacrylamide in water at room temperature. Data of D. V. Boger.

to take the functional dependence explicitly into account:

$$\tau_s = \eta(\Gamma_s)\Gamma_s \tag{2.8}$$

(The function $\eta(\Gamma_s)$ is sometimes called the *apparent viscosity*, reserving the name "viscosity" for the constant that characterizes Newtonian fluids, but there is no clear advantage to this distinction.) Substitution of Eq. (2.8) into Eq. (2.6) then leads to a useful inequality,

$$\frac{d\tau_s}{d\Gamma_s} = \frac{d}{d\Gamma_s}\eta(\Gamma_s)\Gamma_s = \eta(\Gamma_s) + \Gamma_s\frac{d\eta(\Gamma_s)}{d\Gamma_s} > 0 \tag{2.9a}$$

or, equivalently,

$$\frac{d\ln\eta(\Gamma_s)}{d\ln\Gamma_s} > -1 \tag{2.9b}$$

That is, on a log-log scale, the viscosity function cannot drop off faster than a slope of -1.

It is useful to note at this point that shear stress and shear rate are both directional quantities; in the plane of the paper, F and U are positive when they point in a positive coordinate direction and they are negative when they point in a negative coordinate direction. Thus, if the velocity of the plate changes in direction from right to left but retains the same magnitude, then the sign of the force, and hence of the shear stress, must also change with no change in magnitude. According to Eq. (2.8), it then follows that the viscosity function $\eta(\Gamma_s)$ must be a positive function which depends only on the magnitude of Γ_s. This same conclusion can be established more generally and rigorously from the second law of thermodynamics.

Viscosity data for polymeric systems frequently fall on a straight line on a log-log scale for a decade or more in shear rate. This indicates a power relation between viscosity and shear rate that is often represented in the empirical *power law* equation:

$$\text{power law (empirical)}: \eta(\Gamma_s) = K\,|\Gamma_s|^{n-1} \qquad (2.10)$$

K is called the consistency factor and n the power law index. It readily follows from Eq. (2.9) that $n > 0$. The data in Fig. 2.8 show a power law region with $n \approx 1/2$. Molten polystyrene is typically processed in a region in which $n \approx 1/3$, which is about the lowest value of n usually observed. The Newtonian fluid is represented by the limit $n \to 1$. The power law is the most common empirical function used to represent polymer viscosity functions for two reasons: first, most processing does in fact take place in a shear rate range where the power law fits the data quite well. Second, power functions have convenient analytical properties for differentiation and integration. A number of other empirical and semi-empirical functions are sometimes used in the polymer processing literature.

In contrast to the "shear-thinning" (i.e., decreasing) viscosity-shear rate behavior of polymeric systems, slurries and suspensions often tend to be "shear thickening." Figure 2.9 shows typical suspension viscosity data; the apparent growth to an infinite viscosity at a finite shear rate probably indicates the development of internal structure in the suspension and a transition to solid-like properties. Viscosity data on slurries and suspensions are mean-

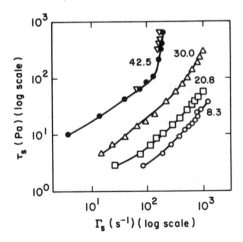

Figure 2-9. Shear stress as a function of shear rate for suspensions of TiO_2 in a 47.1 wt% sucrose solution. The viscosity of the sucrose solution is 0.017 Pa·s. The number next to each curve is the *volume* percent of TiO_2 in the suspension. Data of Metzner and Whitlock, *Trans. Soc. Rheology*, **2**, 239 (1958). Copyright by John Wiley and Sons, Inc; reproduced by permission.

ingful only when the average particle size is very small compared to the plate spacing, H, so that the system can be considered to be a homogenous single phase. In addition, particle migration away from the wall region can invalidate some experiments. Shear thickening materials are often represented over finite shear rate ranges by a power law, Eq. (2.10), with $n > 1$.

Two other types of materials need to be noted briefly, although they are sufficiently limited in scope that we shall not deal with them further in this text. Some liquids, primarily suspensions, are time-dependent, in that the viscosity changes in time with steady shearing, generally because of structural changes. Other liquids, including mayonnaise, blood, and some paints, show a *yield stress;* that is, a shear stress-shear rate curve has a positive intercept at zero shear rate. Typical data are shown in Fig. 2-10. A material with a yield stress acts like a solid and does not flow until a critical stress is exceeded; beyond the critical stress typical liquid behavior prevails. When the shear stress-shear rate function is linear beyond the yield point, such fluids are called *Bingham plastics.*

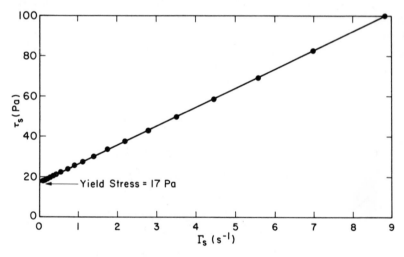

Figure 2-10. Shear stress as a function of shear rate for a meat extract at 77°C. Data of D. V. Boger.

2.4.4 Kinematic Viscosity

The viscosity and the density often enter engineering problems only in the ratio η/ρ. This quantity, which has units of m^2/s, is known as the *kinematic viscosity* and is often given the symbol ν. To avoid confusion in terminology, some authors refer to η as the shear viscosity.

In developing a feeling for orders of magnitude of physical quantities, it is useful to refer again to Figs. 2-5 and 2-6. The viscosity of water is two orders of magnitude greater than that of air at room temperature and atmospheric pressure. The density of water, however, is three orders of magnitude greater than that of air at these conditions. Thus, the kinematic viscosity of air is an order of magnitude *greater* than that of water; the kinematic viscosity of the latter is approximately 10^{-6} m²/s.

The kinematic viscosity is sometimes recorded in the literature in the cgs unit *stoke*, named for the English scientist Stokes. One stoke equals 1 cm²/s. The kinematic viscosity of water is approximately 1 centistoke.

2.4.5 Measurement

Finally, we must deal with the practical implementation of the experiment described in Fig. 2-1. It would be exceedingly difficult to keep two large plates at a constant uniform gap width, and after a short time the plates would have moved relative to one another a sufficient distance to introduce unacceptable error into the determination of the surface area A. Both of these problems are alleviated, however, by noting that infinite parallel plates can be approximated by concentric cylinders in which the gap, H, is very small compared to the inner radius, R, as shown in Fig. 2-11. Suppose that

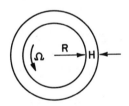

Figure 2-11. Concentric cylinders with a small gap/radius ratio as an approximation to infinite parallel plates.

the inner cylinder is turned with an angular velocity Ω and the outer cylinder is fixed. Then the linear velocity, U, is $R\Omega$, and $\Gamma_s = R\Omega/H$. The area, A, is equal to $2\pi RL$. (The area of the outer cylinder is larger by a factor $1 + H/R$, which is nearly equal to unity in the approximation $H/R \rightarrow 0$.)

Force is a vector quantity, and the direction of the force changes continuously about the circumference in this cylindrical system; indeed the net force is zero, since equal and opposite forces cancel for points 180° apart. In a cylindrical system it is the torque that must be considered, and this requires a very slight alteration in our approach.

Consider the differential area $dA = RL\,d\theta$ in Fig. 2-12. The differential

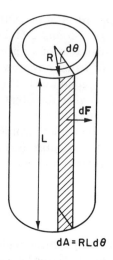

dA = RLdθ

Figure 2-12. Schematic for calculation of torque on rotating concentric cylinders.

force dF acting on this area is the product of the shear stress and the area:

$$dF = \tau_s \, dA = \eta \Gamma_s \, dA = \eta \left(\frac{R\Omega}{H}\right) RL \, d\theta \qquad (2.11)$$

The direction of the force is tangential. The differential torque dG is the product of the force and the lever arm, R:

$$dG = R \, dF = \eta \left(\frac{R^3 L\Omega}{H}\right) d\theta \qquad (2.12)$$

Torque is a vector whose direction is determined by the cross product of the radius and force vectors. The *right-hand rule* shows that the torque vector points in the axial direction; hence, the differential torques at all circumferential positions point in the same direction and can be added together.

The total torque, which is an easily measured experimental quantity, is obtained by summing (integrating) all the small contributions dG to obtain

$$G = \int dG = \int_0^{2\pi} \eta \left(\frac{R^3 L\Omega}{H}\right) d\theta = \frac{2\pi\eta R^3 L\Omega}{H} \qquad (2.13)$$

or

$$\eta = \frac{GH}{2\pi R^3 L\Omega} \qquad (2.14)$$

The length L can be increased until end effects are negligible, or other methods, not appropriately described here, can be employed to remove the effect of a finite length. The data in Figs. 2-2, 2-3, and 2-4 were taken in a device of this type, sometimes called a *Couette viscometer*, with $R = 56.31$, 57.00, and 57.68 mm and $L = 41.0$ mm.

2.5 VISCOELASTICITY

The initial transient in the shearing experiment described in Secs. 2.4.1 and 2.4.5 cannot be observed for a Newtonian fluid, but it can be estimated from molecular arguments to be of the order of 10^{-12} s. Any transient seen experimentally is an artifact which results from the inertia of the instrument, since moving parts cannot be set in motion instantaneously. Thus, for Newtonian fluids the transient can be completely ignored.

The transient cannot be ignored in polymeric systems, where the initial transient can be quite long because of the need of the long polymer chains to rearrange. Figure 2-13 shows data for molten low-density polyethylene; here,

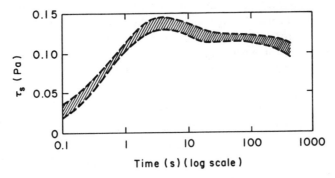

Figure 2-13. Shear stress during initial transient, $\Gamma_s = 1$ s^{-1}. Data for eight samples of a low-density polyethylene (LDPE A) fall within the shaded region. Reported by Meissner, *Pure Appl. Chem.*, **42**, 553 (1975). Copyright by Int. Union of Pure and Applied Chemistry; reproduced by permission.

the time of the transient is of the order of seconds and is comparable to the time that might be required in a processing operation. The transient in a dilute polymer solution will be shorter, perhaps 10^{-3} or 10^{-2} s, and even these time scales can be significant in some flow fields.

In polymeric liquids the time scale of the transient can be shown to be a property of the material and not an artifact of a single experiment. The finite transient time is a consequence of a solidlike response of the material. (Think of silicone play putty, which is sold as a novelty toy. When it is stretched slowly, it appears to be a viscous liquid, but it fractures like a brittle solid when stretched rapidly.) Materials that show both fluid- and solidlike behavior, depending on the time scale of the experiment, are called *viscoelastic*. Viscoelastic liquids are characterized by a viscosity, η, and a *relaxation time*, λ.

The relaxation time can be measured from the initial transient in shear, as described here, but it is more convenient to use other measurements,

23

described in Chapter 19, where viscoelasticity is treated quantitatively. Viscoelastic measurement is a part of the field of *rheology*.

Viscoelasticity should not be confused with the time dependence of the viscosity for some fluids that was described briefly in Sec. 2.4.3. In the latter case, there is no steady state, while for viscoelastic liquids there is a well-defined transient which does go to a final steady state from which a time-invariant viscosity can be determined.

2.6 INTERFACIAL TENSION

The interfacial tension, or *surface tension*, is a property of a liquid–gas or liquid–liquid interface. It is shown in texts on physical chemistry that work is required to create interfacial area, and that the differential change in free energy is equal to $\sigma \, dA$, where the interfacial tension σ has dimensions of energy per area or force per length. It is readily established that the interfacial tension causes a pressure difference across a static curved surface equal to

$$\Delta p = \sigma \left(\frac{1}{R_1} + \frac{1}{R_2} \right) \tag{2.15}$$

Here, R_1 and R_2 are the principal radii of curvature of the surface. In the case of a spherical bubble or droplet, the two radii of curvature equal the sphere radius, R, and the pressure inside the bubble or droplet exceeds the pressure outside by an amount

$$\text{spherical bubble or droplet:} \quad \Delta p = \frac{2\sigma}{R} \tag{2.16}$$

Interfacial tension effects can sometimes be quite important in two-phase flows, such as those discussed in Chapter 18, and in certain free surface flows discussed elsewhere in the text. We shall point out those situations where interfacial tension may play a role, but we shall only deal in this text with physical situations in which interfacial effects may be neglected without serious error.

2.7 CONCLUDING REMARKS

We have introduced a number of concepts in this chapter in order to define the scope of fluid characterization through physical property measurement. The viscosity is the property that is of the greatest concern to us, particularly the constant viscosity that characterizes Newtonian fluids. This is because of the emphasis on Newtonian fluids in this introductory treatment. It is important to understand the meaning of the viscosity and to develop a feeling for relevant magnitudes. In the next two chapters we shall see how this physical property enters into the analysis of fluid flow problems of engineering interest.

BIBLIOGRAPHICAL NOTES

Units and dimensions are discussed in most introductory engineering texts. See, for example,

RUSSELL, T. W. F., and DENN, M. M., *Introduction to Chemical Engineering Analysis*, John Wiley & Sons, Inc., New York, 1972.

For a more complete discussion of SI units, see

MULLIN, J. W., *AIChE J.*, **18**, 222 (1972).

OLDSHUE, J. Y., *Chem. Eng. Progr.*, **73**, No. 8, 135 (August 1977).

Viscosity data can be found in standard references such as *The Chemical Engineers' Handbook*, the *Handbook of Chemistry and Physics*, and the *International Critical Tables*. Molecular theories require advanced study of physical chemistry; the classic treatment is

CHAPMAN, S. and COWLING, T. G., *The Mathematical Theory of Non-uniform Gases*, Cambridge University Press, New York, 1961.

Good introductory discussions of non-Newtonian viscosity and of viscoelasticity may be found in

METZNER, A. B., "Flow of Non-Newtonian Fluids," in *Handbook of Fluid Mechanics*, V. L. Streeter, ed., McGraw-Hill Book Company, New York, 1961.

MIDDLEMAN, S., *The Flow of High Polymers*, John Wiley & Sons, Inc., New York, 1968.

There is a recent review of suspensions in

JEFFREY, D. J., and ACRIVOS, A., *AIChE J.*, **22**, 417 (1976).

PROBLEMS

2.1. The following shear stress-shear rate data were obtained on a 6 weight percent solution of a sodium cellulose sulfate polymer in water at 24.5°C:

Shear rate (s^{-1})	Shear stress (Pa)	Shear rate	Shear stress
0.23	5.5	5.8	52
0.29	7	7.2	57
0.36	8.5	9.1	63
0.46	11	11.5	69
0.58	13	14.5	74
0.72	15	18	80
0.91	17.5	23	85
1.15	20.5	29	92
1.45	24	36	100
1.8	27.5	46	108
2.3	31.5	58	113
2.9	36	72	120
3.6	41	91	130
4.6	46		

a) Plot the viscosity-shear rate function.

b) Is there a shear rate range in which the fluid appears to be Newtonian?

c) Is there a range in which the viscosity follows a power law relation? If so, determine the power law constants.

d) The polymer solution is to be used in a process at a shear rate of 5000 s^{-1}. Estimate the viscosity under these conditions from the available data. Do you think that the estimated value is reliable?

2.2. Viscosity data are sometimes represented by the empirical Ellis model:

$$\frac{\eta_0}{\eta} = 1 + \left(\frac{\tau_s}{\tau_{1/2}}\right)^{\alpha-1}$$

a) What are the physical meanings of η_0 and $\tau_{1/2}$?

b) What is the relation (if any) between α and the power-law index n in Eq. 2.10?

c) How well does the Ellis model fit the data in Problem 2.1? Show results graphically.

d) Consider the data in Fig. 2-8. Why can't the Ellis model be expected to fit the data over the entire shear rate range shown? Can you suggest an empirical modification that might do a better job?

2.3. A crude molecular model of a dilute gas indicates that viscosity should be proportional to the square root of absolute temperature. Are the data in Fig. 2-6 consistent with this relation? Are they better fit by some other power of absolute temperature?

2.4. A crude molecular model of a low viscosity liquid, combined with an empiricism called *Trouton's rule*, indicates that viscosity should be proportional to $\exp(3.8T_b/T)$, where T is the absolute temperature and T_b is the boiling point at atmospheric pressure. Are the data for water and glycerol in Fig. 2-5 consistent with this relation?

2.5. Look up the viscosity of steam as a function of temperature and pressure. At each temperature, estimate the pressure at which the pressure dependence of viscosity causes a change of at least five percent from the low pressure asymptote, and plot this pressure as a function of temperature.

Part II
Dimensional Analysis
and Experimentation

Pipe Flow   3

3.1 INTRODUCTION

Flows in pipes and other straight conduits occur often in process applications. A great deal can be learned about pipeline flow from the results of a small number of experiments. These experiments are motivated by the use of dimensional analysis, which is an important tool in any experimental program.

We begin this chapter with a brief introduction to dimensional analysis. This is followed by an illustration of the use of the technique in order to study the relation between the flow variables and physical properties in pipeline flow of a Newtonian fluid. The remainder of the chapter includes some engineering applications and extensions to related problems.

3.2 DIMENSIONAL ANALYSIS

Dimensional analysis is a tool that is of considerable use in experimental design. The basic principle can be stated as follows:

A physical process involves a relation between $\mathcal{U}$ variables,

$$x_1 = \text{function of } (x_2, x_3, \ldots, x_{\mathcal{U}}) \tag{3-1}$$

If the $\mathcal{U}$ variables require $\mathcal{D}$ dimensions (length, mass, time, etc.), then the variables can be combined into $\mathcal{G} = \mathcal{U} - \mathcal{D}$ independent dimensionless groups* of the form $N_1 = x_1^{a_1} x_2^{a_2} x_3^{a_3} \ldots x_\mho^{a_\mho}$, $N_2 = x_1^{b_1} x_2^{b_2} x_3^{b_3} \ldots x_\mho^{b_\mho}$, and so on. Equation (3-1) can then be written in the form

$$N_1 = \text{function of } (N_2, N_3, \ldots, N_g) \tag{3-2}$$

This statement is sometimes known as the *Buckingham pi theorem*.

The usefulness of dimensional analysis follows from the fact that it reduces the number of variables that must be studied from $\mho$ to g, and it indicates what *groupings* of physical variables affect the process. Thus, it may be possible to limit the scope of an experimental program and, as we shall see subsequently, to design experiments on one physical scale that are appropriate to process problems occurring on an entirely different physical scale.

The application of dimensional analysis will be demonstrated in the next section and again in Chapter 4. We shall not present a proof of the pi theorem here, but simply note that it is a rather elementary consequence of the principle that each term in an equation must have the same dimensions. The proof is a constructive one, in that it provides a procedure for determining the groups $N_1, N_2, \ldots, N_g$. The formal procedure is rather tedious, however, and the dimensionless groups are usually constructed on an *ad hoc* basis.

3.3 SMOOTH PIPE FLOW, NEWTONIAN FLUID

3.3.1 Dimensionless Groups

Consider a long smooth horizontal pipe of length L and inside diameter D. (We call this a smooth pipe to indicate that the inside diameter is perfectly uniform over the entire length.) An incompressible Newtonian fluid flows through the pipe with a volumetric flow rate Q. The flow is caused by a pressure difference Δp over the length of pipe. We wish to determine the relation between the several flow variables and physical property parameters.

The variables and their dimensions are listed in Table 3-1. Mean velocity v is used for convenience in place of flow rate; the two are related through the equation $v = 4Q/\pi D^2$. The absolute value signs about the pressure difference Δp are used to enable us to use the common terminology "pressure drop," which is a positive number; since flow occurs from a higher to a lower pressure, the pressure change Δp over the length L is negative in sign. The Newtonian fluid is characterized by two constant physical properties, density and viscosity; we are assuming that any temperature changes are

---

*There are rare exceptions to this rule. See the specialized texts on dimensioned analysis cited in the Bibliographical Notes for details.

sufficiently small to enable us to neglect the additional parameters describing temperature sensitivity of these properties.

Table 3-1 contains six variables ($\mho = 6$) and three dimensions ($\mathfrak{D} = 3$); thus, there are three independent dimensionless groups ($\mathcal{G} = \mho - \mathfrak{D} = 3$). These can be determined most easily by inspection, as follows:

TABLE 3-1

VARIABLES AND DIMENSIONS FOR PIPE FLOW

Variable	Dimensions	
D	length	L
L	length	L
$|\Delta p|$	force/area	$ML^{-1}\Theta^{-2}$
v	length/time	$L\Theta^{-1}$
ρ	mass/volume	ML^{-3}
η	(force/area) × time	$ML^{-1}\Theta^{-1}$

Clearly, the ratio L/D is dimensionless.

The pressure drop, $|\Delta p|$, is a force per area, which is dimensionally the same as a stress. From the discussion of viscosity and shear stress in Chapter 2 it is clear that $\eta v/D$ has dimensions of stress. Thus, $|\Delta p|$ divided by $\eta v/D$, or $|\Delta p| D/\eta v$, is dimensionless.

The only variable not used thus far is ρ. The quantity $\frac{1}{2}\rho v^2$ is kinetic energy per unit volume, which is dimensionally the same as stress (energy equals force times distance, so energy/volume is the same dimensionally as force/area). Thus, $\rho v^2/|p|$ is dimensionless. So, too, are $\rho v^2 D/\eta v$ and $\rho v^2 L/\eta v$.

Only three of the groups listed above can be independent. All others can be obtained by combinations of three independent groups. For example, $\rho v^2 L/\eta v$ is the product of $\rho v^2/|\Delta p|$ with $|\Delta p| D/\eta v$ and L/D. The three most convenient groups to use are

$$\frac{|\Delta p|}{\rho v^2} \qquad \frac{Dv\rho}{\eta} \qquad \frac{L}{D}$$

In that case, according to Eq. (3.2), we can write

$$\frac{|\Delta p|}{\rho v^2} = \text{function of } \left(\frac{Dv\rho}{\eta}, \frac{L}{D}\right) \tag{3.3}$$

Now, why have we selected these three groups, rather than some others, since we have an infinity of choices? L/D is a natural choice because it involves only geometry, thus placing all flow and physical property variables in the two remaining groups. We will often wish to solve explicitly for the pressure drop, so $|\Delta p|$ should not appear on both sides of Eq. (3.3). The second group must then be $Dv\rho/\eta$ or $Lv\rho/\eta$ if we wish to retain the simplicity of first powers whenever possible. $Dv\rho/\eta$ is preferred over $Lv\rho/\eta$ because of the intuitive belief that D is far more important than L in determining the nature of the flow; a second reason will be introduced subsequently. Finally, the choice $|\Delta p|/\rho v^2$ is arbitrary but is dictated by convention; $|\Delta p|D/\eta v$ would serve equally as well.

Equation (3.3) contains all the information available from dimensional analysis. A more useful result can be obtained by the use of some physical insight. Consider two adjacent sections of pipe of equal length, both far from the entrance or exit of the pipe. According to Eq. (3.3), the pressure drop over each of these identical pipe sections will be the same. Thus, the pressure drop over both sections together will be double that over each individual section [i.e., if L is replaced by $2L$ in Eq. (3.3) the value of the function is doubled]. Equivalently, the function on the right hand of Eq. (3.3) is directly proportional to L. Since L appears only in the ratio L/D, it follows that Eq. (3.3) can be rewritten

$$\frac{|\Delta p|}{\rho v^2} = \frac{L}{D} \times \text{function only of } \frac{Dv\rho}{\eta} \tag{3.4}$$

[This logic would not work if we had used $Lv\rho/\eta$ in Eq. (3.3).]

Equation (3.4) is now a relation between only *two* dimensionless groups, both of which occur often and have names. The group $Dv\rho/\eta$ is known as the *Reynolds number*, after Osborne Reynolds:

$$\text{Reynolds number:} \quad \text{Re} = \frac{Dv\rho}{\eta} = \frac{Dv}{v} \tag{3.5}$$

The group $|\Delta p|D/2\rho v^2 L$ is known as the *friction factor**:

$$\text{friction factor:} \quad f = \frac{|\Delta p|}{2\rho v^2} \frac{D}{L} \tag{3.6}$$

The factor of 2 is introduced into the denominator by convention. Equation (3.4) can then be written

$$f = f(\text{Re}) \tag{3.7}$$

That is, *the friction factor is a unique function of the Reynolds number for smooth pipe flow of all incompressible Newtonian Fluids.* This is a result of the greatest engineering importance, for it demonstrates that only two groups of variables need to be studied experimentally to obtain a relation that is universally valid for a wide class of fluids and geometric and flow parameters.

*The friction factor defined by Eq. (3.6) is sometimes called the *Fanning friction factor*. In some texts, particularly books on hydraulics, the friction factor is defined as $(|\Delta p|/\frac{1}{2}\rho v^2) \cdot (L/D)$, causing a factor-of-4 difference in all results. This latter definition is sometimes called the *Moody friction factor* and sometimes the *resistance coefficient*. Care must always be taken to determine which definition of the friction factor is being used.

3.3.2 Friction Factor-Reynolds Number Data

Dimensional analysis can indicate what dimensionless variables are related, but it cannot provide any information about the form of the relation. This must be determined either from experiment or from a more fundamental analysis of the process.

The relation between friction factor and Reynolds number is shown in Fig. 3-1. The data shown are typical of those of many investigators, and the

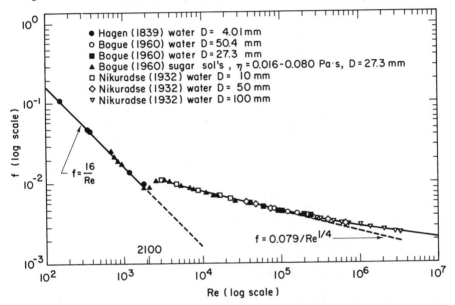

Figure 3-1. Friction factor as a function of Reynolds number for incompressible Newtonian fluids.

available data cover a broad range of viscosities, densities, and pipe diameters.* Note that there is overlap for data from all pipe sizes and over the entire viscosity range, and the data cover more than four decades in Re. Clearly, the friction factor is a unique function of the Reynolds number, as indicated by Eq. (3.7).

The data in Fig. 3-1 suggest two distinct regions of flow behavior, sepa-

---

*Some of the data in the literature were taken with air as the fluid. This is the only way in which a range of densities can be obtained. A gas may be considered to be incompressible in an isothermal flow as long as the pressure drop is less than about 10 to 20% of the mean pressure, as shown in Sec. 5.4.4. Other complications enter unless the velocity is much smaller than the speed of sound in the fluid (Mach number much smaller than unity); see, for example, Sec. 6.11.

rated by Re in the range 2100 to 4000. The region with Re < 2100 is called *laminar flow*; in this region the *f*-Re function is a straight line with a slope of −1. At higher Re, above about 4000, the flow is called *turbulent*; the friction factor-Reynolds number dependence is weaker here. These two terms are quite descriptive and follow from an experiment first carried out by Reynolds in 1883 and reproduced in Fig. 3-2. A thin filament of dye is injected into

Figure 3-2. Re-creation of Reynolds' dye stream experiment. (a) Laminar flow, Re = 1500, dye stream is undisturbed. (b) Turbulent flow, Re = 3000, dye stream is dispersed. Flow is from left to right in both photographs. Photographs by D. Smith, provided courtesy of the Chemical Engineering Department, Monash University.

a fluid stream in a pipe. In the laminar region the die retains its identity and remains at a fixed position relative to the wall, which shows that the fluid moves in concentric lamina through the pipe without motion normal to the wall. In the turbulent region the flow is chaotic and the dye is rapidly mixed, indicating substantial motion normal to the wall.

Turbulent flow is clearly a more favorable regime for some processes because of the intense mixing. The transition from laminar to turbulent flow may vary somewhat from apparatus to apparatus, and the *f*-Re function is not well defined experimentally in the transition region. Laminar flow usually ends at Re = 2100; between Re = 2100 and about 4000, the flow seems to pulsate between laminar and turbulent portions. Fully developed turbulence begins at Re of about 4000. In some very careful experiments, however, where special precautions have been taken to keep disturbances out of the system, laminar flow has been maintained as high as Re = 50,000. This is a laboratory curiosity, and in practice the transition can always be taken as Re = 2100.

The friction factor-Reynolds number relation in the laminar region follows the equation

$$\text{laminar, Re} < 2100: \quad f = \frac{16}{\text{Re}} \tag{3.8}$$

Using the definitions of f and Re we can rewrite this equation as

$$\text{laminar, Re} < 2100: \quad Q = \frac{\pi}{128} \frac{|\Delta p| D^4}{L\eta} \tag{3.9}$$

Equation (3.9) is known as the *Hagen–Poiseuille equation*, and we will derive it from first principles in Sec. 8.4. Note the fourth-power dependence of flow rate on tube diameter and the absence of the density from the equation.

In the range $4000 \leq \text{Re} \leq 10^5$, the data closely follow the *Blasius equation*:

$$\text{turbulent, } 4000 \leq \text{Re} \leq 10^5: \quad f = 0.079\text{Re}^{-1/4} \tag{3.10}$$

or, solving for Q,

$$\text{turbulent, } 4000 \leq \text{Re} \leq 10^5: \quad Q = 2.26 \left(\frac{|\Delta p|}{L}\right)^{4/7} (\rho^3 \eta)^{-1/7} D^{19/7} \tag{3.11}$$

The Blasius equation is empirical and has no theoretical basis, but it is in a convenient form for application. The entire turbulent region is well represented by the *von Kármán–Nikuradse equation*:

$$\text{turbulent, Re} \geq 4000: \quad \frac{1}{\sqrt{f}} = 4.0 \log_{10}(\text{Re}\sqrt{f}) - 0.4 \tag{3.12}$$

This equation has a theoretical basis, which we shall discuss in Sec. 16.4, but the two numerical parameters must be obtained by experiment. The equation is implicit for f, but it is explicit in the velocity, since the product $\text{Re}\sqrt{f}$ is independent of v.

3.3.3 Capillary Viscometry

There is a significance to the success of the friction factor-Reynolds number correlation in Fig. 3-1 which may not be immediately obvious. We defined the viscosity operationally in Chapter 2 in terms of a particular experiment, the shearing of a fluid between two plates. We had no guarantee that the quantity which we were defining was truly a *general property of the fluid*, rather than a specific response to a specific experiment. We find, however, that this measured quantity serves to define behavior in an entirely unrelated flow situation. Thus, we can be considerably more assured that *viscosity is an intrinsic physical property of a Newtonian fluid.*

The Hagen–Poiseuille equation, Eq. (3.9), provides the basis for a particularly simple method of viscosity measurement, the capillary viscometer. The equation is rearranged to the form

$$\eta = \frac{\pi}{128} \frac{|\Delta p|}{L} \frac{D^4}{Q} \tag{3.13}$$

The pressure difference is measured over a length L for a known flow rate and the viscosity is determined directly. Alternatively, the flow rate may be measured for a fixed pressure drop. Thus, if a separate flow meter is available in a process line undergoing laminar flow, a pressure drop measurement can provide continuous on-line measurement of viscosity, which can be useful for control purposes. The equivalent relationship for non-Newtonian fluids is derived in Sec. 19.6.

Example 3.1

A pressure drop of 200 Pa is measured over a 1-m length of 10-mm-diameter tubing. The flow rate is recorded as 60 mm³/s. Compute the viscosity.
From Eq. (3.13),

$$\eta = \frac{\pi}{128} \frac{200 \text{ Pa}}{1 \text{ m}} \frac{(10^{-2} \text{ m})^4}{60 \times 10^{-9} \text{ m}^3/\text{s}} = 0.82 \text{ Pa·s}$$

It is necessary to check that Re < 2100 so that the equation applies. If the density is of order 10^3 kg/m³, Re is readily computed to be of order 10^{-2}.

3.3.4 End Effects

The observation that the pressure drop, $|\Delta p|$, is proportional to length L, which was used to pass from Eq. (3.3) to (3.4), depended on the fact that two lengths of pipe were indistinguishable from one another. That is certainly true away from the entrance or exit of the pipe. There is a short region near the entrance to a pipe where the flow is still adjusting and the relations given in this chapter do not apply. This *entry length*, L_e, is given approximately by the relations

$$\text{Re} < 2100: \quad \frac{L_e}{D} \approx 0.59 + 0.055\text{Re} \tag{3.14a}$$

$$\text{Re} > 2100: \quad \frac{L_e}{D} \approx 40 \tag{3.14b}$$

We will establish a theoretical basis for these values in later chapters. The contribution to the total pressure drop from the entry length is greater than the contribution from a comparable length of pipe in the fully developed flow away from the entrance. The region of flow adjustment near the exit is quite small and usually does not need to be considered; it is about one radius in length at low Reynolds numbers and vanishes for Reynolds numbers greater than about 100.

The entry and exit regions are of little importance in most applications, since they usually comprise only a small portion of the total length of pipe and hence contribute only a minor fraction of the total pressure drop. One important exception is some commercial capillary viscometers for measuring the viscosity of very viscous liquids. The capillary lengths are often short because of the very large pressure drops, and corrections must be made for the entry and exit contributions to $|\Delta p|$. The procedures for doing so are discussed in specialized texts on viscometry.

3.3.5 Physical Meaning of the Reynolds Number

It is helpful to have a physical interpretation of the Reynolds number. This can be done crudely by reference to Fig. 3-3. A bit of fluid of mass m

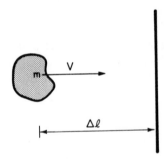

Figure 3-3. Schematic of the motion of a fluid element near a wall.

moves with velocity v. If the fluid were to be decelerated to zero by hitting a wall a distance $\Delta \ell$ away, the fluid would exert a force on the wall and the wall would exert an equal and opposite force on the fluid. This is because the fluid has inertia, and hence a force is required to change its uniform motion. The force is equal to the rate of change of momentum; since the momentum changes from mv to zero in the time Δt required to move distance $\Delta \ell$, the rate of change of momentum is $mv/\Delta t$. We can estimate Δt by assuming that the fluid mass moves at velocity v until it strikes the wall, in which case $\Delta t = \Delta \ell/v$. Thus, the force associated with fluid inertia, F_I, is estimated to be

$$F_I = \frac{mv^2}{\Delta \ell} \tag{3.15}$$

If the same particle is taken to be moving parallel to the wall, on the other hand, it experiences a shear stress $\eta v/\Delta \ell$ and a viscous shear force

$$F_V = \frac{\eta v}{\Delta \ell} A \tag{3.16}$$

A is the area of the fluid element that is acted on by the shear stress.
The ratio of inertial to viscous forces is

$$\frac{F_I}{F_V} = \frac{mv^2/\Delta \ell}{\eta v A/\Delta \ell} = \frac{mv}{\eta A} \tag{3.17}$$

If we take d to be the characteristic dimension of the space occupied by the fluid element, so that A is of order d^2 and the volume of the element is approximately Ad, then $m/A \approx \rho d$, and

$$\frac{F_I}{F_V} = \frac{\rho v d}{\eta} \tag{3.18}$$

Equation (3.18) has the form of the Reynolds number, except that the diameter D is replaced by some characteristic dimension of a fluid element. Thus, *except for a scaling factor* to account for substitution of D in place of the much smaller quantity d, *the Reynolds number can be interpreted as the ratio of the inertial to viscous forces at work in the fluid.*

A similar type of interpretation can be placed on the friction factor. The net imposed force on a fluid particle to cause flow in the axial direction is proportional to $|\Delta p| d^2$, while the inertial force is again proportional to $\rho d^2 v^2$. To within a scaling factor *the friction factor is therefore the ratio of the net imposed external force to the inertial force.*

These interpretations make it possible to explain the friction factor-Reynolds number behavior in the laminar region. If the fluid particles always move at a constant velocity in a straight line parallel to the wall, their momentum is never changed and there are no inertial forces. But Eq. (3.7) can be written

$$|\Delta p| D^2 = \text{proportional to } F_I f\left(\frac{F_I}{F_V}\right) \tag{3.19}$$

Since there are no inertial forces in the laminar region, the right-hand side must be independent of F_I. The only way that $F_I f(F_I/F_V)$ can be independent of F_I is for the function f to be proportional to F_V/F_I or, equivalently, for f to be proportional to the reciprocal of Re. Thus, the absence of inertial forces in the laminar region implies that

$$\text{laminar: } f = \text{proportional to Re}^{-1} \tag{3.20a}$$

or

$$\text{laminar: } f\text{Re} = \text{constant} \tag{3.20b}$$

This is the form taken by the experimental data. The Hagen–Poiseuille equation can therefore be deduced (except for the value of the numerical factor

$\pi/128$) as a direct consequence of the constant linear motion of fluid particles parallel to the pipe wall.

3.4 POWER

3.4.1 Power Input

The power P required to pump an incompressible Newtonian fluid through a straight length of pipe can be computed from the definition of power as the rate of performing work. The force acting upstream (position 1) to push the fluid into the length of pipe is the pressure times the area, $p_1(\pi D^2/4)$. At the downstream location (position 2) the fluid outside the pipe is exerting a force $p_2(\pi D^2/4)$ on the fluid in the pipe. Thus, the net force acting on the fluid in the length L is $|\Delta p|\pi D^2/4$.

The work done to move the fluid in the pipe a distance $\Delta \ell$ is $|\Delta p|(\pi D^2/4)\Delta \ell$. This takes place over a time Δt, so the rate of doing work is $|\Delta p|(\pi D^2/4)(\Delta \ell/\Delta t)$. But $\Delta \ell/\Delta t = v$, and $Q = (\pi D^2/4)v$, so we have the equation for power input,

$$\text{power input:} \quad P = Q|\Delta p| \tag{3.21}$$

3.4.2 Dissipation

The power input is the rate at which work is being done on the flowing system. According to the principle of conservation of energy, this work must go into increasing the energy of the system, or it must be removed as heat by some heat exchange mechanism at the wall.

It is useful to estimate the maximum increase in fluid temperature that can result from pumping power. The maximum will occur in an adiabatic pipe, with no heat removal by cooling at the walls. It is usually reasonable to assume that axial conduction of heat is negligible, in which case the energy balance for an incompressible fluid at steady state simply reduces to an equality between the rate of work done on the system and net rate at which the additional internal energy flows from the system. This equality is expressed as

$$\rho c_v Q \, \Delta T = P = Q|\Delta p| \tag{3.22}$$

or

$$\Delta T = \frac{|\Delta p|}{\rho c_v} \tag{3.23}$$

Here, ΔT is the adiabatic increase in temperature and c_v is the heat capacity per unit mass at constant volume. For liquids, c_v and c_p, the heat capacity at constant pressure, are almost equal numerically.

In steady pipe flow the power input does not increase the kinetic energy of the fluid. In this sense it represents a degradation in the quality of the energy, since the energy put into the system cannot be recovered in the form of useful work. This type of power input is sometimes referred to as *lost work* or *viscous loss*. The power per unit volume that goes only into viscous losses is often referred to as *viscous dissipation*; for steady pipe flow this is simply $P/(\pi D^2 L/4)$. It is useful to have this relation for steady laminar pipe flow, which follows from Eqs. (3.9) and (3.21) as

$$\text{laminar pipe flow:} \quad \text{dissipation} = \frac{P}{\pi D^2 L/4} = \frac{1}{2}\eta\left(\frac{8V}{D}\right)^2 \quad (3.24)$$

The grouping $8V/D$ is the shear rate at the wall in laminar pipe flow.*

Example 3.2

Compute the adiabatic temperature rise for the conditions in Example 3.1, assuming that $\rho = 10^3$ kg/m³ and $c_v = 3.5 \times 10^3$ J/kg·°C.
$|\Delta p|$ is given as 200 Pa. Thus,

$$\Delta T = \frac{200 \text{ Pa}}{(10^3 \text{ kg/m}^3)(3.5 \times 10^3 \text{ J/kg·°C})} = 0.6 \times 10^{-4} \text{ °C}$$

This temperature rise is clearly negligible. Were the tubing 1 mm in diameter, the fourth-power dependence of $|\Delta p|$ on D would have resulted in a temperature rise of 0.6°C over a 1-m length. Were the viscosity of order 10^5 Pa·s, as might occur for polymer melts, but the diameter still 10 mm, ΔT would be of order 7°C. A polymer melt in a smaller diameter and/or with a higher flow rate could clearly experience extremely large adiabatic temperature increases, necessitating efficient heat removal.

3.4.3 Optimal Pipe Diameter

We have sufficient information to estimate the optimal size of pipe to use in a given flow situation. The economic trade-off in this problem is between the increased capital cost of large-diameter pipe and the increased energy cost in pumping through small-diameter pipe.

Figure 3-4 shows the cost per meter in 1973 of standard (Schedule 40) steel pipe for diameters ranging from 19 to 300 mm ($\frac{3}{4}$ to 12 in). The costs are well represented by a straight line on log-log coordinates. Thus, we may

---

*The shear rate is the shear stress divided by the viscosity. In steady pipe flow the fluid in the pipe experiences a net force $|\Delta p|(\pi D^2/4)$ in the direction of flow. The equal and opposite force acting on the fluid is the product of the wall shear stress and the wall area, πDL. Thus,

$$\pi DL\tau_s = \pi DL(\eta\Gamma_s) = (\pi D^2/4)|\Delta p|$$

It then follows from the Hagen–Poiseuille equation, Eq. (3.9), that for laminar pipe flow, $\Gamma_s = 8V/D$.

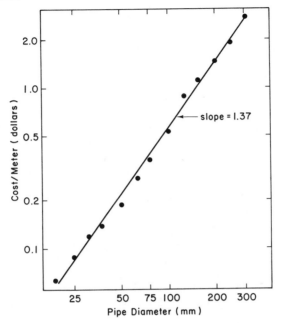

Figure 3-4. Cost per linear meter of standard steel pipe in 1973 U.S. dollars.

write

$$\text{cost of pipe} = K_0 \left(\frac{D}{D_0}\right)^n L \tag{3.25}$$

where K_0 = cost/meter of pipe of diameter D_0
D, D_0 = pipe diameter, reference diameter
L = pipe length
n = parameter, equal to 1.37 for standard steel pipe

If the pipeline is to have a useful life of N years, and we assume straight-line depreciation for simplicity, we have

$$\text{yearly cost of pipe} = \frac{K_0}{N} \left(\frac{D}{D_0}\right)^n L \tag{3.26}$$

The cost of construction will be related to the pipeline size and hence to the cost of pipe. Thus, we can take the depreciated cost of construction as a multiple of the depreciated cost of pipe and write

$$\text{yearly cost of construction} = \Lambda_c \times \text{yearly cost of pipe}$$

Similarly, the cost of fittings will be directly related to the cost of pipe, and we have

$$\text{yearly cost of fittings} = \Lambda_f \times \text{yearly cost of pipe}$$

Yearly maintainance will be taken as some fraction of the total cost of construction,

$$\text{yearly maintainance} = \Lambda_m \times \text{total cost of construction}$$

Thus, the total annual cost of the piping system for capital and maintainance, denoted C_{pipe}, is

$$C_{\text{pipe}} = \frac{1 + \Lambda_c + \Lambda_f}{N}(1 + \Lambda_m)K_0\left(\frac{D}{D_0}\right)^n L \qquad (3.27)$$

For simplicity we combine all parameters into a single dimensional parameter, Λ, and write

$$C_{\text{pipe}} = \Lambda D^n L \qquad (3.28)$$

The yearly cost of power is equal to the product of the power requirement, P; the hours of operation per year, H; the hourly cost of power, K_p; and the reciprocal of the efficiency of the pumping device, E:

$$C_{\text{power}} = \frac{PHK_p}{E} = \frac{Q|\Delta p|HK_p}{E} \qquad (3.29)$$

Here, we have used Eq. (3.21), $P = Q|\Delta p|$.

The pressure drop is related to the operating variables through the friction factor,

$$|\Delta p| = \frac{2f\rho v^2 L}{D} \qquad (3.30)$$

We will use the Blasius equation, Eq. (3.10), which is valid in the range $4 \times 10^3 \leq \text{Re} \leq 10^5$:

$$f = 0.079\text{Re}^{-1/4} \qquad (3.10)$$

in which case we can write

$$|\Delta p| = \frac{0.16L\rho^{0.75}v^{1.75}\eta^{0.25}}{D^{1.25}} \qquad (3.31)$$

Thus, the annual cost of power can be written

$$\begin{aligned}
C_{\text{power}} &= 0.16\left(\frac{4}{\pi}\right)^{1.75}\frac{HK_pL}{E}\frac{\rho^{0.75}\eta^{0.25}Q^{2.75}}{D^{4.25}} \\
&\equiv \lambda L\frac{\rho^{0.75}\eta^{0.25}Q^{2.75}}{D^{4.75}}
\end{aligned} \qquad (3.32)$$

where λ is a dimensional cost coefficient for the power cost.

The total annual cost, C, is the sum of pipe and power costs,

$$C = C_{\text{pipe}} + C_{\text{power}} = \Lambda D^n L + \lambda L\frac{\rho^{0.75}\eta^{0.25}Q^{2.75}}{D^{4.75}} \qquad (3.33)$$

The first term is a monotonically increasing function of D, while the second is a monotonically decreasing function of D. Thus, the function becomes infinite in the limits of D going to zero and infinity, and there is a minimum cost at some intermediate value of D. We can find this point by setting dC/dD

to zero:

$$\frac{dC}{dD} = n\Lambda D^{n-1}L - 4.75\lambda L \frac{\rho^{0.75}\eta^{0.25}Q^{2.75}}{D^{5.75}} = 0$$

or

$$D = \left(\frac{4.75\lambda}{n\Lambda}\rho^{0.75}\eta^{0.25}Q^{2.75}\right)^{1/(4.75+n)} \tag{3.34}$$

It would be unusual for the optimum to correspond exactly to an available pipe size, and in practice the nearest regular pipe size would be used.

It is extremely instructive to invert Eq. (3.34) in order to solve for the velocity when the optimum pipe size is used. Using $v = 4Q/\pi D^2$, we obtain

$$v = \frac{4}{\pi}\left(\frac{n\Lambda}{4.75\lambda}\right)^{1/2.75} \frac{D^{(n-0.75)/2.75}}{\rho^{3/11}\eta^{1/11}} \tag{3.35}$$

The velocity is virtually independent of viscosity and is essentially independent of density for all liquids. For n of the order shown in the figure for steel pipe, the diameter dependence is $D^{0.23}$. Thus, v will change by less than a factor of 2 for an order-of-magnitude change in D. As a result, we expect to find the velocity in all liquid pipelines to lie in a very narrow range for economic reasons. This observation is borne out in practice. The linear velocity in oil pipelines of all sizes, for example, is nearly always approximately 1.5 m/s.

3.5 COMMERCIAL PIPE

3.5.1 Relative Roughness

The friction factor-Reynolds number data in Fig. 3-1 were taken in drawn tubing with a very smooth inner surface. The section of pipe shown schematically in Fig. 3-5 represents an opposite extreme, in which the inner surface varies in a regular manner. In such a case it is evident that the characteristic size of the surface roughness, k, needs to be taken into account in the dimensional analysis. This will lead to one additional dimensionless group, k/D, which is known as the *relative roughness*. The consequence of dimensional

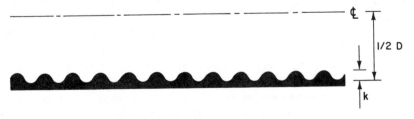

Figure 3-5. Schematic of pipe with walls of uniform, regular roughness.

analysis is then

$$f = f\left(Re, \frac{k}{D}\right) \tag{3.36}$$

The ideal situation shown in Fig. 3-5 was studied experimentally by Nikuradse, who glued sand particles of nearly equal size to the inner walls of smooth tubes in order to obtain uniform and known values of k/D. His data are shown in Fig. 3-6; at the largest value of k/D, the sand particle thickness was 6.7% of the pipe radius.

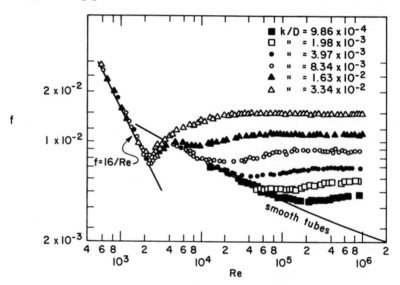

Figure 3-6. Friction factor for sand-roughened pipes as a function of Reynolds number. Data of Nikuradse, *NACA Tech. Memo. 1292,* November 1950 (translation of original German 1933 publication).

In laminar flow the friction factor is independent of k/D over the entire range studied, as is the transition to turbulence. For each relative roughness there is a region of turbulent flow in which the friction factor follows the curve defined for smooth pipes in Fig. 3-1. This region is referred to as *hydraulically smooth*, since the f-Re relation is unaffected by the surface roughness.

For each relative roughness there is a critical Reynolds number at which the friction factor data deviate from the smooth pipe line. Following a short transition, the friction factor becomes independent of Reynolds number. The region of constant friction factor is often called *complete turbulence*, although this terminology has no physical significance. When the friction factor is constant, the viscous forces in the fluid play no measurable role, since they enter only through the Reynolds number, and the pressure drop comes entirely from the inertial forces.

3.5.2 Pipe Roughness

The use of a single roughness factor works well for Nikuradse's sand-roughened pipes because there is only one additional length scale, since both the height and spacing of the roughness are uniform and both are equal to the size of the sand particles. Thus, the dimensional analysis accurately reflects the true experimental situation. The surface of real pipe, however, is quite irregular on a microscopic scale, and the irregularities cannot be characterized by a single number. Nevertheless, motivated by the sand-roughness experiments, data for real commercial pipe are commonly analyzed in terms of a single roughness, and the approach works reasonably well for clean pipes.

Figure 3-7 shows a chart of friction factor versus Reynolds number prepared by Moody for commercial pipe. Typical values of k given by Moody for various types of pipe are shown in Table 3-2. The curves in Fig. 3-7 represent experimental data to within about $\pm 10\%$, and they follow the

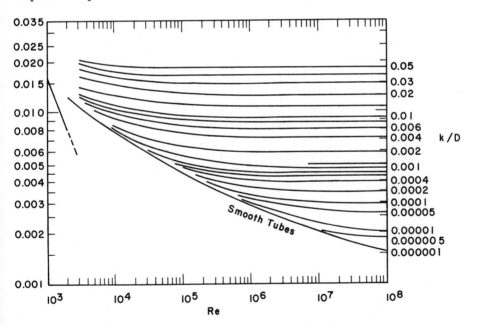

Figure 3-7. Friction factor as a function of Reynolds number for rough pipe. The lines are a graphical representation of the empirical Colebrook formula, Eq. (3.37).

empirical *Colebrook formula*,

$$\frac{1}{\sqrt{f}} = -4.0 \log_{10}\left(\frac{k}{D} + \frac{4.67}{\text{Re}\sqrt{f}}\right) + 2.28 \qquad (3.37)$$

A rough explicit expression for typical commercial pipe in the range $4 \times 10^3 \leq \text{Re} \leq 2 \times 10^7$ with typical velocities is

$$4000 \leq \text{Re} \leq 2 \times 10^7: \quad f \approx 0.04\text{Re}^{-0.16} \qquad (3.38)$$

TABLE 3-2
SURFACE ROUGHNESS FOR VARIOUS MATERIALS

Material	k (mm)
Drawn tubing (brass, lead, glass, etc.)	1.5×10^{-3}
Commercial steel or wrought iron	0.05
Asphalted cast iron	0.12
Galvanized iron	0.15
Cast iron	0.46
Wood stave	0.2–0.9
Concrete	0.3–3
Riveted steel	0.9–9

If the optimal pipe diameter calculation is repeated using Eq. (3.38) in place of the Blasius equation, the result is only slightly different from that given previously.

The curves for commercial pipe show regions of hydraulic smoothness and complete turbulence, as in the sand-roughened experiments, but the transition between the two regions is much more gradual. The relative roughness given in Table 3-2 is determined by comparison of the completely turbulent friction factor with the corresponding sand-roughened pipe. Thus, because of the gross irregularities, the equivalent k may (and usually will) be quite different from the average value that would be estimated from careful observation of the surface.

Example 3.3

Water at 20°C is pumped through a 50-mm commercial steel pipe at a velocity of 1.5 m/s. Determine the pressure drop per unit length.

The physical properties of water are approximately $\eta = 10^{-3}$ Pa·s and $\rho = 10^3$ kg/m³. Thus,

$$\text{Re} = \frac{Dv\rho}{\eta} = \frac{(50 \times 10^{-3} \text{ m})(1.5 \text{ m/s})(10^3 \text{ kg/m}^3)}{10^{-3} \text{ Pa·s}} = 7.5 \times 10^4$$

From Table 3.2 we have $k = 0.05$ mm, so $k/D = 10^{-3}$. Thus, from Fig. 3-7 or Eq. (3.37), $f = 0.0058$. From the definition of f, we obtain

$$\frac{|\Delta p|}{L} = \frac{2\rho v^2 f}{D} = \frac{2(10^{-3} \text{ kg/m}^3)(1.5 \text{ m/s})^2}{50 \times 10^{-3} \text{ m}}(5.8 \times 10^{-3}) = 520 \text{ Pa/m}$$

If we use the rough approximation, Eq. (3.38), instead, we have

$$f \approx 0.04(7.5 \times 10^4)^{-0.16} = 0.0067$$

In that case we estimate $|\Delta p|/L$ to be 600 Pa/m. The lower value is probably slightly more reliable, although the two estimates do not differ significantly within the $\pm 10\%$ uncertainty on the curves in Fig. 3-7.

The corresponding friction factor in a smooth pipe is 0.0047, giving a much lower pressure drop of $|\Delta p|/L = 420$ Pa/m.

3.5.3 Nominal and Real Diameter

There is one additional factor with regard to commercial pipe that needs to be taken into account in doing practical calculations. The diameter of commercial pipe is a nominal size. The actual inner pipe diameter will depend on the designed strength of the pipe, as manifested by the wall thickness. Table 3-3 shows the diameter for various types of "50-mm" pipe. All have an

TABLE 3-3
WALL THICKNESS AND INSIDE DIAMETER OF NOMINAL
50-mm ("2-in") PIPE

Schedule number	Wall thickness (mm)	Inner diameter (mm)
5S	1.65	57.02
10S	2.77	54.79
40ST, 40S	3.91	52.50
80ST, 80S	5.54	49.25
160	8.74	42.85
xx	11.07	38.18

outer diameter of 60.33 mm. (At the time of writing, pipe is still sold in the United States in inches. The numbers given here are for nominal 2-in pipe.) Complete data are contained in handbooks.

3.6 NONCIRCULAR CROSS SECTIONS

There is an empirical method for predicting flow rates and pressure drops in conduits with noncircular cross sections that is quite effective. The method defines a pipe diameter that is "equivalent" to the noncircular conduit and then makes use of the friction factor-Reynolds number relation for smooth, round pipes.

A cylinder of diameter D and length L has a volume $\pi D^2 L/4$ and a surface area wetted by the fluid of πDL. Thus, we can write

$$D = \frac{4 \times \text{volume of liquid}}{\text{surface wetted by liquid}}$$

For any other conduit we define the *hydraulic diameter*, D_H, by analogy:

$$\text{hydraulic diameter:} \quad D_H = \frac{4 \times \text{volume of liquid}}{\text{surface wetted by liquid}} \tag{3.39}$$

For channels of constant but noncircular area, an equivalent and somewhat more common definition is

$$D_H = \frac{4 \times \text{cross-sectional area}}{\text{wetted perimeter}} \tag{3.40}$$

For example, for a rectangular channel with sides a and b, the area is ab and the perimeter is $2(a + b)$. Thus,

$$\text{rectangle:} \quad D_H = \frac{4(ab)}{2(a + b)} = 2\left(\frac{1}{a} + \frac{1}{b}\right)^{-1} \tag{3.41}$$

For conduits whose cross sections are "nearly circular," such as rectangles of small aspect ratio and triangles, the friction factor-Reynolds number correlation developed for turbulent flow in round pipes can be applied with little error when D_H is substituted for D. This is demonstrated in Fig. 3-8 with data for a conduit with an equilateral triangular cross section. The laminar–turbulent transition also occurs at a Reynolds number in the range 2100 to 4000. The data in the laminar regime follow the relation $f\,\text{Re} = \text{constant}$, but the constant may differ a great deal from the value 16 characteristic of round pipes. For the triangle, the constant does have a value equal to about 16.

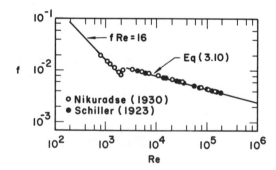

Figure 3-8. Friction factor as a function of Reynolds number for incompressible Newtonian fluid flow in a conduit with an equilateral triangular cross section. f and Re are calculated using the hydraulic radius. Solid lines are for smooth, round pipes.

Other cross sections have values both greater and smaller than 16; a theoretical basis is described in Sec. 8.6.

The hydraulic diameter concept, although based on a slight conceptual foundation for turbulent flow, is an empirical tool of great use and is widely employed. It provides a consistent means of introducing a unique conduit size for analysis purposes when the actual flow situation might be extremely complex.

There is one point of confusion in the use of the hydraulic diameter concept. The *hydraulic radius*, R_H, is often used in the engineering literature. This is defined, somewhat strangely, as

$$R_H = \frac{\text{cross-sectional area}}{\text{wetted perimeter}} \qquad (3.42)$$

Note that $R_H = D_H/4$. Thus, the hydraulic radius is equal to one-half the cylinder radius!

3.7 CONCLUDING REMARKS

It is important to keep in mind that the relations developed in this chapter apply only to incompressible Newtonian fluids. Polymeric liquids, for example, have an additional material property with dimensions of time, so an additional dimensionless group will occur. Thus, polymeric liquids should not be expected to follow the f–Re relations developed in this chapter, and indeed, deviations of an order of magnitude are sometimes observed.

The use of the friction factor-Reynolds number chart, with allowance for pipe roughness and noncircular cross section, is one of the most important tools for the solution of process flow problems. The definitions in this chapter, the Hagen–Poiseuille equation, and the Blasius equation should be committed to memory to enable rapid estimation of orders of magnitude in practical problems.

BIBLIOGRAPHICAL NOTES

Dimensional analysis is discussed in *The Chemical Engineers' Handbook* and in

RUSSELL, T. W. F., and DENN, M. M., *Introduction to Chemical Engineering Analysis*, John Wiley & Sons, Inc., New York, 1972, Chap. 3.

The classical treatments are in

BRIDGMAN, P. W., *Dimensional Analysis*, Yale University Press, New Haven, Conn., 1931; paperback edition, 1963.

LANGHAAR, H. W., *Dimensional Analysis and Theory of Models*, John Wiley & Sons, Inc., New York, 1951.

Citations of the original sources of round pipe and noncircular conduit data, as well as data not included here, can be found in

PRANDTL, L., and TIETJENS, O. G., *Applied Hydro- and Aeromechanics*, Dover Publications, New York, 1957.

SCHLICHTING, H., *Boundary Layer Theory*, 6th ed., McGraw-Hill Book Company, New York, 1968.

The data by Bogue in Fig. 3-1 were taken from

BOGUE, D. C., *Velocity Profiless in Turbulent Non-Newtonian Flow*, Ph.D. dissertation, University of Delaware, Newark, Del., 1960.

See *The Chemical Engineers' Handbook* for further discussion of the optimal-pipe-size problem, including more realistic economics.

PROBLEMS

3.1. Flow rate is often expressed in mass/time (kg/s), w. Express the Reynolds number in terms of w, η, and D.

3.2. Water at 93°C flows through a 100 mm diameter smooth pipe at a flow rate of 100,000 kg/hr. Find the pressure drop per meter of pipe.

3.3. Molten poly (ethylene terephthalate), the polymer used to make polyester fibers, is approximately a Newtonian liquid at processing conditions (about 280°C) with $\eta = 600$ Pa·s, $\rho = 1300$ kg/m³. Find the pressure drop per meter if 5×10^{-4} kg/s is transported in a 25 mm diameter smooth pipe.

3.4. An incompressible Newtonian fluid is to be pumped a distance L at flow rate Q, and the available pressure difference is $|\Delta p|$. Show that the capital investment is always smaller by use of one large pipe than two smaller ones, regardless of whether the flow is laminar or turbulent.

3.5. Repeat the optimal pipe diameter and velocity calculations in Sec. 3.4.3 for laminar flow.

3.6. Crude oil of specific gravity 0.86 ($\rho = 860$ kg/m³) is to be pumped at 20°C. What is the minimum pipe size that can be used in order to operate in the turbulent regime at a velocity of 1.5 m/s?

3.7. The following data were obtained by R. Metzner for water at 22°C flowing through precision bore glass tubing having an inside diameter of 0.289 mm and length 57.6 mm:

Run	Pressure Drop (Pa)	Volumetric Flowrate (m³/s)
A	3.6×10^4	1.024×10^{-7}
B	1.9×10^4	0.564×10^{-7}
C	8.0×10^4	2.066×10^{-7}
D	11.5×10^4	2.763×10^{-7}

a) Compute the Reynolds number for each run, and show that all data are in the laminar regime.

b) Compute the expected pressure drop for the given flow rates and compare to the experimental values.

c) The extra pressure drop in laminar flow caused by flow rearrangement at the entrance and exit is believed to be proportional to the wall shear stress $(8\eta v/D)$ at low Re and to inertial stresses (ρv^2) at high Re, with coefficients of order unity in both cases. (The exit correction is unimportant at high Re, as shown in Sec. 6.7.) Show that the ρv^2 term will dominate for these experiments, and compute the entry correction K in the equation

$$|\Delta p| = |\Delta p|_{\text{Poiseuille}} + K \rho v^2$$

How does this compare to the "best" value of $K = 1.17$?

3.8. Repeat Problem 3.2 when the pipe is

a) steel

b) galvanized iron

3.9. The calculation of flow rate given diameter, or of diameter given flow rate, for a specified pressure gradient is iterative with a friction factor-Reynolds number plot. Describe a way of replotting f-Re data so that the following calculations can be carried out without iteration:

a) Calculate the flow rate when pressure drop and pipe diameter are specified.

b) Calculate the pipe diameter required to provide a specified mass flow rate under a given pressure drop.

3.10. Water at 20°C flows at 3 m/s in the annulus of a concentric tube heat exchanger. The inner wall of the annulus has a diameter of 25 mm and the outer wall a diameter of 50 mm. What is the pressure drop over 30 meters if

a) the walls are taken as smooth?

b) the walls are made of commercial steel?

Flow of Particulates $\quad$ **4**

4.1 INTRODUCTION

In many practical applications we need to know the force required to move a solid object through a surrounding fluid or, equivalently, the force that a moving fluid exerts on a solid as the fluid moves past. These problems can be studied at a first level by dimensional analysis and experimentation, as we did in Chapter 3 for pipe flow, and many results of practical importance can be obtained.

It is helpful to keep the following physical principle in mind when studying particulate flow: *Physical phenomena do not depend on the frame of reference of the observer.* In its simplest and most important application, this principle means that a problem can be analyzed by use of the most convenient coordinate system, and the results can then be converted to any other system. When a particle moves through a fluid, for example, we might choose to utilize a coordinate system that is fixed in the laboratory. It will sometimes be more convenient, however, to use a coordinate system that is fixed on the *particle*. In that case, it would appear to an observer making measurements that the particle was fixed and the fluid was moving past. (This is the phenomenon often experienced on railroad trains, in which the train appears to be motionless and the station moving past.) Results can be transformed from one frame of reference to the other as long as we know the relative motion of the two coordinate systems; the transformation is most straightforward when the relative velocity is a constant. If one coordinate frame is accelerating relative to the other, the forces measured by the two observers will be different.

4.2 FLOW PAST A SPHERE

4.2.1 Drag Coefficient

We will consider the steady motion of a spherical particle in an infinite expanse of an incompressible Newtonian fluid. We think of the surrounding fluid as infinite in order to eliminate the effect of nearby particles and container walls. As we shall see subsequently, we can consider a fluid to be infinite if the nearest wall is at least 20 sphere diameters away. A suspension of spheres can be considered to consist of single spheres in an infinite fluid if the volume fraction of spheres is less than 0.1 (10%); at higher volume fractions, the spheres interact with one another.

We will first determine the relation between the velocity of the sphere and the force on the sphere. The fluid is assumed to be stationary except for motion that might be induced near the sphere. The net force on the sphere will include viscous forces from the fluid and external forces such as gravity. The problem is completely equivalent, except for a change in frame of reference, to the uniform motion of an infinite fluid past a stationary sphere.

The force is denoted F_D (for "drag force") and the sphere velocity V_p (p is for "particle"). Force and velocity are vector quantities, and we assume that both vectors point in the same direction. The relevant variables and their dimensions are then as follows:

$$F_D = \text{force}, \; ML\,\Theta^{-2}$$

$$V_p = \text{velocity}, \; L\,\Theta^{-1}$$

$$D_p = \text{sphere diameter}, \; L$$

$$\rho \;\; = \text{fluid density}, \; ML^{-3}$$

$$\eta \;\; = \text{fluid viscosity}, \; ML^{-1}\,\Theta^{-1}$$

The mass of the sphere is not included. It seems reasonable to presume that a moving fluid will exert the same force on a stationary sphere whether the sphere is solid or hollow and regardless of the material of construction.

There are five variables and three dimensions, so there will be two independent dimensionless groups. Two groups are readily found by inspection as follows:

$$\text{Reynolds number:} \quad \text{Re} = \frac{D_p V_p \rho}{\eta} \qquad (4\text{-}1a)$$

$$\text{drag coefficient:} \quad C_D = \frac{8}{\pi}\frac{F_D}{\rho V_p^2 D_p^2} \qquad (4\text{-}1b)$$

The factor $8/\pi$ in the drag coefficient is included by convention, as discussed in Sec. 4.3. The Reynolds number should not be confused with the Reynolds number in pipe flow; Re is used to refer to any dimensionless group which is the product of a length, velocity, and density, divided by a viscosity. The Reynolds number can always be interpreted as a ratio of inertial to viscous forces, but particular regimes established in one flow field have no relevance to any other flow.

The drag coefficient will be a unique function of the Reynolds number,

$$C_D = C_D(\text{Re}) \tag{4.2}$$

Drag coefficient data are shown in Fig. 4-1. The data were obtained using the setting velocity method discussed in Sec. 4.2.2. Most process applications take place at Re much less than 10^5, and there are no sharp transitions between flow regimes.

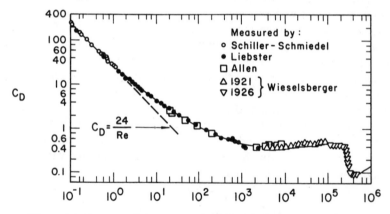

Figure 4-1. Drag coefficient as a function of Reynolds number for flow past a sphere. Reproduced from H. Schlichting, *Boundary Layer Theory*, 6th ed., McGraw-Hill Book Company, New York, 1968, by permission.

The region Re < 1 is inertialess, as can be seen from the fact that C_D varies as Re^{-1}. The existence of an inertialess region is not as obvious intuitively for flow about a sphere as it is for pipe flow, because the flow lines must bend as fluid passes around the sphere, and hence there is a change in fluid momentum. Evidently, the effect does not influence the drag force. This inertialess region is known as *Stokes flow*, and here the drag coefficient follows the equation

$$\text{Re} < 1: \quad C_D = \frac{24}{\text{Re}} \tag{4.3}$$

The result was obtained theoretically by Stokes in 1851, and we will derive it from first principles in Chapter 12. The corresponding relation for the force,

$$\text{Re} < 1: \quad F_D = 3\pi\eta V_p D_p \tag{4.4}$$

is usually known as *Stokes' law*.

The intermediate region, $1 \leq \text{Re} \leq 10^3$, can be roughly approximated for computational purposes by

$$1 \leq \text{Re} \leq 10^3: \quad C_D \approx 18\text{Re}^{-0.6} \tag{4.5}$$

The *Newton regime*, $10^3 \leq \text{Re} \leq 2 \times 10^5$, is a region of approximately constant drag coefficient,

$$10^3 \leq \text{Re} \leq 2 \times 10^5: \quad C_D \approx 0.44 \tag{4.6}$$

Here, the drag force is approximately proportional to the square of the velocity, and the drag force is independent of the viscosity. A quadratic drag–velocity relation was postulated by Newton. The sharp drop in C_D at a Reynolds number of approximately 2×10^5 is discussed in Sec. 15.8.

4.2.2 Settling Velocity

When a sphere is dropped in a viscous fluid and allowed to experience the acceleration of gravity, there is a brief transient period, discussed in Appendix 4.2.B, after which the sphere falls with a constant terminal (or *settling*) velocity. The drag coefficient-Reynolds number data provide sufficient information to compute this settling velocity.

The situation is shown schematically in Fig. 4-2. There are three forces acting on the sphere: gravity, F_G; buoyancy, F_B; and drag, F_D. The forces are collinear, so we need not be concerned with the vectorial nature of force. Since

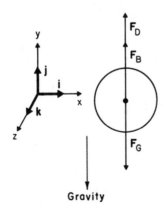

Figure 4-2. Forces acting on a sphere falling in a viscous fluid.

the sphere is moving in a straight line with no change in velocity, there is no change in linear momentum. Thus, Newton's second law reduces to the requirement that the forces on the sphere sum algebraically to zero. These forces are as follows:

gravity: $$F_G = -\rho_p \frac{\pi D_p^3 g}{6} \qquad (4.7)$$

buoyancy: $$F_B = +\rho \frac{\pi D_p^3 g}{6} \qquad (4.8)$$

drag: $$F_D = \frac{\pi}{8} \rho V_p^2 D_p^2 C_D \qquad (4.9)$$

ρ_p is the density of the solid sphere and g is the gravitational acceleration (9.8 m/s²). The gravitational force is equal to the weight of the sphere, and the sign is negative because it is directed downward. The buoyancy force, which points upward, is equal to the weight of displaced fluid; if the concept of buoyancy is unfamiliar, see Appendix 4.2.A. The drag force, which is directed upward, is written in terms of the drag coefficient. Thus, we can write

$$0 = -\rho_p \frac{\pi D_p^3 g}{6} + \rho \frac{\pi D_p^3 g}{6} + \frac{\pi}{8} \rho V_p^2 D_p^2 C_D \qquad (4.10)$$

For a given velocity, diameter, and fluid and sphere properties, Eq. (4.10) can be solved for C_D. This is the way that Fig. 4-1 was constructed.

The greatest interest regarding terminal velocity is in the Stokes regime, Re < 1. Substituting $C_D = 24/\text{Re}$ in Eq. (4.10) and solving for V_p, we obtain an expression for the Stokes settling velocity:

$$\text{Re} < 1: \quad V_p = \frac{g D_p^2 (\rho_p - \rho)}{18\eta} \qquad (4.11)$$

In the *intermediate regime*, $1 \le \text{Re} \le 10^3$, the drag coefficient is roughly approximated by $C_D \approx 18\text{Re}^{-0.6}$, in which case Eq. (4.10) can be solved for V_p to give

$$1 \le \text{Re} \le 10^3: \quad V_p \approx \left[\frac{2g}{27} \left(\frac{\rho_p}{\rho} - 1 \right) \right]^{5/7} D_p^{8/7} \left(\frac{\rho}{\eta} \right)^{3/7} \qquad (4.12)$$

Finally, in the Newton regime, $10^3 \le \text{Re} \le 2 \times 10^5$, the drag coefficient is a constant, $C_D \approx 0.44$. Then Eq. (4.10) gives

$$V_p \approx \left[3 D_p g \left(\frac{\rho_p}{\rho} - 1 \right) \right]^{1/2} \qquad (4.13)$$

The different diameter dependence of the velocity in the three regimes is of some interest. The velocity is quadratic in diameter in the Stokes regime, nearly linear in the intermediate regime, and varies with the one-half power in the Newton regime.

Example 4.1

Estimate the maximum spherical particle that will fall in Stokes flow in a given liquid.

Stokes flow requires $Re < 1$, or, from Eq. (4.1a),

$$D_p V_p < \frac{\eta}{\rho}$$

Multiplying Eq. (4.11) by D_p gives

$$D_p V_p = \frac{g D_p^3 (\rho_p - \rho)}{18 \eta} < \frac{\eta}{\rho}$$

or

$$D_p < \left[\frac{18 \eta^2}{g \rho (\rho_p - \rho)} \right]^{1/3}$$

For air, $\rho \sim 1.3$ kg/m³ and $\eta \sim 2 \times 10^{-5}$ Pa·s. If we assume that $\rho_p \sim 10^3$ kg/m³, we obtain

$$D_p < \left[\frac{(18)(2 \times 10^{-5} \text{ Pa·s})^2}{(9.8 \text{ m/s}^2)(1 \text{ kg/m}^3)(1.3 \times 10^3 \text{ kg/m}^3)} \right]^{1/3} \sim 10^{-4} \text{ m}$$

The settling velocity of this particle in air is

$$V_p \sim \frac{(9.8 \text{ m/s}^2)(10^{-4} \text{ m})^2(10^3 \text{ kg/m}^3)}{(18)(2 \times 10^{-5} \text{ Pa·s})} = 0.3 \text{ m/s}$$

4.2.3 Falling-Sphere Viscometer

The falling-sphere viscometer is one of the practical applications of the results of the preceding section. Falling sphere viscometry requires the measurement of the terminal velocity of a sphere, usually by measuring the time required to fall between two marks a distance L apart. The time, t_p, is related to the terminal velocity by $V_p = L/t_p$. Measurements are made in the Stokes regime, so Eq. (4.11) can be rewritten in the form

$$\eta = \frac{g D_p^2 (\rho_p - \rho) t_p}{18 L} \tag{4.14}$$

Falling-sphere devices are in common use, as are similar viscometers that utilize the rise time of a bubble in a viscous fluid.

Example 4.2

A steel ball, with diameter $D_p = 3$ mm and density $\rho_p = 7.6 \times 10^3$ kg/m³ is dropped in a liquid with density $\rho = 1.2 \times 10^3$ kg/m³. The average time for the ball to drop a distance of 0.5 m is 10.0 s. What is the viscosity?

From Eq. (4.14),

$$\eta = \frac{(9.8 \text{ m/s}^2)(3 \times 10^{-3} \text{ m})^2(7.6 \times 10^3 - 1.2 \times 10^3 \text{ kg/m}^3)(10.0 \text{ s})}{(18)(0.5 \text{ m})} = 0.6 \text{ Pa·s}$$

A check shows that $Re = 0.3$, so Eq. (4.14) does apply.

4.2.4 Separation of Particulates

Another practical application of the results of this section is the analysis of devices for removing solids from fluid streams. The simplest such device, which we shall analyze here, is the *gravity settling chamber*, in which the gravitational force acts on the particles. *Electrostatic precipitators* operate on a similar principle, but the particles are charged and an electric field is used in place of gravity. *Cyclone separators* also operate in a similar manner, but here the fluid is forced into a spiraling motion and the centrifugal force operates on the particles in place of gravity.

The gravity settling chamber is sometimes used on natural draft exhausts from kilns and furnaces and is effective in removing particles larger than 0.042 mm in diameter ("325 mesh"). We will compute the maximum particle size that can pass through a chamber without being removed, under the assumption that the volume fraction of particles in the gas stream is less than 0.1, so that particles do not interact with one another.

The chamber is shown schematically in Fig. 4-3. It is simply a box of

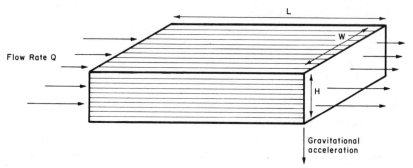

Figure 4-3 Schematic of a gravity settling chamber.

length L, height H, and width W. The top and bottom planes are horizontal and, therefore, perpendicular to the direction of gravitational acceleration. Gas flows through the chamber with volumetric flow rate Q. Particles are carried in the gas stream, but they settle out because of the gravitational force, as shown in Fig. 4.4.

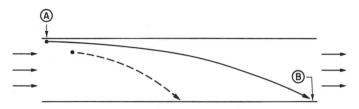

Figure 4-4. Particle trajectories in a gravity settling chamber.

Consider a particle at position A in the figure. For that particle to be trapped, it must drop the height H before it has traveled the length L. We are interested in the particle starting at A and ending exactly at B. A smaller particle starting at A will fall more slowly and cannot be trapped. A larger particle will fall to the bottom before B. Thus, the particle starting at A and ending at B represents the smallest particle normally removed by the settling chamber. Smaller particles entering below the top will also be caught, but we cannot determine these with any certainty.

The fluid velocity through the chamber is the flow rate divided by the cross-sectional area of the flow,

$$V_f = \frac{Q}{WH} \tag{4.15}$$

The time required for horizontal fluid motion from left to right is

$$t_H = \frac{L}{V_f} = \frac{WHL}{Q} \tag{4.16}$$

If the particle falls with a settling velocity V_p, the time required for vertical movement from top to bottom is

$$t_V = \frac{H}{V_p} \tag{4.17}$$

To start at A and stop at B, the times t_H and t_V must be equal. Thus, we have

$$V_p = \frac{Q}{WL} \tag{4.18}$$

This result assumes that the particle can fall freely in the gas stream and that it is not reentrained by large-scale eddying motions. In practice, the linear velocity V_f must then be less than 3 m/s. For a given flow rate and chamber width, this places a design restriction on H through Eq. (4.15). Because V_p is independent of H, this restriction is the only way in which the height enters explicitly into the analysis.

We assume that the particle Reynolds number is sufficiently small for Stokes' law to apply. This is usually an excellent assumption for particulates in a gas stream, but if it is not valid in any particular case, it can easily be relaxed by using the drag coefficient-Reynolds number correlation in the appropriate range. Equating the Stokes terminal velocity Eq. (4.11) to V_p in Eq. (4.18), we obtain

$$V_p = \frac{gD_p^2(\rho_p - \rho)}{18\eta} = \frac{Q}{WL} \tag{4.19}$$

This is then rearranged to obtain the smallest particle removed by the gravity settling chamber:

$$D_p = \left[\frac{18\eta Q}{gWL(\rho_p - \rho)} \right]^{1/2} \tag{4.20}$$

Example 4.3

This example is adapted from Lapple. A gas stream with $Q = 0.55$ m³/s, consisting mostly of air at 70°C, is passed through a dust collector of dimensions $L = 6$ m, $W = 3.6$ m, and $H = 3$ m. The gas contains a mist of 62°Be sulfuric acid. What is the smallest diameter mist that can be removed?

Small mist droplets behave very much like small solid particles. Degrees Baumé is a scale for measuring specific gravity, and conversions can be found in standard references such as *The Chemical Engineers' Handbook*. The density of 62°Be H_2SO_4 is 1.75×10^3 kg/m³. The physical properties of air at 70°C are $\eta \approx 2 \times 10^{-5}$ Pa·s and $\rho \approx 1$ kg/m³ $\ll \rho_p$. Equation (4.20) then gives

$$D_p = \left[\frac{(18)(2 \times 10^{-5} \text{ Pa·s})(0.55 \text{ m}^3/\text{s})}{(9.8 \text{ m/s}^2)(3.6 \text{ m})(6 \text{ m})(1.75 \times 10^3 \text{ kg/m}^3)} \right]^{1/2}$$

$$= 2.3 \times 10^{-5} \text{ m} = 0.023 \text{ mm}$$

This is slightly below the nominal 0.042-mm diameter for efficient performance, so the device will probably operate just a bit less efficiently than designed.

In any calculation of this type it is important to calculate the Reynolds number to ensure that the hypothesis that Stokes' law applies is indeed valid. From Eq. (4.18),

$$V_p = \frac{0.55 \text{ m}^3/\text{s}}{(3.6 \text{ m})(6 \text{ m})} = 0.038 \text{ m/s}$$

Then

$$\text{Re} = \frac{D_p V_p \rho}{\eta} = \frac{(2.3 \times 10^{-5} \text{ m})(3.8 \times 10^{-2} \text{ m/s})(1 \text{ kg/m}^3)}{2 \times 10^{-5} \text{ Pa·s}} = 0.04 \ll 1$$

4.2.5 Sizing a Distillation Column

The diameter of a plate distillation column is chosen to provide an upward gas velocity that will not entrain excessive amounts of liquid. The liquid falls through the gas from tray to tray, and entrainment is the process by which liquid droplets are carried up with the gas stream.

To avoid entrainment it is necessary that the settling velocity of the droplet be at least equal to the upward gas velocity. The droplets will typically have a diameter of about 0.1 mm. We assume that the droplets behave more or less like solid spheres, and the usual assumption is that the liquid- and gas-phase properties are such that the particles fall in the Newton regime. The settling velocity is therefore given by Eq. (4.13), and this must equal the maximum upward gas velocity. Setting $C_D = 0.44$, $D_p = 10^{-4}$ m, and $g = 9.8$ m/s², we therefore obtain the design equation,

$$V_{\text{gas}} = 0.055 \left(\frac{\rho}{\rho_g} - 1 \right)^{1/2} \tag{4.21}$$

The coefficient in Eq. (4.21) used for actual design is affected by the plate spacing and the liquid level on the tray, but the value is relatively insensitive to both of these parameters for tray spacings greater than 0.75 m. The empirical values used in practice for bubble plates with wide spacings range from 0.052 to 0.059. Detailed values for bubble-plate and sieve-tray columns for a range of conditions are given in *Perry's Handbook*.

4.2.6 Wall Effects

The drag coefficient data in Sec. 4.2.1 are based on a sphere falling in an infinite fluid medium, and the same assumption, therefore, applies to all the subsequent results. In any real situation we must account for the possible influence of the container walls. Let us suppose, for example, that the sphere is falling along the axis of a cylinder of diameter D_c, as shown in Fig. 4-5.

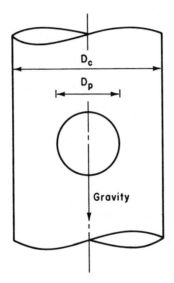

Figure 4-5. Schematic of a sphere falling along the axis of a cylinder.

Dimensional analysis would then lead to the relation

$$C_D = C_D\left(\text{Re}, \frac{D_p}{D_c}\right) \tag{4.22}$$

In the Stokes (inertialess) regime, this relation can be written

$$C_D = \frac{24}{\text{Re}}\phi\left(\frac{D_p}{D_c}\right) \tag{4.23}$$

The function $\phi(D_p/D_c)$ is shown in Fig. 4-6, together with the linear asymptotic relation for small D_p/D_c. When the sphere diameter is only one-tenth of the cylinder diameter, the wall correction is more than 20%, and the effect increases dramatically with increasing D_p/D_c. It is evident from the figure that the sphere must be at least 20 or so diameters from the nearest wall

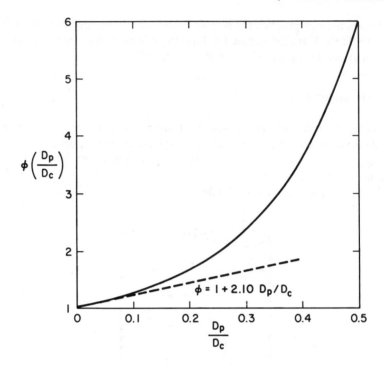

Figure 4-6. Correction to Stokes' law for a sphere falling in a cylinder of finite diameter.

$(D_p/D_c = 0.025)$ in order to be able to treat the fluid as unbounded with an error of no more than 5%.

Example 4.4

The experiment in Example 4.2 was carried out in a tube with a 25-mm diameter. What is the true fluid viscosity?

In that experiment we had $D_p = 3$ mm, so $D_p/D_c = \frac{3}{25} = 0.12$. From Fig. 4-6, then, $\phi \approx 1.3$. It is readily established that the correct viscosity is the viscosity computed from Eq. (4.14) divided by ϕ. Thus, $\eta = 0.6/1.3 = 0.5$ Pa·s.

APPENDIX 4.2.A

Buoyant Force

The buoyant force exerted on a submerged particle by the surrounding fluid is equal to the weight of fluid displaced by the solid. This is known as *Archimedes' theorem* and is usually proved in introductory courses in physics. We present an elementary proof here for completeness.

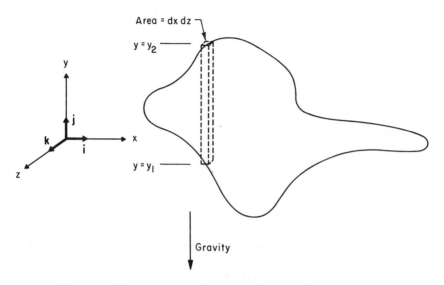

Figure 4-7. Breaking a body up into small prisms for calculation of the bouyant force.

The configuration is shown in Fig. 4-7. At any elevation y in a fluid of constant density, the pressure is

$$p = p_0 - \rho g y \tag{4.24}$$

where p_0 is the pressure at $y = 0$. (The minus sign indicates that the positive y direction is opposite to the direction of gravitational acceleration.) Following the usual approach in the calculus we divide the solid object up into a large number of very small rectangular prisms of upper and lower surface area $dx\,dz$ and height $y_2 - y_1$. One such prism is shown in the figure. The pressure at the upper face is $p_0 - \rho g y_2$, so the small force pushing down on that face (i.e., in the negative y direction) is

$$dF_2 = -(p_0 - \rho g y_2)\, dx\, dz \tag{4.25a}$$

Similarly, the force pushing up on the bottom face is

$$dF_1 = +(p_0 - \rho g y_1)\, dx\, dz \tag{4.25b}$$

The net force is the algebraic sum of the forces on the top and bottom:

$$dF_B = dF_1 + dF_2 = \rho g(y_2 - y_1)\, dx\, dz \tag{4.26}$$

But $(y_2 - y_1)\, dx\, dz$ is simply the volume $d\mho$ of the prism, so we may write

$$dF_B = \rho g\, d\mho \tag{4.27}$$

The total force is the sum (i.e., the integral) over all infinitesimal prisms making up the particle:

$$F_B = \int_{\mho} dF_B = \rho g \int_{\mho} d\mho = \rho g \mho \qquad (4.28)$$

where $\rho g \mho$ is the weight of fluid equal in volume to the volume of the submerged object. Note that the force is exerted in the positive y direction, opposite to the direction of gravitational acceleration. By an identical proof it can be shown that there is no contribution to the buoyant force directed in the xz plane.

APPENDIX 4.2.B

Transient Settling of a Sphere

In calculating the settling velocity of a sphere using, say, Stokes' law, it is helpful to have an estimate of the time or distance required for the initial transient to vanish. This can be done provided that we made one assumption and accept one additional relation without proof.

The configuration is as shown in Fig. 4-2. At time $t = 0$, the sphere is placed in the fluid with an initial velocity of zero. For $t > 0$, the sphere accelerates until it reaches the terminal velocity. Fluid around the sphere must also be accelerated from rest, resulting in a change in fluid momentum. This rate of change of fluid momentum shows up as an additional force acting on the sphere, pointing in the direction opposite to the motion of the sphere. Without proof we will use the result that this additional force has a magnitude equal to $(\pi \rho D_p^3/12)(dV_p/dt)$. Note that this is one-half the rate of change of momentum of a sphere of liquid moving at the same velocity as the solid sphere, but in the opposite direction. The quantity $\pi \rho D_p^3/12$ is sometimes called the *virtual mass*, since the net contribution of this term is to appear to increase the mass of the sphere.

Newton's second principle, that the rate of change of momentum of the sphere equals the sum of the imposed forces, can be written in words as

rate of change of momentum $=$ force of gravity $+$ buoyancy force
(negative direction) $\quad$ (negative direction) $\quad$ (positive direction)

$+$ drag force $+$ fluid inertia force
(positive direction) $\quad$ (positive direction)

Each of these terms can be expressed quantitatively except F_D. We will *assume* that the drag force on an accelerating sphere is the same as the drag force on a nonaccelerating sphere that is moving with the same velocity. Thus, F_D can be replaced by $\pi \rho V_p^2 D_p^2 C_D/8$, where C_D is given as a function of Re in Fig. 4-1. We can then write Newton's second principle as

$$-\frac{\pi\rho_p D_p^3}{6}\frac{dV_p}{dt} = -\frac{\pi\rho_p D_p^3}{6}g + \frac{\pi\rho D_p^3}{6}g + \frac{\pi}{8}\rho V_p^2 D_p^2 C_D + \frac{\pi\rho D_p^3}{12}\frac{dV_p}{dt} \quad (4.29)$$

For simplicity we will assume that $Re < 1$ and use Stokes' law, $C_D = 24/Re$. With some additional algebra, we can then write Eq. (4.29) as

$$\frac{dV_p}{dt} = \frac{18\eta}{(\rho_p + \frac{1}{2}\rho)D_p^2}\left[\frac{gD_p^2(\rho_p - \rho)}{18\eta} - V_p\right] \quad (4.30)$$

The first term in brackets is the steady-state settling velocity for $Re < 1$, given by Eq. (4.11); we will refer to it here as $V_{p,\infty}$. It is convenient to introduce a dimensionless time, $\theta = V_{p,\infty}t/D_p$, so that each unit of dimensionless time corresponds to the time required to move one sphere diameter at terminal velocity. Equation (4.30) is then written

$$\frac{dV_p}{d\theta} = \frac{18}{\left(\dfrac{\rho_p}{\rho} + \dfrac{1}{2}\right)Re}(V_{p,\infty} - V_p) \quad (4.31)$$

where $Re = D_p V_{p,\infty}\rho/\eta$ is based on the terminal velocity. The solution is

$$\frac{V_p}{V_{p,\infty}} = 1 - \exp\left[-\frac{18\theta}{\left(\dfrac{\rho_p}{\rho} + \dfrac{1}{2}\right)Re}\right] \quad (4.32)$$

The sphere will be within 5% of its terminal velocity when the argument of the exponential is -3; we will call this time θ_∞. Thus,

$$\theta_\infty = Re\left(\frac{\rho_p}{6\rho} + \frac{1}{12}\right) \quad (4.33)$$

For a solid sphere falling in a liquid, the ratio $\rho_p/6\rho$ will usually be of order unity. Since Re is required to be less than unity, the terminal velocity will be reached within a few sphere diameters at most, and if $Re \ll 1$, the terminal velocity will be attained almost instantaneously. In a gas, where $\rho_p/\rho \gg 1$, a long time may be required, depending on the value of the Reynolds number.

Example 4.5

Estimate the transient period for the particles in Example 4.3.
 In that example, $Re = 0.04$, $\rho_p = 1.75 \times 10^3$, $\rho = 1$, $V_{p,\infty} = 3.8 \times 10^{-2}$, and $D_p = 2.3 \times 10^{-5}$. Hence,

$$\theta_\infty = 0.04\left(\frac{1.75 \times 10^3}{6} + \frac{1}{12}\right) \approx 12$$

$$t_\infty = \frac{D_p}{V_{p,\infty}}\theta_\infty = \frac{2.3 \times 10^{-5}}{3.8 \times 10^{-2}} \times 12 \approx 10^{-2}\text{ s}$$

Note that the virtual mass term $(\frac{1}{12})$ is negligible when the fluid is a gas.

4.3 OTHER SUBMERGED OBJECTS

The ideas developed in the preceding sections can be readily extended to other particle geometries. The general definition of the drag coefficient for an object of any shape is

$$C_D = \frac{F_D}{\frac{1}{2}\rho V_p^2 A_p} \tag{4.34}$$

where A_p is the projection of the solid object on a plane normal to the direction of flow. For a sphere, A_p is a circle of diameter D_p, hence the factor $8/\pi$ in the definition (4.1b). For a long cylinder of diameter D_p and length L_p, with flow normal to the axis, the projected area is a rectangle of height D_p and length L_p. Thus, we have

$$\text{long cylinder:} \quad C_D = \frac{F_D/L_p}{\frac{1}{2}\rho V_p^2 D_p} \tag{4.35}$$

The drag coefficient for circular disks and infinite cylinders is shown in Fig. 4-8, with the sphere curve included for comparison. The flow is normal to the flat side of the disk. An "infinite" cylinder is a cylinder for which $L_p \gg D_p$; clearly, the force per unit length is the relevant quantity, which is the reason that Eq. (4.35) is expressed in terms of F_D/L_p. It is interesting to note that there is no inertialess region at low Reynolds number for infinite cylinders; the low-Reynolds-number asymptote is

$$\text{long cylinder, Re} < 1: \quad C_D = \frac{8\pi}{\text{Re}}\frac{1}{\ln(8/\text{Re}) - 0.077} \tag{4.36}$$

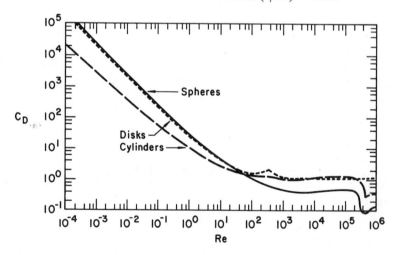

Figure 4-8. Drag coefficient as a function of Reynolds number for spheres, disks, and cylinders. Reproduced from Lapple and Shepherd, *Ind. Eng. Chem.*, **32**, 605 (1940), copyright by The American Chemical Society, by permission.

Drag coefficients have been computed for a number of other shapes. Equations and relevant references can be found in *Perry's Handbook*.

4.4 BEDS OF PARTICLES

4.4.1 Porous Media

Flow of a fluid through a porous solid bed occurs often in process applications, ranging from packed tubular reactors to the removal of oil from porous rock. We will concentrate here on the special case of porous beds that are made up of unconsolidated spherical particles. This includes many of the cases of practical interest, and the extension to other types of porous media is straightforward.

The most important parameters that characterize the bed are the particle diameter, D_p, and the *void fraction*, ϵ. The void fraction is the ratio of empty volume available for fluid to pass to the total volume of the bed. The void fraction depends on the manner of packing. All that follows assumes that the particles are all spheres of the same diameter. If there is a mixture of diameters, a mean diameter is defined by

$$D_p = \left(\sum \frac{X_n}{D_{pn}} \right)^{-1} \tag{4.37}$$

where X_n is the weight fraction of spheres with diameter D_{pn}. If the particles are nonspherical, an "equivalent sphere" is defined in a manner similar to the hydraulic diameter concept for pipes, with

$$D_p = 6 \times \frac{\text{volume of particle}}{\text{area of particle}} \tag{4.38}$$

4.4.2 Friction Factor-Reynolds Number Relation

The flow problem is shown schematically in Fig. 4-9. We wish to determine the pressure drop $|\Delta p|$ across a bed of cross-sectional area A and depth L when the flow rate of an incompressible Newtonian fluid is Q. Two observations can be made at once to simplify the analysis. First, it is evident that, all other things being equal, the flow rate is directly proportional to the bed area. Hence, only the flow rate per unit area of bed is relevant. This quantity is called the *superficial velocity* and is denoted v_∞:

$$v_\infty = \frac{Q}{A} \tag{4.39}$$

The superficial velocity is the velocity that would exist in the absence of particles. Second, all other things being equal, the pressure drop will increase in direct proportion to the bed depth. Hence, only $|\Delta p|/L$ is relevant.

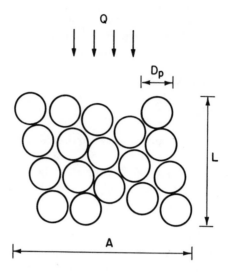

Figure 4-9. Schematic of flow through a bed of uniform spheres.

For purposes of dimensional analysis we must consider $|\Delta p|/L$, v_∞, D_p, ρ, η, and ϵ. We can then write

$$\frac{D_p}{\rho v_\infty^2}\frac{|\Delta p|}{L} = \text{function of} \left(\frac{D_p v_\infty \rho}{\eta}, \epsilon\right) \qquad (4.40)$$

This equation is not particularly useful because it says nothing about the dependence on ϵ. Thus, a series of experiments would have to be carried out over a range of void fractions. This is all the information that can be obtained from dimensional analysis.

We can achieve some insight into the dependence on ϵ by a less direct approach in which we construct an idealized model of a packed bed. The fluid follows a tortuous path through the bed, but we might visualize this tortuous path as being replaced by a cylinder with the same resistance to the motion. In that case we would treat the packed bed as though it were as shown in Fig. 4-10, a solid traversed by N_c cylinders of diameter D_{eff} (for "effective"). The velocity in each cylinder is v_{eff}, and we can write

$$\frac{D_{\text{eff}}}{\rho v_{\text{eff}}^2}\frac{|\Delta p|}{L} = \text{function of} \left(\frac{D_{\text{eff}} v_{\text{eff}} \rho}{\eta}\right) \qquad (4.41)$$

The problem now is to obtain D_{eff} and v_{eff} in terms of the measurable variables.

The total cross-sectional area of the N_c cylinders is ϵA, so the volumetric flow rate is

$$Q = \epsilon A v_{\text{eff}} \qquad (4.42a)$$

or

$$v_{\text{eff}} = \frac{1}{\epsilon}\frac{Q}{A} = \frac{v_\infty}{\epsilon} \qquad (4.42b)$$

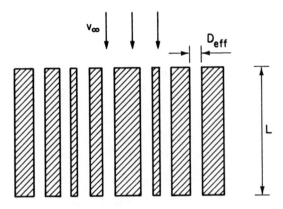

Figure 4-10. Model of a packed bed as a set of parallel cylinders of diameter D_{eff}.

The effective diameter can be computed by use of the *hydraulic diameter* concept for noncircular conduits, Sec. 3.6. From Eq. (3.39),

$$D_{eff} = \frac{4 \times \text{volume of fluid}}{\text{surface area wetted by fluid}} \qquad (4.43)$$

The volume of fluid is the void volume, while the wetted surface is the total surface area of the spheres. We note that

$$\frac{\epsilon}{1 - \epsilon} = \frac{\text{void volume/total volume}}{\text{solids volume/total volume}} = \frac{\text{void volume}}{\text{solids volume}} \qquad (4.44)$$

If there are N_p spherical particles, then the solids volume is $N_p \pi D_p^3/6$, and Eq. (4.44) can be rearranged to

$$\text{void volume} = \frac{\epsilon}{1 - \epsilon} N_p \frac{\pi D_p^3}{6} \qquad (4.45)$$

The surface area of the void space is

$$\text{surface area} = N_p \pi D_p^2 \qquad (4.46)$$

so that Eq. (4.43) can be written

$$D_{eff} = \frac{4 \dfrac{\epsilon}{1 - \epsilon} N_p \dfrac{\pi D_p^3}{6}}{N_p \pi D_p^2} = \frac{2}{3} D_p \frac{\epsilon}{1 - \epsilon} \qquad (4.47)$$

Substituting into Eq. (4.41) then gives

$$\frac{\frac{2}{3} D_p \dfrac{\epsilon^3}{1 - \epsilon}}{\rho v_\infty^2} \frac{|\Delta p|}{L} = \text{function of} \left(\frac{\dfrac{2}{3} \dfrac{D_p}{1 - \epsilon} v_\infty \rho}{\eta} \right) \qquad (4.48)$$

Comparison with Eq. (4.40) shows how the conceptual model incorporates the void fraction.

It is customary to define the *packed bed friction factor*, f_p, and *packed bed*

Reynolds number, Re_p, as

$$f_p = \frac{D_p \epsilon^3}{\rho v_\infty^2 (1 - \epsilon)} \frac{|\Delta p|}{L} \qquad (4.49a)$$

$$Re_p = \frac{D_p v_\infty \rho}{(1 - \epsilon)\eta} \qquad (4.49b)$$

Then Eq. (4.49) simply states that

$$f_p = f_p(Re_p) \qquad (4.50)$$

Experimental data are shown in Fig. 4-11. There is considerable scatter, but

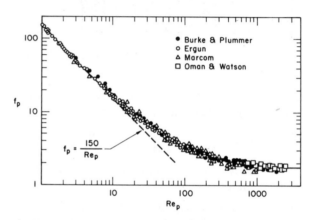

Figure 4-11. Friction factor as a function of Reynolds number for flow through granular packed beds. The solid line through the data is the Ergun equation, Eq. (4.51). Reproduced from Ergun, *Chem. Eng. Progr.*, **48**, 93 (1952), copyright from the American Institute of Chemical Engineers, by permission.

the data follow a single curve, indicating that the conceptual model does work quite well. The data correlate well over the entire range with the *Ergun equation*,

$$\text{Ergun equation:} \quad f_p = \frac{150}{Re_p} + 1.75 \qquad (4.51)$$

The second term can be neglected in the inertialess region, $Re_p \leq 10$, while the first term can be neglected in the inviscid Newton region, $Re_p \geq 1000$. It should be noted that the experimental scatter introduces an uncertainty of at least $\pm 20\%$ into pressure drop calculations.

Example 4.6

A sand pack is used to filter impurities from molten polyester downstream of the extruder and prior to spinning into filaments (see Fig. 1-2). The polymer at 280°C is approximately a Newtonian liquid with $\eta = 600\,\text{Pa·s}$ and $\rho = 1300\,\text{kg/m}^3$.

The sand pack is 38 mm in diameter and 16 mm in depth, and the mass flow rate is 5×10^{-4} kg/s. Estimate the pressure drop through the pack if the particles have a mean diameter of 0.7 mm and $\epsilon = 0.38$.

$$v_\infty = \frac{4(5 \times 10^{-4} \text{ kg/s})/(1300 \text{ kg/m}^3)}{\pi(38 \times 10^{-3} \text{ m})^2} = 3.4 \times 10^{-4} \text{ m/s}$$

Re_p will be much less than 10, so the last term in Eq. (4.51) can be neglected, and we have

$$|\Delta p| = \frac{150 L v_\infty \eta (1 - \epsilon)^2}{D_p^2 \epsilon^3} = \frac{150(16 \times 10^{-3} \text{ m})(3.4 \times 10^{-4} \text{ m/s})}{(0.7 \times 10^{-3} \text{ m})^2(0.38)^3}$$

$$= 7 \times 10^6 \text{ Pa}$$

4.4.3 Fluidized Beds

Fluidized bed reactors are used quite commonly in a number of processing applications. In a fluidized bed the solid particles move about chaotically in the gas (or liquid) stream. This includes a substantial amount of mixing and particle–particle and particle–wall contact. A fluidized bed is thus an efficient device for heat transfer, and reasonable temperature uniformity can be maintained. This is important in highly exothermic chemical reactions.

The manner in which a bed becomes fluidized can be understood by reference to Figs. 4-12 and 4-13. Gas or liquid is passed upward through a bed of solids at ever-increasing superficial velocity, v_∞. The pressure drop across the bed, $|\Delta p|$, is given by Eq. (4.49). For simplicity we assume that

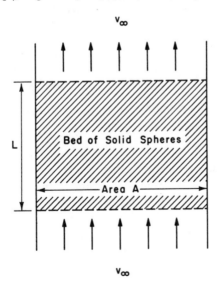

Figure 4-12. Schematic of a fluidized bed.

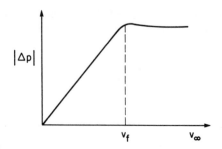

Figure 4-13. Typical curve of pressure drop as a function of superficial velocity for upward flow through a granular bed. v_f is the point of incipient fluidization.

$Re_p \leq 10$, which is generally true, in which case we can use the Ergun equation in the form

$$Re_p \leq 10: \quad \frac{|\Delta p|}{L} = \frac{150 v_\infty \eta (1 - \epsilon)^2}{D_p^2 \epsilon^3} \tag{4.52}$$

The pressure drop thus increases linearly with v_∞.

Now there is a net upward force on the bed of particles equal to $|\Delta p| A$. The volume of solid particles is $(1 - \epsilon)AL$, so the net gravitational and buoyant force on the solid particles is equal to $(1 - \epsilon)(\rho_p - \rho)ALg$. When these two forces are equal there is no net force on the solid particles; they are in a state of "weightlessness" and are free to move about unhindered by gravity. Thus, there will be no further increase in $|\Delta p|$ as v_∞ is increased. Rather, the bed will tend to expand, or become fluidized, and $|\Delta p|$ will level off.

We can calculate the minimum superficial velocity, v_f, at which the bed becomes fluidized by equating the two forces:

$$|\Delta p| A = (1 - \epsilon)(\rho_p - \rho)ALg \tag{4.53}$$

From Eq. (4.52) we can eliminate $|\Delta p|$ and solve for v_f,

$$v_f = \frac{(\rho_p - \rho)g D_p^2 \epsilon^3}{150 \eta (1 - \epsilon)} \tag{4.54}$$

This is known as the point of *incipient fluidization*. The void fraction at incipient fluidization is a function of the material and the particle size. If data are not available, then $\epsilon^3/(1 - \epsilon)$ may be roughly approximated by 0.091.

There is also an upper limiting velocity at which a fluidized bed can be operated. Individual particles will be carried up in the gas stream. Unless they settle back into the bed at a faster speed than the upward speed of the gas stream, the particles will have a net upward velocity and will be carried out of the bed. Thus, the maximum superficial gas velocity without particle entrainment, v_{max}, is equal to the settling velocity for a sphere. If we assume

Stokes' flow for simplicity, the maximum velocity is given by Eq. (4.11),

$$v_{\max} = \frac{(\rho_p - \rho)gD_p^2}{18\eta} \qquad (4.55)$$

It is readily established that $v_{\max}/v_f > 1$ for all physically possible values of ϵ.

The behavior of fluidized beds beyond the point of fluidization is quite complex. The bed no longer remains homogeneous, and "bubbles" of particle-free gas move through the system. For analytical purposes in reaction engineering it is often sufficient to treat the fluidized bed as a well-stirred tank containing a single homogeneous fluid phase.

Example 4.7

Pulverized coal is to be burned at atmospheric pressure in a fluidized bed. The density of the coal is approximately 1000 kg/m³. The mean particle diameter is 0.074 mm and the gas, mostly air, has a viscosity $\eta = 10^{-4}$ Pa·s. Estimate the minimum fluidization velocity.

The void fraction is not given, so we use the approximation $\epsilon^3/(1 - \epsilon) = 0.091$. The gas density can be neglected. In that case, Eq. (4.54) is

$$v_f = \frac{(1000)(9.8)(7.4 \times 10^{-5})^2(0.091)}{(150)(10^{-4})} = 3.2 \times 10^{-4} \text{ m/s}$$

The entrainment velocity, from Eq. (4.55), is

$$v_{\max} = \frac{(1000)(9.8)(7.4 \times 10^{-5})^2}{(18)(10^{-4})} = 3.0 \times 10^{-2} \text{ m/s}$$

The Reynolds numbers for the packed bed at the point of incipient fluidization and the free entrained particles are both less than unity.

4.5 CONCLUDING REMARKS

This concludes the introduction to the use of dimensional analysis and experimentation for the solution of engineering problems. The presentation should suggest the broad applicability of the methods to problems far beyond the scope of this chapter and the preceding one.

The drag coefficient is a quantity that appears frequently in process applications, and its definition should be committed to memory. So, too, should Stokes' law in any of its forms, and the limiting value $C_D = 0.44$ for a sphere, for it is important to be able to estimate orders of magnitude rapidly.

We turn next to the use of more fundamental principles in order to obtain quantitative descriptions that include a better understanding of the underlying physical processes. In later sections of the book we will return to some of the problems considered in this chapter and the preceding one, and we will show how certain of the results shown here as the outcome of experiments are, in fact, obtainable directly from first principles.

BIBLIOGRAPHICAL NOTES

There is a good discussion of much of the material in this chapter, with extensions and applications, in

LAPPLE, C. E., and coworkers, *Fluid and Particle Mechanics*, University of Delaware Press, Newark, Del., 1956.

There is also a very detailed discussion with much data in

ZENZ, F. A., and OTHMER, D. A., *Fluidization and Fluid-Particle Systems*, Van Nostrand Reinhold, New York, 1960.

Portions of Zenz and Othmer are beyond the scope of the presentation here, but most of the material can be read and appreciated. A more difficult text with some sections of interest to readers of this book is

KUNII, D., and LEVENSPIEL, O., *Fluidization Engineering*, John Wiley & Sons, Inc., New York, 1969.

One of the important applications of the material of this chapter is to filtration. There is a good discussion in Chapter 15 of

BENNETT, C. O., and MYERS, J. E., *Momentum, Heat, and Mass Transfer*, 2nd ed., McGraw-Hill Book Company, New York, 1974.

PROBLEMS

4.1. Design an experiment that will enable you to construct Fig. 4.1. You may need to choose several fluids, sphere sizes, and sphere materials in order to cover the entire Reynolds number range. Note that other investigators have used spheres made of nylon, steel, and ruby.

4.2. Find experimental data on drag coefficients for gas bubbles and immiscible Newtonian droplets in a Newtonian liquid, and determine the parameter range over which the solid sphere correlation can be used.

4.3. The use of a drag coefficient-Reynolds number plot to calculate either the terminal velocity or particle diameter of a falling sphere involves trial and error. Determine a way of plotting C_D-Re data in order to carry out the following calculations without trial and error, given physical properties of the fluid and the sphere:
a) Compute the particle diameter, given the terminal velocity.
b) Compute the terminal velocity, given the particle diameter.

4.4. A hollow steel sphere, 5.0 mm in diameter, with a mass of 5.0×10^{-5} kg, is released in a liquid of density $\rho = 900$ kg/m³ and attains a terminal velocity of 5.0 mm/s. Compute the viscosity of the liquid.

4.5. An incompressible Newtonian fluid of density ρ and viscosity η flows with velocity V normal to the axis of an "infinite" cylinder of diameter D. Denote the force exerted by the fluid per unit length of cylinder as F_D/L.
a) Use dimensional analysis to establish the general form between F_D/L and the other variables.

b) Prove that there is no analogue to Stokes' Law for this problem; i.e., it is impossible to find a region in which C_D is proportional to $1/\text{Re}$.

4.6. The process of filtration is shown schematically in Fig. 4P6. In *constant pressure* filtration the pressure difference $p_s - p_l$ is kept constant. Suppose that a slurry contains a volume fraction ϕ of essentially spherical particles of diameter D_p which pack into a filter cake with void fraction ϵ. Obtain a design equation for the height of the filter cake as a function of time. You should assume that the filter cloth is a porous medium whose pressure drop-velocity equation is

$$\Delta p = -K_f \eta v_\infty$$

and that the low Reynolds number asymptote applies at all times in the filter cake.

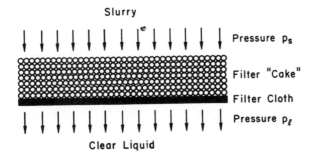

Slurry

Pressure p_s

Filter "Cake"

Filter Cloth

Pressure p_l

Clear Liquid

4.7. A bed of spheres of identical diameter D_p and density ρ_p is fluidized by a Newtonian liquid of density ρ and viscosity η. The fluidization is uniform; i.e., the void fraction ϵ is the same everywhere. Estimate the void fraction of the fluidized bed. The empirical Richardson-Zaki equation for the velocity V_{sw} of a uniform swarm of particles with respect to the surrounding fluid is expressed in terms of the Stokes settling velocity of a single sphere, V_{st}, as

$$V_{sw} = V_{st} \epsilon^n$$

$n = 4.65$ for $\text{Re} < 0.1$. Compare your result to the rule of thumb $\epsilon^3/(1 - \epsilon) = 0.091$.

4.8. A bypass-reactor system is shown in Fig. 4P8. The reactor is a tube of diameter D_R, packed with void fraction ϵ with uniform spherical catalyst of diameter D_p. All pipes have diameter D. Compute the fraction of the flow passing through the reactor. The Reynolds number is low in all cases, and you may neglect excess pressure drops associated with flow splits, bends, etc.

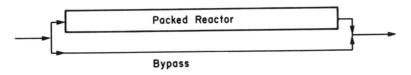

Packed Reactor

Bypass

Part III
Macroscopic Problems

Macroscopic Balances   **5**

5.1 INTRODUCTION

We now turn to a more fundamental analysis of fluid mechanics problems in order to obtain analytical relationships between the process variables. We will use the principles of conservation of mass, energy, and momentum to construct *mathematical models* of the flow. A mathematical model is a quantitative description of a physical process, and the development of an appropriate model requires an understanding of the relative importance of the many factors that can influence the process response. The outcome of a modeling study is a set of mathematical relations that describes the behavior of the process and which can be tested against experiment.

In this section of the text we will consider flows of process interest for which we can construct *macroscopic* models. These are flows in which we are interested only in overall process performance, and not in the detailed structure of the flow field. One important characteristic of such flows is a well-defined flow direction, usually in a conduit or a free jet; we are rarely interested in variations in any other spatial direction, so we deal with averages in directions normal to the flow. This is sometimes referred to as *lumping*.

This chapter is mostly concerned with development of the model equations, and Chapter 6 with applications.

5.2 CONTROL VOLUME AND CONSERVATION PRINCIPLE

The application of a conservation principle requires that we define a *control volume*. This is a region of space with well-defined boundaries where we can (in principle) monitor the flow in and out of the quantity that is being conserved. The conservation principle is then expressed as follows: *The rate of change of the conserved quantity within the control volume equals the rate at which the conserved quantity enters the control volume minus the rate at which the conserved quantity leaves the control volume.*

We restrict ourselves here to *one-dimensional* flows such as that shown schematically in Fig. 5-1. There is a single surface 1 at which fluid enters the control volume and a single surface 2 at which fluid leaves. The control volume consists of the entire conduit, bounded by the walls, between 1 and 2. The mean flow direction is parametrized by a distance variable z; z might be arc length along the centerline, for example. (It will be apparent in all that follows that the control volume need not be bounded by conduit walls in reality, as long as the boundary is a surface over which no fluid flows. Thus, the conduit walls could be replaced by the outer surface of a free liquid jet, for example.)

Let $\mathcal{C}Q$ refer to the conserved quantity (say, energy), and $\mathcal{C}q$ refer to the conserved quantity on a unit mass basis (say, energy per unit mass). Then $\rho\mathcal{C}q$ is the amount of the conserved quantity on a unit volume basis. We now compute the rate at which $\mathcal{C}Q$ enters and leaves the control volume by reference to Fig. 5-2.

Consider a small element of surface 1 or 2 with area dA. Let $d\mathbf{A}$ denote a vector* with magnitude dA and direction normal to the surface, pointing in

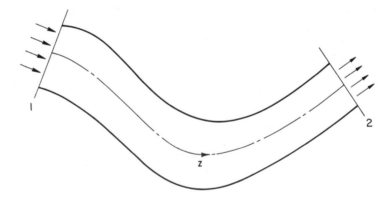

Figure 5-1. One-dimensional flow.

*Boldface type will always be used to denote vectors.

80

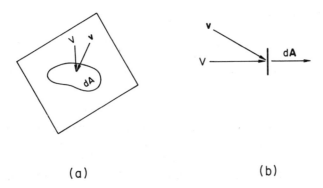

(a)   (b)

Figure 5-2. Differential area with velocity vector **v** and normal component V.

the direction of flow.† The fluid velocity near the surface element dA is **v**. **v** may vary from point to point over surface 1 or 2, and it need not be orthogonal to the surface. **v** can be decomposed into two orthogonal vectors, one normal to the surface and one parallel to it. The component of **v** normal to the surface is denoted V.

The differential volumetric flow rate through the small surface element dA is

$$d(\text{volumetric flow rate}) = \mathbf{v} \cdot d\mathbf{A} = V\,dA \qquad (5.1)$$

This relation follows from the fact that volumetric flow rate is simply velocity times area, and only that portion of the velocity vector that is normal to the surface can carry fluid across the surface.

The differential mass flow rate through the small surface element dA is the volumetric flow rate multiplied by the density:

$$d(\text{mass flow rate}) = \rho\mathbf{v} \cdot d\mathbf{A} = \rho V\,dA \qquad (5.2)$$

The differential flow rate of eQ across the small surface element dA is then the amount per unit mass, eq, times the differential mass flow rate:

$$d(\text{flow rate of } eQ) = \rho(eq)\mathbf{v} \cdot d\mathbf{A} = \rho(eq)V\,dA \qquad (5.3)$$

The total flow rate of eQ over the surface is the sum (integral) of the flows through all the differential areas:

$$\text{flow rate of } eQ = \int_{\text{surface}} \rho(eq)V\,dA \qquad (5.4)$$

It is often useful to define the *surface average*. Let ψ be any quantity that varies from position to position on the surface. We define the surface average,

---

†Note that this is not the normal vector usually used in courses in integral calculus, where the normal always points out from the volume. Here, the vector $d\mathbf{A}$ points *into* the control volume at surface 1. The corresponding vector at surface 2 points out of the control volume.

$\langle \psi \rangle$, by the relation

$$\langle \psi \rangle = \frac{1}{A} \int_{surface} \psi \, dA \qquad (5.5)$$

where A is the total cross-sectional area of the surface. Thus, Eq. (5.4) can be written

$$\text{flow rate of } \mathcal{C}\mathcal{Q} = \langle \rho(\mathcal{c}q)V \rangle A \qquad (5.6)$$

In writing the conservation equations it will also be necessary to have an expression for the total amount of conserved quantity in the control volume. The amount of $\mathcal{C}\mathcal{Q}$ in a differential volume element of area dA and length dz is $\rho(\mathcal{c}q) \, dA \, dz$. Thus, the total amount of $\mathcal{C}\mathcal{Q}$ in the control volume from surface 1 to surface 2 is

$$\mathcal{C}\mathcal{Q} \text{ in control volume} = \int_{z_1}^{z_2} \int_{surface} \rho(\mathcal{c}q) \, dA \, dz = \int_{z_1}^{z_2} \langle \rho(\mathcal{c}q) \rangle A \, dz \quad (5.7)$$

Here, z_1 and z_2 denote the positions of surfaces 1 and 2, respectively.

Example 5.1

A quantity ψ is distributed over a circular cross section of radius R according to the equation

$$\psi = \psi_m \left(1 - \frac{r^2}{R^2} \right)$$

where r is the radial coordinate and ψ_m is a constant. Compute $\langle \psi \rangle$.

Refer to Fig. 5-3. The differential area in cylindrical coordinates is $dA =$

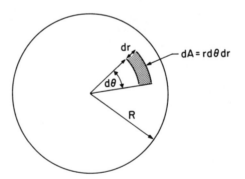

Figure 5-3. Differential area in polar coordinates.

$r \, d\theta \, dr$, and $A = \pi R^2$. Thus, from Eq. (5.5), we have

$$\langle \psi \rangle = \frac{1}{\pi R^2} \int_{r=0}^{R} \int_{\theta=0}^{2\pi} \psi_m \left(1 - \frac{r^2}{R^2} \right) r \, d\theta \, dr = \frac{2\pi \psi_m}{\pi R^2} \int_0^R \left(1 - \frac{r^2}{R^2} \right) r \, dr$$

$$= 2\psi_m \int_0^1 (1 - \xi^2) \xi \, d\xi = \tfrac{1}{2}\psi_m$$

The θ integration was included for generality, but for this problem we could have noted that ψ was independent of θ and used the differential ring $dA = 2\pi r \, dr$ as the starting point.

Example 5.2

A quantity ψ is distributed over the rectangle shown in Fig. 5-4 according to the

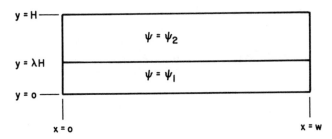

Figure 5-4. Rectangular cross section with different values of ψ in the two parts.

relation

$$\psi = \psi_1 \qquad 0 \leq y < \lambda H$$
$$\psi = \psi_2 \qquad \lambda H < y \leq H$$

where ψ_1 and ψ_2 are constants. Compute $\langle \psi \rangle$.

There is no variation in the x direction, so we may take the differential area to be $dA = W \, dy$ and write

$$\langle \psi \rangle = \frac{1}{WH} \int_{y=0}^{y=H} \psi W \, dy = \frac{W}{WH} \left\{ \int_{y=0}^{\lambda H} \psi_1 \, dy + \int_{\lambda H}^{H} \psi_2 \, dy \right\}$$

$$= \frac{W}{WH} [\psi_1 \lambda H + \psi_2 (H - \lambda H)] = \lambda \psi_1 + (1 - \lambda)\psi_2$$

5.3 CONSERVATION OF MASS

5.3.1 Basic Equation

When the conserved quantity is mass, $\mathscr{e}q = 1$ (mass/mass). We can then use Eqs. (5.4) and (5.7) to write

$$\frac{d}{dt} \int_{z_1}^{z_2} \langle \rho \rangle A \, dz = \langle \rho V \rangle_1 A_1 - \langle \rho V \rangle_2 A_2 \qquad (5.8)$$

Subscripts 1 and 2 refer to surfaces 1 and 2, respectively, so Eq. (5.8) simply states that the rate of change of mass in the control volume equals the rate at which mass enters over surface 1 minus the rate at which mass leaves over surface 2.

We sometimes use the symbol w for the mass flow rate over a surface,

$$w = \langle \rho V \rangle A \tag{5.9}$$

Equation (5.8) can then be written

$$\frac{d}{dt} \int_{z_1}^{z_2} \langle \rho \rangle A \, dz = w_1 - w_2 \tag{5.10}$$

Equation (5.8) is sometimes called the *continuity equation*.

Many flows of processing interest take place at *steady state*, when there is no change with time in the control volume. In this case, $d/dt = 0$ and Eq. (5.8) simplifies to

$$\text{steady state:} \quad \langle \rho V \rangle_1 A_1 = \langle \rho V \rangle_2 A_2 = w \tag{5.11a}$$

$$\text{steady state:} \quad w_1 = \quad\quad w_2 = w \tag{5.11b}$$

At steady state the mass flow rate is the same over all surfaces in the flow direction; in that case, we do not need a subscript on w to distinguish between positions.

5.3.2 Single Fluid

When we are dealing with a single fluid phase we can usually assume that the density does not change significantly over a cross section of the conduit; indeed, we shall restrict ourselves to such situations. We can then write

$$\text{single fluid:} \quad \int_{\text{area}} \rho(\mathbf{e}q) V \, dA \simeq \rho \int_{\text{area}} (\mathbf{e}q) V \, dA \tag{5.12a}$$

and

$$\text{single fluid:} \quad \langle \rho(\mathbf{e}q) V \rangle \simeq \rho \langle (\mathbf{e}q) V \rangle \tag{5.12b}$$

It further follows that $\langle \rho \rangle \simeq \rho$, and we can write the continuity equation, Eq. (5.8), as

$$\text{single fluid:} \quad \frac{d}{dt} \int_{z_1}^{z_2} \rho A \, dz = \rho_1 \langle V \rangle_1 A_1 - \rho_2 \langle V \rangle_2 A_2 \tag{5.13}$$

(Note that ρ might change with position along the flow direction, even though it is a constant over any cross section.)

If the fluid is *incompressible*, which means that the density does not change at all, then $\rho_1 = \rho_2 = \rho$. For cases in which the volume is a constant, we then have $d/dt \left(\int_{z_1}^{z_2} A \, dz \right) = 0$, and the continuity equation simplifies to

$$\text{incompressible:} \quad \langle V \rangle_1 A_1 = \langle V \rangle_2 A_2 \tag{5.14a}$$

or

$$\text{incompressible:} \quad \frac{\langle V \rangle_1}{\langle V \rangle_2} = \frac{A_2}{A_1} \tag{5.14b}$$

This is simply a quantitative expression of the intuitive notion that if the area decreases, the velocity must increase to move a comparable amount of fluid.

5.4 CONSERVATION OF ENERGY

5.4.1 Basic Equation

The principle of conservation of energy states that the rate of change of total energy in the control volume equals the rate at which energy enters the control volume by flow, minus the rate at which energy leaves by flow, plus the rate at which heat is added through the boundaries, plus the rate at which work is done on the fluid in the control volume (the power input).* This is sometimes called the *first law of thermodynamics* for a flowing system.

We take the total energy to be the sum of internal energy, potential energy, and kinetic energy. Other forms of energy may be important in some applications and must be included. Internal energy per unit mass is denoted e. Kinetic energy per unit mass is $v^2/2$, where v is the magnitude of the velocity vector, **v**. Note that the fluid kinetic energy does not depend on the direction of flow, and v does not equal V except in the special case in which the flow is directed normal to the surface at every point. Potential energy per unit mass is gh, where h is the height above an arbitrary datum and g is the gravitational acceleration. Thus,

$$eq = e + \tfrac{1}{2}v^2 + gh \tag{5.15}$$

It is convenient to divide the rate of doing work into two parts. *Flow work* is the work required to move the fluid into and out of the control volume, and *shaft work* is all other work done on the fluid, such as the work required to turn a stirrer. We have already briefly discussed the flow work in Sec. 3.4.1, and the development is identical here except that it is applied to the differential area dA and then integrated over the entire surface. The rate of doing flow work to move fluid into the control volume is $\langle pV \rangle_1 A_1$, and the rate of doing work to move fluid out is $-\langle pV \rangle_2 A_2$. The minus sign is required on the latter term because the fluid is doing work on the surroundings. Thus, the equation of conservation of energy is

$$\frac{d}{dt} \int_{z_1}^{z_2} \langle \rho(e + \tfrac{1}{2}v^2 + gh) \rangle A\, dz$$
$$= \langle \rho(e + \tfrac{1}{2}v^2 + gh)V \rangle_1 A_1 - \langle \rho(e + \tfrac{1}{2}v^2 + gh)V \rangle_2 A_2 \tag{5.16}$$
$$+ \langle pV \rangle_1 A_1 - \langle pV \rangle_2 A_2 + \dot{Q}_H + \dot{W}_S$$

*There are two conventions in common use regarding the work term in the energy equation. We will use the convention that work is positive when done on the system by the surroundings.

where $\dot{Q}_H$ and $\dot{W}_S$ represent the rate at which heat is added and the rate at which shaft work is done, respectively. $\dot{Q}_H$ could, in principle, include a contribution from heat conduction across surfaces 1 and 2, but such terms are rarely important and we will presume that all heat transfer is across the surfaces of the control volume through which there is no flow.

5.4.2 Simplifying Assumptions

For most applications, Eq. (5.16) can be simplified considerably. It is important to note each of the simplifying assumptions carefully, because situations might arise in which one of the assumptions made here will not be applicable.

Assumption 1: Steady State. We assume that nothing changes with time, in which case $d/dt = 0$. Eq. (5.16) then simplifies to

$$\langle \rho e V \rangle_1 A_1 + \tfrac{1}{2}\langle \rho v^2 V \rangle_1 A_1 + \langle \rho g h V \rangle_1 A_1$$
$$- \langle \rho e V \rangle_2 A_2 - \tfrac{1}{2}\langle \rho v^2 V \rangle_2 A_2 - \langle \rho g h V \rangle_2 A_2 \qquad (5.17)$$
$$+ \langle p V \rangle_1 A_1 - \langle p V \rangle_2 A_2 + \dot{Q}_H + \dot{W}_S = 0$$

We have expanded some of the terms for ease of future manipulation by noting that $\langle a + b \rangle = \langle a \rangle + \langle b \rangle$ (i.e., the average of a sum equals the sum of the averages).

It should be noted that the steady-state assumption can be applied under *quasi-steady-state* conditions, in which the rate of change term is small compared to other terms in the equation but is not identically zero.

Assumption 2: Single phase, uniform properties. If there is one fluid phase with uniform properties, the density does not change over a cross section. For example,

$$\langle \rho v^2 V \rangle \simeq \rho \langle v^2 V \rangle \qquad (5.18)$$

The internal energy per unit mass in a nonreacting system depends only on the density and temperature. If we further assume that the temperature is uniform at each cross section, it follows that e does not vary over the cross section; thus,

$$\langle \rho e V \rangle \simeq \rho e \langle V \rangle \qquad (5.19)$$

Assumption 3: Uniform equivalent pressure. We assume that the pressure is the same over the entire cross section, except for a variation resulting from the weight of fluid above:

$$\langle (p + \rho g h) V \rangle \simeq (p + \rho g h) \langle V \rangle \qquad (5.20)$$

We refer to the sum $p + \rho g h$ as the *equivalent pressure*. This is the same as assuming that the pressure has the same properties in motion as in a static fluid. We will show later that the assumption is exact when all fluid motion is parallel to straight conduit walls. The assumption may break down if the sides of a conduit have significant curvature, because the curvilinear motion of the fluid will have a centrifugal contribution to the pressure that is not uniform in all directions but is directed toward the center of curvature.

It is important to note that we have *not* assumed that the pressure is a constant over the cross section. In open channel flows with an air–liquid interface at the top and a conduit wall at the bottom, the pressure variation with height at each cross section is a dominant factor in the flow behavior. In closed conduits there will rarely be any error in taking both p and h as constants at any cross section, thus ignoring small changes in each term from the top to the bottom of the conduit. *All equations that follow do assume that there is a single pressure and a single height that are characteristic of each cross section; the equations must be modified for open channel flow by always considering only the sum $p + \rho g h$, not the individual terms.*

With the three simplifying assumptions, and incorporating Eqs. (5.18) to (5.20) into Eq. (5.17), we can rewrite the energy balance as

$$e_1(\rho_1\langle V\rangle_1 A_1) + \frac{1}{2}\frac{\langle v^2 V\rangle_1}{\langle V\rangle_1}(\rho_1\langle V\rangle_1 A_1) + gh_1(\rho_1\langle V\rangle_1 A_1)$$

$$- e_2(\rho_2\langle V\rangle_2 A_2) - \frac{1}{2}\frac{\langle v^2 V\rangle_2}{\langle V\rangle_2}(\rho_2\langle V\rangle_2 A_2) - gh_2(\rho_2\langle V\rangle_2 A_2) \quad (5.21)$$

$$+ \frac{p_1}{\rho_1}(\rho_1\langle V\rangle_1 A_1) - \frac{p_2}{\rho_2}(\rho_2\langle V\rangle_2 A_2) + \dot{Q}_H + \dot{W}_S = 0$$

Some terms have been multiplied and divided by the same quantity in order to have each of the flow terms multiplied by $\rho_1\langle V\rangle_1 A_1$ or $\rho_2\langle V\rangle_2 A_2$. At steady state these quantities are both equal to the mass flow rate, w; see Eq. (5.11). Thus, we may divide each term by w to obtain the considerably simplified form:

$$e_1 + \frac{1}{2}\frac{\langle v^2 V\rangle_1}{\langle V\rangle_1} + gh_1 + \frac{p_1}{\rho_1} - e_2 - \frac{1}{2}\frac{\langle v^2 V\rangle_2}{\langle V\rangle_2} - gh_2 - \frac{p_2}{\rho_2} = -\frac{\dot{Q}_H}{w} - \frac{\dot{W}_S}{w}$$

$$(5.22)$$

One further simplification in nomenclature is useful. $\dot{Q}_H/w$ and $\dot{W}_S/w$ represent the heat and shaft work added to the system at steady state per unit mass of fluid contained in the control volume. We define new symbols for these ratios as follows:

$$\delta Q_H = \frac{\dot{Q}_H}{w} \quad (5.23)$$

$$\delta W_S = \frac{\dot{W}_S}{w} \quad (5.24)$$

Equation (5.22) is then written, finally, as

$$\Delta\left(e + \frac{1}{2}\frac{\langle v^2 V\rangle}{\langle V\rangle} + gh + \frac{p}{\rho}\right) = \delta Q_H + \delta W_S \quad (5.25)$$

The operator Δ means "evaluated at z_2 minus evaluated at z_1."

Example 5.3

Compute the temperature rise for adiabatic flow of a nonreacting incompressible fluid in a horizontal pipe of uniform cross section, and the heat removal required to keep the flow isothermal.

We have done this calculation in a sketchy manner in Sec. 3.4.2. If the pipe is horizontal, $\Delta h = 0$. If the cross section is uniform, we may expect that the velocity distribution will be essentially the same at both ends and $\Delta(\langle v^2 V \rangle / \langle V \rangle) = 0$. For a nonreacting incompressible fluid, $\Delta e = c_V \Delta T$, where c_V is heat capacity at constant volume per unit mass and T is temperature. If we assume that no shaft work is done, Eq. (5.25) becomes

$$c_V \Delta T + \frac{\Delta p}{\rho} = \delta Q_H$$

If the flow is adiabatic, $\delta Q_H = 0$, and we recover Eq. (3.23),

$$\Delta T = \frac{-\Delta p}{\rho c_V}$$

(Δp is negative, since the pressure at z_2 is less than at z_1.)

If we wish to keep the system isothermal, $\Delta T = 0$ and the heat that must be removed is

$$\delta Q_H = \frac{\Delta p}{\rho}$$

Heat removal is negative in algebraic sign.

5.4.3 Velocity Averages

Each term in Eq. (5.25) is expressed in a form that is convenient for measurement and computation except the velocity term. It is conventional to introduce one further piece of nomenclature,

$$\alpha = \frac{\langle v^2 V \rangle}{\langle V \rangle^3} \tag{5.26}$$

Equation (5.25) can then be written in the simpler form

$$\Delta \left(e + \frac{\alpha}{2} \langle V \rangle^2 + gh + \frac{p}{\rho} \right) = \delta Q_H + \delta W_S \tag{5.27}$$

This form is somewhat more compact since it involves only the average velocity, but unless we can estimate α we cannot obtain a solution.

In most applications, it is acceptable to presume that the flow is normal to the surface at the entrance and exit of the control volume, in which case $\mathbf{v}$ and $d\mathbf{A}$ are collinear and $v = V$. In that case Eq. (5.26) becomes

$$v = V: \quad \alpha = \frac{\langle V^3 \rangle}{\langle V \rangle^3} \tag{5.28}$$

The ratio of the average of the cube to the cube of the average is not unity unless V is uniform over the entire cross section. This is approximately true for turbulent pipe flow, and here $\alpha \simeq 1.07$, for which we usually substitute unity. For laminar flow in a round pipe, $\alpha = 2.0$.

Example 5.4

The velocity distribution in laminar pipe flow is found empirically in Sec. 6.6 to be

$$V = V_m\left(1 - \frac{r^2}{R^2}\right)$$

Compute α.
   We have already shown in Example 5.1 that $\langle V \rangle = V_m/2$, so $\langle V \rangle^3 = V_m^3/8$.

$$\langle V^3 \rangle = \frac{1}{\pi R^2} \int_0^R \left[V_m\left(1 - \frac{r^2}{R^2}\right)\right]^3 2\pi r \, dr = V_m^3 \int_0^1 (1 - \xi^2)^3 \xi \, d\xi = \frac{V_m^3}{4}$$

Thus,

$$\alpha = \frac{V_m^3/4}{V_m^3/8} = 2$$

Example 5.5

A liquid flows with the following velocity distribution in a pipe:

$$V = V_1 \qquad 0 \le r < \lambda R$$
$$V = V_2 \qquad \lambda R < r \le R$$

Compute α.

$$\langle V \rangle = \frac{1}{\pi R^2}\left\{\int_0^{\lambda R} V_1 2\pi r \, dr + \int_{\lambda R}^R V_2 2\pi r \, dr\right\} = \lambda^2 V_1 + (1 - \lambda^2)V_2$$

$$\langle V^3 \rangle = \frac{1}{\pi R^2}\left\{\int_0^{\lambda R} V_1^3 2\pi r \, dr + \int_{\lambda R}^R V_2^3 2\pi r \, dr\right\} = \lambda^2 V_1^3 + (1 - \lambda^2)V_2^3$$

$$\alpha = \frac{\langle V^3 \rangle}{\langle V \rangle^3} = \frac{\lambda^2 V_1^3 + (1 - \lambda^2)V_2^3}{[\lambda^2 V_1 + (1 - \lambda^2)V_2]^3}$$

5.4.4 Engineering Bernoulli Equation

The most useful form of the energy equation is one that replaces the internal energy term in Eq. (5.27) with a thermodynamic equivalent. We will require the use of some basic thermodynamics, including the notion of *entropy*. Entropy is associated with irreversibility, and the transformation of energy into forms which cannot be converted to useful work. Those who have not had a course in engineering thermodynamics or physical chemistry may wish to skip the next few steps and continue with Eq. (5.34), which contains a nonnegative term l_v that accounts for the viscous losses in the system. The viscous losses were first introduced in Sec. 3.4.2. We shall show subsequently that the same equation can be derived directly from the equation for conservation of momentum, although the meaning of the losses term is not as clearly brought out as it is when the energy equation is used.

Because thermodynamic relationships are usually given in differential form, it is helpful to consider the case in which the surfaces 1 and 2 are only a differential distance apart. In that case, the difference operator Δ is replaced by the differential operator d, and Eq. (5.27) is written

$$de + \tfrac{1}{2} d(\alpha\langle V\rangle^2) + g\,dh + d\left(\frac{p}{\rho}\right) = dQ_H + dW_S \qquad (5.29)$$

(We have replaced δQ_H and δW_S by dQ_H and dW_S, respectively, since the heat and shaft work are now also differential quantities.)

The internal energy per unit mass can be written in terms of the *entropy per unit mass, s,* as

$$de = T\,ds - pd\left(\frac{1}{\rho}\right) \qquad (5.30a)$$

or, equivalently,

$$de = T\,ds - d\left(\frac{p}{\rho}\right) + \frac{1}{\rho}\,dp \qquad (5.30b)$$

Thus, Eq. (5.29) can be written

$$(T\,ds - dQ_H) + \tfrac{1}{2}d(\alpha\langle V\rangle^2) + gdh + \frac{1}{\rho}\,dp = dW_S \qquad (5.31)$$

A fundamental postulate of thermodynamics, sometimes called the *second law of thermodynamics,* states that the entropy differential in any real (irreversible) process satisfies the inequality

$$T\,ds - dQ_H \equiv dl_V \geq 0 \qquad (5.32)$$

The losses per unit mass, l_V (V is for "viscous losses"), stem from small-scale friction within the fluid, and they are positive for all real fluids. dl_V equals zero only for a reversible process; flow can be truly reversible only if the fluid has no viscosity.

Introducing the viscous losses, Eq. (5.31) becomes

$$\tfrac{1}{2}d(\alpha\langle V\rangle^2) + g\,dh + \frac{1}{\rho}\,dp = dW_S - dl_V \qquad (5.33)$$

Integrating from z_1 to z_2, we then obtain the final working equation, sometimes known as the *engineering Bernoulli equation,*

$$\frac{\alpha_2}{2}\langle V\rangle_2^2 + gh_2 = \frac{\alpha_1}{2}\langle V\rangle_1^2 + gh_1 - \int_{p_1}^{p_2} \frac{dp}{\rho} + \delta W_S - l_V \qquad (5.34)$$

p_1 and p_2 are the pressures at z_1 and z_2, respectively. Equation (5.34) is also sometimes called the *mechanical energy balance,* since every term is associated with a mechanical (i.e., nonthermal) form of energy. The name is an unfortunate one, because mechanical energy is not conserved; the losses term represents the nonconservative nature of the equation. We return to this point in Sec. 5.5.4.

Two special cases are of interest. For an ideal gas, $\rho = M_w p / R_g T$, where M_w is the molecular weight and R_g is the gas constant. If the temperature is constant between z_1 and z_2, then

$$\text{isothermal, ideal gas:} \quad \int_{p_1}^{p_2} \frac{dp}{\rho} = \frac{R_g T}{M_w} \int_{p_1}^{p_2} \frac{dp}{p} = \frac{R_g T}{M_w} \ln \frac{p_2}{p_1} \qquad (5.35)$$

If the fluid is incompressible, ρ does not change between z_1 and z_2 and

$$\text{incompressible:} \quad \int_{p_1}^{p_2} \frac{dp}{\rho} = \frac{1}{\rho} \int_{p_1}^{p_2} dp = \frac{p_2 - p_1}{\rho} \qquad (5.36)$$

Example 5.6

Estimate the conditions under which isothermal flow of an ideal gas can be treated as incompressible.

The pressure term in Eq. (5.35) can be written

$$\int_{p_1}^{p_2} \frac{dp}{\rho} = \frac{R_g T}{M_w} \ln \left(1 + \frac{p_2 - p_1}{p_1} \right) = \frac{R_g T}{M_w} \left[\frac{p_2 - p_1}{p_1} - \frac{1}{2} \left(\frac{p_2 - p_1}{p_1} \right)^2 + \cdots \right]$$

$$= \frac{R_g T}{M_w p_1} (p_2 - p_1) \left(1 - \frac{p_2 - p_1}{2 p_1} + \cdots \right) = \frac{p_2 - p_1}{\rho_1} \left(1 - \frac{p_2 - p_1}{2 p_1} + \cdots \right)$$

Here, we have made use of the fact that $\ln (1 + x) = x - \frac{1}{2} x^2 + \ldots$, and $\rho_1 = M_w p_1 / R_g T$. We thus recover the pressure change term for an incompressible fluid, Eq. (5.36), with a correction term $1 - (p_2 - p_1)/2p_1$. The correction term will be negligible as long as $|p_2 - p_1| \ll 2p_1$. Thus, pressure changes that are 10 to 20% of the mean pressure can be ignored, and the fluid can be treated as incompressible.

5.4.5 Equivalent Heads

The Bernoulli equation is sometimes written in terms of *equivalent heads*. Consider, for simplicity, the special case of an incompressible fluid. If we divide Eq. (5.34) by g, we have

$$\frac{\alpha_2}{2g} \langle V \rangle_2^2 + h_2 + \frac{p_2}{\rho_2 g} = \frac{\alpha_1}{2g} \langle V \rangle_1^2 + h_1 + \frac{p_1}{\rho_1 g} + \frac{\delta W_s}{g} - \frac{l_v}{g} \qquad (5.37)$$

Each term in Eq. (5.37) has units of height, and hence each is equivalent to a head of liquid. The quantity $\langle V \rangle^2 / 2$ is often called a *velocity head*. A good rule of thumb is that the losses in turbulent flow in a pipe 50 diameters long are approximately equal to one velocity head.

Inspection of Eq. (5.37) makes it clear that fluid motion can be carried out by a number of equivalent mechanisms; there is an elevation change that has the same effect as a given pressure change, for example, and there is an equivalent input of shaft work, perhaps through a pump.

5.4.6 Pipeline Losses

The losses in a piping network can be broken down into losses in the straight lengths of pipe and losses in the fittings (expansions, curves, etc.). The losses in the straight pipe can be computed from information which we already have available. Consider a horizontal pipe ($h_2 = h_1$) of constant cross section. If the fluid is incompressible, then from the continuity equation, $\langle V \rangle_2 = \langle V \rangle_1$. We assume that no shaft work is done in the segment of interest, so the engineering Bernoulli equation simplifies to

$$\frac{p_1 - p_2}{\rho} = l_V \tag{5.38}$$

The pressure difference can be written in terms of the friction factor, Eq. (3.6):

$$f = \frac{p_1 - p_2}{2\rho\langle V \rangle^2} \frac{D}{L} \tag{3.6}$$

so that the losses can be expressed as

$$l_V = \frac{2\langle V \rangle^2 L f}{D} \tag{5.39}$$

Losses in fittings are usually recorded in terms of velocity heads. A small number of fittings losses can be computed theoretically, as we shall see subsequently, but most are determined by experiment. If K_f is the number of velocity heads lost by flow through the fitting, the fittings loss can be written

$$l_V = \tfrac{1}{2}\langle V \rangle^2 K_f \tag{5.40}$$

A small number of fittings losses for turbulent flow are recorded in Table 5-1. These are probably good to about $\pm 30\%$. More extensive tabulations may be found in handbooks such as *Perry's Handbook* and in the Crane Co. publication cited in the Bibliographical Notes.

If we wish to compute the complete losses term for the Bernoulli equation in a pipeline network, we can now simply write

$$l_V = \sum_{\substack{\text{lengths} \\ \text{of pipe}}} \frac{2\langle V \rangle_i^2 L_i f_i}{D_i} + \sum_{\text{fittings}} \tfrac{1}{2}\langle V \rangle_i^2 K_{fi} \tag{5.41}$$

The subscript i refers to the lengths of pipe and fittings which are being summed over. If there is a change in cross section at the fitting, then $\langle V \rangle_i$ refers to the *downstream* velocity.

TABLE 5-1

LOSSES IN FITTINGS AND VALVES FOR TURBULENT FLOW[a]

Fitting or valve	Velocity heads lost, K_f
90° elbow, standard	0.75
90° elbow, square	1.3
Coupling	0.04
Gate valve	
Open	0.17
Half-open	4.5
Globe valve, bevel seat	
Open	6.4
Half-open	9.5
Sudden expansion	$\left(\dfrac{A_2}{A_1} - 1\right)^2$

Sudden contraction	$\left(\dfrac{2}{m} - \dfrac{A_2}{A_1} - 1\right)^2$

m is the root of the quadratic

$$\frac{1 - m(A_2/A_1)}{1 - (A_2/A_1)^2} = \left(\frac{m}{1.2}\right)^2$$

Rounded entrance	0.05

[a]The result for the sudden expansion is derived in Sec. 6.2. The result for the sudden contraction is from Martin, *Chem. Eng. Educ.*, Summer 1974, p. 138. Other values are from *Perry's Handbook*.

Example 5.7

A liquid is pumped through a 50-mm-diameter smooth pipe between two tanks at a rate of 3 kg/s in the section of the process stream shown in Fig. 5-5. The liquid has properties $\rho = 10^3$ kg/m³, $\eta = 10^{-3}$ Pa·s. The pressure above the liquid is the same in each tank. Compute the pumping power required.

First consider the sections of straight pipe.

$$\langle V \rangle = \frac{4w}{\pi D^2 \rho} = \frac{(4)(3 \text{ kg/s})}{\pi(50 \times 10^{-3} \text{ m})^2(10^3 \text{ kg/m}^3)} = 1.5 \text{ m/s}$$

$$\text{Re} = \frac{D\langle V \rangle \rho}{\eta} = \frac{(50 \times 10^{-3} \text{ m}) (1.5 \text{ m/s})(10^3 \text{ kg/m}^3)}{10^{-3}\text{Pa·s}} = 7.5 \times 10^4$$

From Fig. 3-1 or Eq. (3.10),

$$f = 0.079\text{Re}^{-1/4} = 4.8 \times 10^{-3}$$

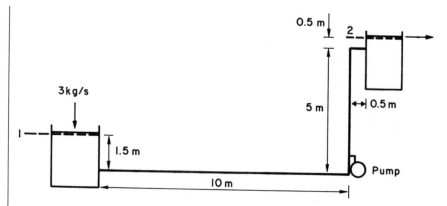

Figure 5-5. Schematic of a process stream. The pipe is smooth with a 50 mm diameter. The pressure above the liquid in each tank is the same.

There are four fittings terms. At the entrance to the first pipe, $A_2/A_1 \rightarrow 0$ and $K_f \cong 0.5$. For each right-angle bend, $K_f = 0.75$. At the exit from the second pipe, $A_2/A_1 \rightarrow \infty$ and $\langle V \rangle_2 \rightarrow 0$, leading to an indeterminate form. This can be resolved by noting that, for an incompressible fluid,

$$\left(\frac{A_2}{A_1} - 1\right)^2 \langle V \rangle_2^2 = \left(1 - \frac{A_1}{A_2}\right)^2 \langle V \rangle_1^2 \xrightarrow[A_1/A_2 \rightarrow 0]{} \langle V \rangle_1^2$$

Thus, in terms of the velocity in the *pipe*, $K_f = 1$ at the final expansion.

The pressure is the same at planes 1 and 2, since the pressure above the fluid is the same in each tank. The net elevation difference is 4 m. The velocity can be neglected at planes 1 and 2 relative to the flow in the pipe, so we have, from Eq. (5.34),

$$W_S = w\,\delta W_S = wg(h_2 - h_1) + wl_v$$

$$= wg(h_2 - h_1)$$

$$+ w\left[\frac{2\langle V \rangle^2 (L_1 + L_2 + L_3)f}{D} + \tfrac{1}{2}\langle V \rangle^2 (K_{f1} + K_{f2} + K_{f3} + K_{f4})\right]$$

$$= (3 \text{ kg/s})(9.8 \text{ m/s}^2)(4 \text{ m}) +$$

$$(3 \text{ kg/s})\left[\frac{2(1.5 \text{ m/s})^2(10 + 5 + 0.5 \text{ m})(4.8 \times 10^{-3})}{50 \times 10^{-3} \text{ m}}\right.$$

$$\left. + \tfrac{1}{2}(1.5 \text{ m/s})^2(0.5 + 0.75 + 0.75 + 1.0)\right] = \underset{\text{(elevation)}}{118} + \underset{\text{(pipe)}}{20} + \underset{\text{(fittings)}}{10}$$

$$= 148 \text{ W}$$

Note that the losses satisfy the rule of thumb of one velocity head for each 50 diameters of length, and that the major effect is the elevation change. (148 W is about 0.2 hp, for those more accustomed to the latter unit.) A somewhat larger pump would be needed to account for the fact that the pump will not operate at 100% efficiency.

5.5 CONSERVATION OF LINEAR MOMENTUM

5.5.1 Basic Equation

The principle of conservation of linear momentum states that the rate of change of linear momentum in the control volume equals the rate at which linear momentum enters by flow, minus the rate at which linear momentum leaves by flow, plus the sum of all forces acting on the system. *Linear momentum is a vector quantity.*

In writing the momentum equation, it is helpful to restrict the surfaces 1 and 2 over which fluid flows to be planes. We can then define the vector $\mathbf{A}$ as

$$\mathbf{A} = \int_{\text{surface}} d\mathbf{A} \tag{5.42}$$

That is, $\mathbf{A}$ is a vector with magnitude equal to the total surface area and direction normal to the planar surface in the direction of mean flow. We will retain Assumptions 2 and 3 of Sec. 5.4.2: a single phase with uniform physical properties and uniform equivalent pressure* at each cross section.

Linear momentum per unit mass is simply the velocity vector, $\mathbf{v}$. The conservation equation for a single fluid phase is therefore

$$\frac{d}{dt} \int_{z_1}^{z_2} \rho \langle \mathbf{v} \rangle A \, ds = \rho_1 \langle \mathbf{v} V \rangle_1 A_1 - \rho_2 \langle \mathbf{v} V \rangle_2 A_2 + p_1 \mathbf{A}_1$$
$$- p_2 \mathbf{A}_2 - \mathbf{F} + \left(\int_{z_1}^{z_2} \rho A \, dz \right) \mathbf{g} \tag{5.43}$$

$\mathbf{F}$ is the net force exerted *by the fluid* on the surrounding solid surface; hence, it appears with a negative sign. $p_1 \mathbf{A}_1$ is the force exerted by fluid outside the the control volume on fluid inside the control volume at z_1, and $-p_2 \mathbf{A}_2$ is the corresponding force at z_2. The negative sign associated with the latter stems from the fact that the force points into the control volume (to the left), while the positive $\mathbf{A}_2$ direction is out of the control volume (to the right). $\mathbf{g}$ is the gravitational acceleration, directed toward the center of the earth. The coefficient of $\mathbf{g}$ is simply the total mass of fluid in the control volume.

5.5.2 Simplifying Assumptions

The most important form of the momentum equation for application is when $\mathbf{v}$ can be taken as normal to the cross-sectional plane over the entire entrance and exit. In that case, $\mathbf{v}$ has magnitude V and the same direction as

---

*We will, in fact, neglect vertical variations and take the pressure as being uniform over a cross section. This must be modified for open-channel flows.

A, so we can write Eq. (5.43) as

$$\frac{d}{dt} \int_{z_1}^{z_2} \rho \langle \mathbf{v} \rangle A\, dz = \rho_1 \langle V^2 \rangle_1 A_1 - \rho_2 \langle V^2 \rangle_2 \mathbf{A}_2$$
$$+ p_1 \mathbf{A}_1 - p_2 \mathbf{A}_2 - \mathbf{F} + \left(\int_{z_1}^{z_2} \rho A\, dz \right) \mathbf{g} \tag{5.44}$$

As with the energy equation, it is often convenient to define the ratio between $\langle V^2 \rangle$ and $\langle V \rangle^2$,

$$\beta \equiv \frac{\langle V^2 \rangle}{\langle V \rangle^2}$$

β is nearly unity for turbulent flow and equals $\frac{4}{3}$ for laminar pipeline flow. An equivalent form for Eq. (5.44) is then

$$\frac{d}{dt} \int_{z_1}^{z_2} \rho \langle \mathbf{v} \rangle A\, dz = \beta_1 \rho_1 \langle V \rangle_1^2 \mathbf{A}_1 - \beta_2 \rho_2 \langle V \rangle_2^2 \mathbf{A}_2$$
$$+ p_1 \mathbf{A}_1 - p_2 \mathbf{A}_2 - \mathbf{F} + \left(\int_{z_1}^{z_2} \rho A\, dz \right) \mathbf{g} \tag{5.45a}$$

Since $\rho \langle V \rangle A = w$ and $\langle V \rangle \mathbf{A} = \langle \mathbf{v} \rangle A$, another way of writing Eq. (5.45) is

$$\frac{d}{dt} \int \rho \langle \mathbf{v} \rangle A\, dz = \beta_1 w_1 \langle \mathbf{v} \rangle_1 - \beta_2 w_2 \langle \mathbf{v} \rangle_2$$
$$+ p_1 \mathbf{A}_1 - p_2 \mathbf{A}_2 - \mathbf{F} + \left(\int_{z_1}^{z_2} \rho A\, dz \right) \mathbf{g} \tag{5.45b}$$

At steady state, $w_1 = w_2 = w$ and $d/dt = 0$, so Eq. (5.45) simplifies to

$$\text{steady state:} \quad \mathbf{0} = w(\beta_1 \langle \mathbf{v} \rangle_1 - \beta_2 \langle \mathbf{v} \rangle_2) + p_1 \mathbf{A}_1 - p_2 \mathbf{A}_2 - \mathbf{F} + \left(\int_{z_1}^{z_2} \rho A\, dz \right) \mathbf{g} \tag{5.46}$$

5.5.3 Engineering Bernoulli Equation, Second Derivation

We noted in Sec. 5.4.4 that the engineering Bernoulli equation can be derived directly from the principle of conservation of momentum, without the need to use any results from thermodynamics. We shall carry out such a derivation here, with a small number of assumptions that make the development easier without seriously restricting the range of applicability. Specifically, we assume that there is no shaft work; that the velocity is essentially uniform over the cross section ($\alpha = \beta = 1$); and, to facilitate a geometrical calculation, that the conduit cross section at any position is circular or two-dimensional with symmetry about a center plane.

The starting point is Eq. (5.46). We write the equation for the case in which points 1 and 2 are differentially close to one another, so that differences become differentials and the integral equals the integrand. With the uniform velocity assumption the equation is then written

$$0 = -wd\langle \mathbf{v} \rangle - d(p\mathbf{A}) - d\mathbf{F} + \rho A \, dz \, \mathbf{g} \qquad (5.47)$$

We write $d\mathbf{F}$ because the force will be differential in magnitude. We now take the inner (dot) product of this equation with the unit vector in the direction of flow, $\mathbf{i}$. Note that

$$\mathbf{i} \cdot \langle \mathbf{v} \rangle = \langle V \rangle$$

$$\mathbf{i} \cdot \mathbf{A} = A$$

$$\mathbf{i} \cdot \mathbf{g} \, dz = -g \, dh$$

The last term is $-dh$ because h is measured up from a datum, while $\mathbf{F}$ points down. Thus, we have

$$0 = -wd\langle V \rangle - d(pA) - \mathbf{i} \cdot d\mathbf{F} - \rho A g \, dh \qquad (5.48a)$$

or, equivalently,

$$0 = \rho A \langle V \rangle d\langle V \rangle + (p \, dA + A \, dp) + \mathbf{i} \cdot d\mathbf{F} + \rho A g \, dh \qquad (5.48b)$$

Here, we have substituted $\rho A \langle V \rangle$ for w and expanded the differential of pA. Finally, dividing by ρA and using the fact that $\langle V \rangle d\langle V \rangle = \tfrac{1}{2} d\langle V \rangle^2$, we obtain

$$\tfrac{1}{2} d\langle V \rangle^2 + g \, dh + \frac{dp}{\rho} + \left(\frac{p}{\rho A} dA + \frac{1}{\rho A} \mathbf{i} \cdot d\mathbf{F} \right) = 0 \qquad (5.49)$$

Equation (5.49) is quite close to Eq. (5.33), the differential form of the engineering Bernoulli equation, since we have neglected shaft work and set $\alpha = \beta = 1$. We need only examine the term in parentheses further.

It is convenient to break the force $d\mathbf{F}$ into two parts. One part, which we denote $d\mathbf{F}_p$, is the force exerted by the fluid on the surroundings because of the pressure acting on a changing cross section. This force will be present even in an inviscid liquid. The other part, which we denote $d\mathbf{F}_v$, is the force exerted by a viscous fluid and should be related to the viscous losses. The total force $d\mathbf{F} = d\mathbf{F}_p + d\mathbf{F}_v$.

For the differential segment shown in Fig. 5-6 it suffices to take the conduit

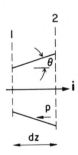

Figure 5-6. Differential segment of a conduit. Note that the pressure exerts a force parallel to the direction of flow because of the changing cross section.

sides as straight over the differential length, with angle θ to the direction of flow. If the fluid pressure is p, there is a net force on the conduit wall of

$$d\mathbf{F}_p = -p \sin \theta \, d\mathbb{S}\mathbf{i} \qquad (5.50)$$

where $d\mathbb{S}$ is the differential surface area. For a circular conduit cross section $d\mathbb{S} = \pi D \, dz$, and

$$\mathbf{i} \cdot \mathbf{F}_p = -\pi D p \sin \theta \, dz \qquad (5.51)$$

We can also compute the area differential,

$$dA = d\left(\frac{\pi D^2}{4}\right) = \frac{\pi D}{2} dD = \frac{\pi D}{2} \frac{dD}{dz} dz$$

For a circular cross section we have $dD/dz = 2 \sin \theta$. It follows then by direct substitution that $p \, dA + \mathbf{i} \cdot d\mathbf{F}_p = 0$; the same result is obtained for a two-dimensional symmetric conduit.

We now let $\mathfrak{F}(z)$ be the viscous force per unit length exerted by the fluid in the flow direction; then $\mathbf{i} \cdot d\mathbf{F}_v = \mathfrak{F}(z) \, dz$ and Eq. (5.49) becomes

$$\tfrac{1}{2}d\langle V \rangle^2 + g \, dh + \frac{1}{\rho} dp + \frac{\mathfrak{F}}{\rho A} dz = 0 \qquad (5.52)$$

This is simply the differential form of the Bernoulli equation if we identify $(\mathfrak{F}/\rho A) \, dz$ with dl_V. A flow to the right exerts a positive force on the conduit, so (if we accept experience in lieu of proof) we have $dl_V \geq 0$.

The fact that the Bernoulli equation can be derived from either the energy or momentum equation is of some interest. For a nondissipative system (i.e., one without losses) the energy equation contains no additional information about the motion not already contained in the momentum equation, so thermodynamic and flow calculations are uncoupled and can be carried out independently. For a viscous system we cannot expect independent information unless we can calculate either forces or losses independently, in which case the equations will provide the other.

5.5.4 Aside on Springs and Dashpots

Some useful insight into the relation between energy and momentum equations, and mechanical and thermal forms of energy, can be obtained by consideration of the mass–spring–dashpot system in Fig. 5-7. To a first approximation the spring force can usually be taken as proportional to the displacement from equilibrium, $-kx$, and the dashpot damping force as proportional to velocity, $-\mu \, dx/dt$. The momentum balance is then

$$m \frac{d^2x}{dt^2} + \mu \frac{dx}{dt} + kx = 0 \qquad (5.53)$$

We can multiply each term in Eq. (5.53) by dx/dt, to obtain

$$m \frac{dx}{dt} \frac{d^2x}{dt^2} + kx \frac{dx}{dt} + \mu \left(\frac{dx}{dt}\right)^2 = 0 \qquad (5.54a)$$

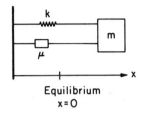

Figure 5-7. Schematic of a mass–spring–dashpot system. The spring constant is k and the dashpot damping coefficient is μ.

or, equivalently,

$$\tfrac{1}{2}m\frac{d}{dt}\left(\frac{dx}{dt}\right)^2 + \tfrac{1}{2}k\frac{dx^2}{dt} + \mu\left(\frac{dx}{dt}\right)^2 = 0 \qquad (5.54b)$$

Equation (5.54) can be integrated, and we have

$$\tfrac{1}{2}m\left(\frac{dx}{dt}\right)^2 + \tfrac{1}{2}kx^2 + \mu\int_0^t \left(\frac{dx}{dt}\right)^2 dt + \text{constant} = 0 \qquad (5.55)$$

Equation (5.55) is a "mechanical energy" equation; the first term on the left is the kinetic energy, while the second term is the potential energy of the extended spring. In the absence of viscous damping by the dashpot ($\mu = 0$), this equation simply states that the sum of potential and kinetic energy is a constant. The dissipation term, $\mu\int_0^t (dx/dt)^2\, dt$, is positive and monotonically increasing with time, indicating a continual loss of "useful" (i.e., kinetic and potential) energy as it is converted to thermal energy and heating of the dashpot. This is the analog of the losses term, l_v, in the engineering Bernoulli equation.

5.6 STREAMLINES AND STREAM TUBES

We have been applying the macroscopic balances to flow in the complete conduit. This is not necessary, and it is sometimes useful to consider flow over only part of a cross section.

The concept of a *streamline* is needed here. At each point in the flow field we draw the velocity vector,* as shown in Fig. 5-8. These vectors are tangent to a family of lines, known as streamlines, and by drawing the "arrows" close enough to one another the streamlines can be sketched in. At steady state, a particle of fluid will move along a streamline.

*In a turbulent flow the velocity fluctuates rapidly at each point in the flow field about a mean value. It is the mean velocity vector that we refer to throughout this section. A steady-state turbulent flow is one in which the mean does not change with time.

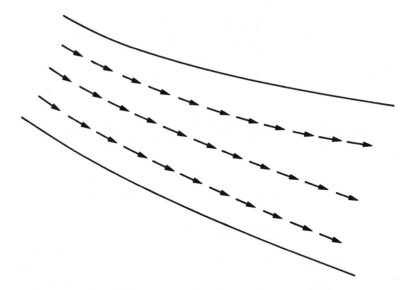

Figure 5-8. Velocity vector at each point in the flow field.

Now consider a steady-state flow field as shown in Fig. 5-9. The fluid that enters the region enclosed by the family of streamlines shown in the figure must remain in that region until it exits at z_2. Such a flow region is called a *stream tube*. Clearly, all the derivations of the macroscopic equations apply equally well to a stream tube, except that we must change our references to a "conduit" to "boundary of the stream tube."

Let us now return to the engineering Bernoulli equation, Eq. (5.34). This equation must apply to a stream tube. If we take the area of the stream tube as small enough, then $\langle V \rangle$ will not differ from V along the center streamline, and $\alpha = 1$. Indeed, we can take the limit as the stream tube shrinks to a single streamline and write the *Bernoulli equation along a streamline,*

$$\tfrac{1}{2}V_2^2 + gh_2 = \tfrac{1}{2}V_1^2 + gh_1 - \int_{p_1}^{p_2} \frac{dp}{\rho} - l_V \qquad (5.56)$$

Here, we have dropped the work term since the equation will not apply in situations where work is done. The l_V term is also usually dropped, since the useful applications are to problems where losses can be neglected. This final form, without l_V, is the equation often referred to in the published literature as the *Bernoulli equation.*

5.7 CONCLUDING REMARKS

This chapter has consisted almost entirely of derivations of the macroscopic balance equations. These equations model flows that are essentially one-

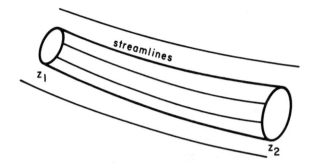

Figure 5-9. Streamlines and a stream tube.

dimensional. As we shall see in Chapter 6, the equations derived here are useful in solving a variety of interesting and important process fluid mechanics problems. It is important to understand the derivations completely, because it may be necessary in practice to apply the equations to situations in which one or more of the assumptions made in this chapter do not apply.

The engineering Bernoulli equation, Eq. (5.34), and the steady-state momentum equation, Eq. (5.46), are the equations used with the greatest frequency in applications. These two equations are so fundamental, and so important, that they should be committed to memory.

BIBLIOGRAPHICAL NOTES

We have assumed a familiarity with the principles of conservation of mass, energy, and momentum, and some previous experience in applying balance equations to flowing systems. Most students will have done so in a prior course in "mass and energy balances."

The fundamentals of mathematical modeling, selection of a control volume, and use of the principles of conservation of mass and energy are the subjects of

RUSSELL, T. W. F., and DENN, M. M., *Introduction to Chemical Engineering Analysis*, John Wiley & Sons, Inc., New York, 1972.

The thermodynamic principles used here are covered in all books on thermodynamics; see, for example,

BALZHISER, R. E., SAMUELS, M. R., and ELIASSEN, J. D., *Chemical Engineering Thermodynamics*, Prentice-Hall, Inc., Englewood Cliffs, N.J., 1972.

SANDLER, S. I., *Chemical and Engineering Thermodynamics*, John Wiley & Sons, Inc., New York, 1977.

There is an extensive treatment of pipe flow and fitting losses in a publication by the Crane Company,

Flow of Fluids Through Values, Fittings, and Pipe, Technical Paper No. 410, Crane Co., 4100 S. Kedzie Ave., Chicago, Ill., sixteenth revised printing, 1976.

Entry and exit losses for laminar pipe flow are reported in

BOGER, D. V., and BINNINGTON, R., *J. Rheol.*, **21**, 515 (1977).
KESTIN, J., SOKOLOV, M., and WAKEHAM, W., *Appl. Sci. Res.*, **27**, 241 (1973).

PROBLEMS

5.1. Experimental data shown in the next chapter, Figs. 6.7 to 6.9, indicate that the velocity varies with radius in pipe flow as follows:
  laminar flow: $v(r) = 2\langle v \rangle[1 - (r/R)^2]$
  turbulent flow: $v(r) = 1.22\langle v \rangle[1 - r/R]^{1/7}$
  a) Show that $\beta = \frac{4}{3}$ for laminar flow.
  b) Compute α and β for turbulent flow.

5.2. Combined entrance and exit losses for laminar flow in a pipe are given by the empirical equation
$$K_f = 2.3(1 + 9.4/\text{Re})$$

If you have not previously solved Problem 3.7, use the given flowrate data to predict pressure drop and compare to the measured values.

5.3. A liquid flow system is shown in Fig. 5P3. Compute the power required for the pump to deliver 1500 kg/hr of fluid with $\rho = 950$ kg/m³, $\eta = 9 \times 10^{-4}$ Pa·s. The pipe is 100 mm in diameter and is constructed of steel.

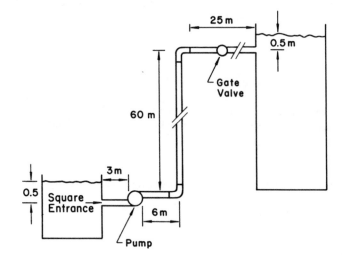

5.4. Derive the rule-of-thumb that the losses in turbulent flow in a pipe that is 50 diameters long are approximately equal to one velocity head.

5.5. The following data were obtained by T.-H. Nguyen for overall pressure drop of a dilute solution of polyacrylamide in maltose syrup, $\rho = 1400 \text{ kg/m}^3$, flowing from a large reservoir into tubes of various L/D and then exiting to the atmosphere:

L/D	$8\langle v\rangle/D \text{ (s}^{-1})$	Δp (Pa)	L/D	$8\langle v\rangle/D \text{ (s}^{-1})$	Δp (Pa)
52.6	5.7	3.5×10^4	7.25	11.6	1.1×10^4
	7.7	4.9		23	2.5
	14.2	8.3		31	3.2
	20	1.2×10^5		39	3.9
	26	1.5		44	4.6
	40	2.2		50	5.3
	50	2.9		57	6.0
	60	3.6		62	6.6
	72	4.0		75	8.0
	80	4.3		83	9.4
				108	1.2×10^5
26.1	4.0	1.3×10^4		133	1.5
	11.2	3.4		200	2.2
	15	4.8		228	2.9
	23	6.8		336	4.2
	32	9.6			
	41	1.2×10^5	3.98	16.6	1.0×10^4
	51	1.5		38	2.4
	60	1.9		47	3.1
	73	2.2		57	3.9
	86	2.5		63	4.5
	120	3.6		70	5.2
	160	5.0		80	5.8
				90	6.5
10.0	10	1.2×10^4		125	9.3
	20	2.6		160	1.2×10^5
	28	3.5		200	1.5
	32	4.1		266	2.2
	37	4.6		380	3.1
	40	5.3			
	50	6.7	Sharp-	25	3.9×10^3
	60	8.1	edged	31	5.3
	70	9.5	orifice	38	8.5
	86	1.2×10^5	$(L/D = 0)$	47	1.1×10^4
	108	1.4		124	4.6
	130	1.8		203	8.0
	180	2.5		360	1.5×10^5
	250	3.6		500	2.2

a) Cross-plot the data as Δp versus L/D at constant $8\langle v\rangle/D$, and extrapolate to $L/D = 0$. How does this limit compare to the pressure drop for flow through the sharp-edged orifice?

b) Use the results in part (a) to obtain a plot of the combined entry and exit losses in the form of an excess pressure drop, Δp_e, as a function of $8\langle v\rangle/D$.

c) Use the result in part (b) to correct the data and plot the shear stress for fully-developed flow as a function of $8\langle v\rangle/D$ for each L/D. Do the data overlap for the various L/D?

   When a plot of τ_s versus $8\langle v\rangle/D$ is linear, the viscosity is a constant and $8\langle v\rangle/D$ is the shear rate at the wall; see Sec. 19.6. Compute the viscosity.

d) Excess entry and exit losses are sometimes written in the form $\Delta p_e = 2e\tau_s$. Compute e as a function of shear rate and compare with the value $e \sim 0.85 - 1.1$ usually found for Newtonian liquids. Is the polymer solution Newtonian?

Applications of **6**
Macroscopic Balances

6.1 INTRODUCTION

In this chapter we will apply the macroscopic balances to the solution of a variety of flow problems. We have three equations available: conservation of mass, conservation of energy, and conservation of momentum. A typical problem would seem to be one in which we are given the flow geometry and the flow rate and asked to calculate the pressure drop, viscous losses, and force. A bit of reflection, however, will reveal that such a problem is underdetermined, in that the number of equations is not sufficient to determine the unknown variables.

This point can be illustrated by considering the specific problem of computing the losses in an expansion, as in Fig. 6-1. The losses are given in Table 5-1 as $(A_2/A_1 - 1)^2$ velocity heads, and we shall derive this result below. We can note here that, with the flow rate and geometry fixed, the equation for conservation of mass provides no information other than a formula for calculating the velocities $\langle V \rangle_1$ and $\langle V \rangle_2$. The energy equation contains two unknowns, the pressure change, $p_1 - p_2$, and the sought-after losses term, l_V. It follows, then, that the pressure difference must be obtained from the equation of conservation of momentum. But this equation involves not only the unknown pressure difference but also the unknown force, $\mathbf{F}$. Thus, in order to calculate the losses we must know or measure either the pressure difference or the force, since the number of unknowns is one more than the number of equations.

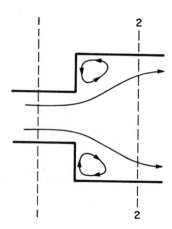

Figure 6-1. Schematic of flow through an expansion.

The simpler applications to be considered below generally fall into one of three classes, with geometry and flow taken as given:

- Class I: Calculate viscous losses for a given flow.
- Class II: Calculate the force for a given flow.
- Class III: Calculate the pressure change for a given flow.

In class I problems, the losses must be obtained from the energy equation, for which the pressure difference is required. The pressure difference must be obtained from the momentum equation. Thus, some prior information must be available about the force.

Class II problems are just the inverse of class I problems. The force must be calculated from the momentum equation. The pressure difference is required for this calculation, so it must be obtained from the energy equation. To obtain the pressure difference from the energy equation it is necessary to have some prior information about the viscous losses. Problems of this type usually yield a functional relationship between the variables which includes a constant that must be measured experimentally.

Class III problems are qualitatively different from the other two. The pressure difference appears in both the momentum and energy equations. It can be calculated from the former if information about the force is available, or from the latter if information about the viscous losses is available. Which equation we use will depend on the extent of our physical understanding of the magnitudes of forces and losses. Different estimates of pressure change are usually obtained from the two equations because of imperfect estimates of forces and losses. The solution to class III problems is also frequently expressed in terms of a constant that must be measured experimentally.

These remarks are intended to emphasize the approximate nature of the calculations that follow. In some cases rather good estimates are obtained, but in others we can obtain from the macroscopic balances only the likely functional forms, orders of magnitude, and guides to experiment. In many engineering situations these results will be enough, but in others more precision is required.

6.2 LOSSES IN EXPANSION

As a first example of the use of the macroscopic balances we will estimate the losses in flow of an incompressible fluid through an expansion. The experimentally measured losses are given in Table 5-1 as $(A_2/A_1 - 1)^2$ velocity heads. This value represents available data to within about 20%.

The flow configuration is shown in Fig. 6-1. This is a class I problem, so we must obtain the pressure difference between points 1 and 2 by use of the momentum equation. The losses are then computed from the energy equation. We will assume that the flow is turbulent and that the velocity at points 1 and 2 is uniform over the cross section, so that $\alpha_1 = \alpha_2 = \beta_1 = \beta_2 = 1$. This requires that plane 2 be sufficiently downstream to be beyond the region of eddying.

Because of the assumption of uniform incompressible flow we can replace $\langle V \rangle$ by V and write the continuity equation, Eq. (5.14), as simply

$$A_1 V_1 = A_2 V_2 \tag{6.1}$$

The component of the momentum equation, Eq. (5.46), in the flow direction is

$$0 = \rho A_1 V_1^2 - \rho A_2 V_2^2 + p_1 A_1 - p_2 A_2 - F \tag{6.2}$$

where F denotes the component of $\mathbf{F}$ in the flow direction. The only conceptual difficulty here is in writing down the force, F. As shown in Fig. 6-2, there are two contributions to the force which the fluid exerts on the surroundings. One, denoted F_t, is the result of the tangential frictional drag along the walls.

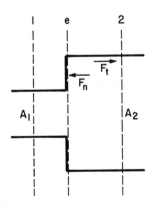

Figure 6-2. Forces on an expansion.

The other, denoted F_n, is the result of fluid pressure on the expansion surface at plane e.

We can calculate F_n as

$$F_n = -p_e(A_2 - A_1) \tag{6.3}$$

where p_e is the pressure at plane e. This pressure acts on the annular surface with area $A_2 - A_1$. The pressure p_e is a gage pressure, with the outside pressure taken as zero.* The negative sign is required because the force acts to the left, while the positive direction is to the right. The momentum equation is then

$$0 = \rho A_1 V_1^2 - \rho A_2 V_2^2 + (p_1 - p_e)A_1 + (p_e - p_2)A_2 - F_t \tag{6.4}$$

It is evident that to go any further we need to say something about F_t and p_e. The force F_t causes little difficulty. The frictional drag varies directly with distance of straight pipe length. The distance is small here, so we can safely assume that $F_t \approx 0$ and can be neglected. (If we wish a quantitative estimate of the error involved in neglecting F_t, we can calculate the force from the friction factor for fully developed pipe flow and surely be within an order of magnitude.) p_e causes more of a problem, but it can be approximated. There is no change in velocity between 1 and e, since all flow at the expansion surface is through the central core. Thus, since there are negligible losses or friction in this short segment, it follows from both the energy and momentum equations that $p_e = p_1$ in the region where there is flow. We assume that the pressure does not change over a cross section, so we can substitute $p_1 = p_e$ in Eq. (6.4) and obtain, together with the Eq. (6.1),

$$\frac{p_1 - p_2}{\rho} = V_2^2 \left(1 - \frac{A_2}{A_1}\right) \tag{6.5}$$

The engineering Bernouilli equation, Eq. (5.34), for horizontal flow ($h_1 = h_2$) of an incompressible fluid with $\alpha_1 = \alpha_2 = 1$ in the absence of shaft work is

$$l_V = \frac{p_1 - p_2}{\rho} + \tfrac{1}{2}(V_1^2 - V_2^2) \tag{6.6}$$

Substituting Eqs. (6.1) for the velocity and (6.5) for the pressure difference then gives

$$l_V = V_2^2 \left[1 - \frac{A_2}{A_1} + \frac{1}{2}\left(\frac{A_2}{A_1}\right)^2 - \frac{1}{2}\right]$$

---

*The outside pressure can be taken as zero, and usually is, whenever we are concerned with the flow in a closed conduit. Atmospheric pressure acts on the outside of the control volume from the left and from the right; the area available for atmospheric pressure is the same to the left and to the right, so the net force exerted by atmospheric pressure is zero and cancels from the momentum equation. See Sec. 6.7 for a case in which there is a liquid-air interface and atmospheric pressure must be accounted for.

or

$$l_V = \frac{V_2^2}{2}\left(\frac{A_2}{A_1} - 1\right)^2 \tag{6.7}$$

Equation (6.7) is known as the *Borda–Carnot equation*. It gives the number of velocity heads lost in terms of the downstream velocity as $(A_2/A_1 - 1)^2$, which is in good agreement with experiment.

Careful pressure measurements at points along the wall show that the assumption that p_e is uniform is a poor one, but the error seems to cancel out. Interestingly, a good estimate of the losses in a contraction cannot be obtained by an analogous procedure, probably because of a greater effect of the uncertainty regarding the pressure near the contraction plane.

6.3 FORCE ON A REDUCING BEND

We now consider the problem of computing the force necessary to maintain a 90° reducing bend in place. The configuration is shown in Fig. 6-3, with the xy plane taken to be horizontal. This is a class II problem. The force is computed from the momentum equation, which requires an expression for the pressure change. The latter must be computed from the energy equation, which will require us to make some statement about the losses.

We assume incompressible turbulent flow with a flat velocity profile ($\alpha = \beta = 1, \langle V \rangle = V$) and no shaft work. The continuity (5.14) and energy (5.34) equations are then readily written down, as follows:

$$A_1 V_1 = A_2 V_2 \tag{6.8}$$

$$\tfrac{1}{2}V_2^2 + \frac{p_2}{\rho} = \tfrac{1}{2}V_1^2 + \frac{p_1}{\rho} - l_V \tag{6.9}$$

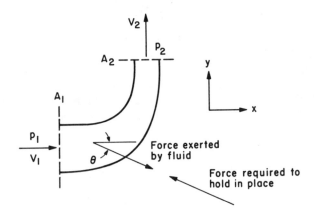

Figure 6-3. Schematic of horizontal reducing bend.

The inflow is entirely in the x direction and the outflow is entirely in the y direction. The x and y components of the momentum equation, Eq. (5.46), are then, respectively,

$$x: \quad 0 = \rho A_1 V_1^2 + p_1 A_1 - F_x \tag{6.10a}$$

$$y: \quad 0 = -\rho A_2 V_2^2 - p_2 A_2 - F_y \tag{6.10b}$$

There is no gravity term because the bend is in the horizontal plane, so the gravity force points entirely in the z direction. It should be remembered that F_x and F_y are the components of the force exerted *by the fluid*. The components of the force needed to hold the bend in place are the negative of these quantities.

We need to make some statement about the losses. We could try to get an order-of-magnitude estimate by neglecting them, but we know from Table 5-1 that there is usually a loss of about three-fourths of a velocity head in a right-angle bend. We will express the losses as K velocity heads,

$$l_V = \tfrac{1}{2} K V_2^2 \tag{6.11}$$

where we expect K to be approximately $\tfrac{3}{4}$. Combining Eqs. (6.8) to (6.11) then leads to the equation for the force components:

$$F_x = A_1 (p_1 + \rho V_1^2) \tag{6.12a}$$

$$F_y = -A_2 \left\{ p_1 + \tfrac{1}{2} \rho V_1^2 \left[1 + (1 - K) \left(\frac{A_1}{A_2} \right)^2 \right] \right\} \tag{6.12b}$$

The magnitude of the net force is

$$|\mathbf{F}| = (F_x^2 + F_y^2)^{1/2} \tag{6.13a}$$

and the direction is

$$\theta = \arctan \left(\frac{F_y}{F_x} \right) \tag{6.13b}$$

Several observations are in order here. First, the force depends explicitly on the magnitude of the upstream (or downstream) pressure. (It should be noted that p_1 is a *gage pressure*, with the pressure in the surroundings taken to be zero.) Second, unless A_1/A_2 is fairly large *and* p/ρ is only of the order of 1 or so velocity heads, the force calculation will be relatively insensitive to the value of K, so a crude approximation for the losses will be adequate. For $p_1/\rho \ll V_1^2/2$, we get

$$\frac{p_1}{\rho} \ll \frac{V_1^2}{2}: \quad |\mathbf{F}| \longrightarrow p_1 (A_1^2 + A_2^2)^{1/2} \tag{6.14b}$$

$$\theta \longrightarrow \arctan \left(-\frac{A_2}{A_1} \right) \tag{6.14b}$$

6.4 JET EJECTOR

The *jet ejector*, or *jet pump*, is a device with no moving parts that is widely used for such operations as moving liquids between tanks and lifting corrosive or abrasive liquids. The basic configuration is shown in Fig. 6-4. The high-velocity "jet" line is introduced concentrically with the low-velocity "suction" line, both at pressure p_1. The mixing will be rather chaotic, but a few diameters downstream the velocity will be relatively uniform and the pressure will be increased, $p_2 > p_1$. We wish to determine the pressure at p_2 in order to compute the pumping capacity of the line. This is therefore a class III problem.

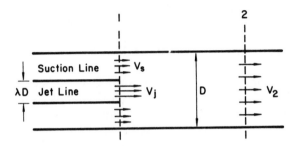

Figure 6-4. Schematic of a jet ejector.

We will assume that both lines contain the same incompressible liquid, and that the flow within each line is turbulent with a uniform velocity. Thus, the velocity in the suction line is V_s everywhere, and the velocity in the jet line is V_j everywhere. We also assume that the velocity is uniform at plane 2, so $\langle V \rangle_2 = V_2$. The calculation of $\langle V \rangle_1$ has already been carried out in Example 5.5 and is given by

$$\langle V \rangle_1 = \lambda^2 V_j + (1 - \lambda^2) V_s \qquad (6.15)$$

where λ is the ratio of the jet diameter to the diameter of the entire tube. Since the upstream and downstream areas are the same, the continuity equation, Eq. (5.14), simply requires that $\langle V \rangle_1 = \langle V \rangle_2$, or

$$V_2 = \lambda^2 V_j + (1 - \lambda^2) V_s \qquad (6.16)$$

Because of the chaotic mixing the losses will be substantial. Since we have no way of estimating these losses, it is unlikely that we can compute the pressure change between 1 and 2 from the energy equation. It should be possible to employ the momentum equation, however. Friction should be small in the relatively short distance from 1 to 2, so the force term in the momentum equation will probably not be important. This means that the momentum equation can be used to compute the pressure change.

111

The component of the momentum equation, Eq. (5.46), in the flow direction, with the gravity and force terms equal to zero, is

$$(p_2 - p_1)A = \rho A \beta_1 \langle V \rangle_1^2 - \rho A V_2^2 \tag{6.17}$$

The area is the same at both cross sections. β_1 is computed as in Example 5.5,

$$\beta_1 = \frac{\langle V^2 \rangle_1}{\langle V \rangle_1^2} = \frac{\lambda^2 V_j^2 + (1 - \lambda^2)V_s^2}{\langle V \rangle_1^2} \tag{6.18}$$

Combination of Eqs. (6.15) through (6.18) then gives the desired result,

$$p_2 - p_1 = \lambda^2 (1 - \lambda^2)\rho(V_j - V_s)^2 \tag{6.19}$$

This is the relation needed for pumping. Note that as long as there is a velocity difference between the two streams, we must have $p_2 > p_1$.

6.5 FLOW THROUGH AN ORIFICE

We will now compute the relation between flow rate and pressure drop across an orifice plate (a plate with a small, sharp-edged hole) placed in a flowing stream. The configuration for this class III problem is shown in Fig. 6-5. This arrangement is sometimes used as a flow-measuring system, since the measured pressure difference can be directly related to the flow rate.

The fluid will emerge from the orifice in a jet and will then expand to fill the downstream section. If the downstream section of the control volume is placed where the jet has expanded and is uniformly filling the pipe, no useful information can be obtained from the balances. We therefore place control surface 2 at the downstream side of the orifice plate. Then $A_2 = A_o$, the orifice area, and the continuity equation, assuming uniform velocity ($\alpha = \beta = 1$) and incompressibility, is

$$AV_1 = A_o V_2 \tag{6.20}$$

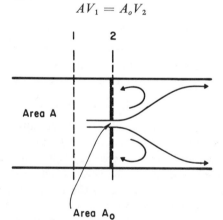

Figure 6-5. Schematic of flow through an orifice.

The pressure drop must be computed from either the momentum or the energy equation. The forces acting on the orifice plate in this complex flow field do not lend themselves to easy estimates of magnitude, so if a solution is to be obtained, it must come from use of the energy equation. This path, too, will be difficult, for the losses are likely to be large and must be accounted for.* The engineering Bernoulli equation, Eq. (5.34), with the assumption of horizontal flow, no shaft work, and $\alpha_1 = \alpha_2 = 1$, is

$$\tfrac{1}{2}V_2^2 = \tfrac{1}{2}V_1^2 + \frac{p_1 - p_2}{\rho} - l_V \qquad (6.21)$$

This can be combined with Eq. (6.20) and rearranged to

$$\frac{p_1 - p_2}{\rho} = \tfrac{1}{2}V_2^2\left[1 - \left(\frac{A_o}{A}\right)^2\right] + l_V \qquad (6.22)$$

The first term on the right of Eq. (6.22) is approximately one velocity head, based on the downstream velocity. This is the order of magnitude of expansion and contraction losses in Table 5.1, so it is clear that l_V must be accounted for in order to relate velocity uniquely to pressure drop. The fact that losses in fittings do correlate well with downstream velocity heads suggests that the correct *form* for the solution will be obtained if we write

$$l_V \approx \tfrac{1}{2}V_2^2 K \qquad (6.23)$$

Then Eq. (6.22) can formally be written as

$$\frac{p_1 - p_2}{\rho} = \tfrac{1}{2}V_2^2\left[1 + K - \left(\frac{A_o}{A}\right)^2\right] \qquad (6.24)$$

or, equivalently,

$$Q = A_o V_2 = A_o\sqrt{\frac{1}{1 + K - (A_o/A)^2}}\sqrt{\frac{2(p_1 - p_2)}{\rho}} \qquad (6.25)$$

Equation (6.25) will be useful only if K is a constant or a unique function of Reynolds number for a given geometry. Data in *Perry's Handbook* for sharp-edged orifices with area ratios from 0.05 to 0.70 correlate well at high orifice Reynolds numbers ($D_o V_2 \rho/\eta > 3 \times 10^4$, $D_o = \sqrt{4A_o/\pi}$) with the relation

$$K = 1.6\left[1 - \left(\frac{A_o}{A}\right)^2\right] \qquad (6.26)$$

Note that the factor 1.6 means that the losses are the cause of more than 60% of the pressure drop (1.6/2.6) in Eq. (6.22). Introducing K from Eq. (6.26)

---

*In many published solutions of this problem and others like it, the losses are taken to be zero. The solution is then multiplied by a coefficient between zero and unity in order to account for the losses and produce agreement with experiment. Such an approach is philosophically unsatisfying and scientifically unsound, for there is no guarantee that even the correct functional form will be obtained.

into Eq. (6.25) leads to the final working equation,

$$Q = 0.62 A_o \sqrt{\frac{2(p_1 - p_2)}{[1 - (A_o/A)^2]\rho}} \tag{6.27}$$

The $1 - (A_o/A)^2$ term is referred to as the *velocity of approach* term and can be set equal to unity with negligible error as long as $A_o/A < 0.25$. The factor 0.62 is the most commonly reported *orifice coefficient*. This value is based on measuring downstream pressure approximately at the location of the *vena contracta*, the minimum diameter of the emerging jet. This point is at about one pipe diameter downstream of the orifice, but the location varies somewhat with A_o/A. The location of the vena contracta is given in *Perry's Handbook*. Furthermore, the result obtained here is independent of downstream diameter as long as the downstream area is large compared to the orifice area. Thus, we would obtain the same flow rate-pressure drop relation for flow through an orifice to the surroundings.

6.6 PITOT TUBE

One of the most common devices for measuring detailed velocity profiles in a flowing stream is the *pitot tube*. We analyze here a version of the pitot tube known as a pitot-static tube. The tube is shown schematically in Fig. 6-6. It consists of an inner tube which is open at the end and an outer tube which is sealed at the end but contains several small openings along the side. Each tube is filled with the flowing fluid, and the tubes are connected to opposite

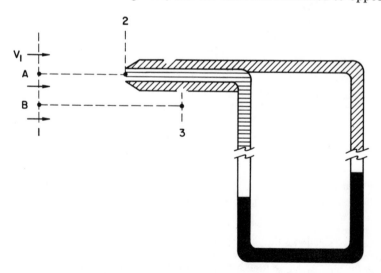

Figure 6-6. Schematic of a pitot-static tube.

ends of a manometer. The analysis is identical for a slightly different arrangement in which the outer tube is replaced by a pressure tap at the conduit wall. The pitot tube must be much smaller in diameter than the conduit in which it is placed, and it is assumed that the presence of the tube does not affect the upstream flow field.

The pitot tube is an instrument for measuring velocity by means of a pressure measurement. Thus, the analysis is a class III problem. We have no choice between the energy and momentum equations, however, because we are now attempting to study behavior of the fluid on individual streamlines, and the only equation available to us for such an analysis is the Bernoulli equation along a streamline, Eq. (5.56).

We now focus on two streamlines in Fig. 6-6, A and B. A is the streamline which goes directly into the tip of the inner tube. Since there is no flow in the tube, the fluid must decellerate from velocity V_1 at point 1 to $V = 0$ at point 2, known as the *stagnation point*. In a horizontal system, noting that $V_2 = 0$ and neglecting losses over the short distance of relatively undisturbed flow from 1 to 2, the Bernoulli equation is simply

$$\tfrac{1}{2}V_1^2 + \frac{p_1 - p_2}{\rho} = 0 \tag{6.28}$$

p_2 is called the *stagnation pressure*.

Now consider streamline B. If the distance from 1 to 3 is short, then in the absence of losses the velocity at 3 is essentially the same as the velocity at 1. That is, since we are flowing past the tube, the flow is not interfered with and the velocity is unchanged. The Bernoulli equation is then

$$\tfrac{1}{2}V_3^2 = \tfrac{1}{2}V_1^2 + \frac{p_1 - p_3}{\rho} \tag{6.29}$$

or, since $V_1 = V_3$,

$$p_1 = p_3 \tag{6.30}$$

Thus, we can write Eq. (6.28) as

$$V_1 = \sqrt{\frac{2(p_2 - p_3)}{\rho}} \tag{6.31}$$

The pressure p_3 is the pressure acting on the side holes in the outer tube, so it is the pressure of the fluid in the outer tube. Thus, the manometer directly measures the pressure difference and hence the velocity. Equation (6.31) is accurate to within a few percent for velocity measurements at high Reynolds numbers, but at low Reynolds numbers a correction must be applied to account for the neglected viscous terms.

Figure 6-7 shows typical pitot tube data for turbulent flow of water in a 50-mm (nominal) pipe. A slight effect of probe diameter is evident. The pitot tube responds slowly to changing velocities, so it filters out the fluctuations in the turbulent flow and records only the mean velocity. Note that the devia-

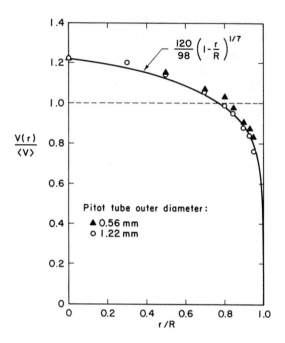

Figure 6-7. Pitot tube data, water in a 50.4-mm diameter pipe. $\langle V \rangle =$ 2.2 m/s, Re $= 1.2 \times 10^5$. Data of D. C. Bogue.

tion in the velocity from the area average, $\langle V \rangle$, is less than 20% over most of the pipe cross section. This relatively flat profile is the reason we can obtain reasonable accuracy with the macroscopic balances in turbulent flow while taking α and β equal to unity. The line shown with the data is an empirical relationship which is sometimes used for turbulent velocity profiles. The one-seventh power curve has no theoretical significance, and it represents turbulent data only approximately, but it is easy to manipulate analytically.

The contrast between laminar and turbulent pipe flow is nicely illustrated by comparison of Fig. 6-7 with Fig. 6-8. The latter shows pitot tube data in laminar flow of a sugar solution in a 25-mm (nominal) diameter pipe at Re $=$ 800. There is substantial deviation from the mean velocity over most of the cross section, and the centerline velocity is twice $\langle V \rangle$. These data are in a region where a correction must be applied to Eq. (6.31) to account for viscous effects, but only the raw uncorrected data are shown here. The data should be right on the parabola

$$\frac{V(r)}{\langle V \rangle} = 2\left(1 - \frac{r^2}{R^2}\right) \tag{6.32}$$

which is drawn in the figure. The viscous correction brings the data closer to the line, which is derived theoretically in Sec. 8.4. More precise data verify

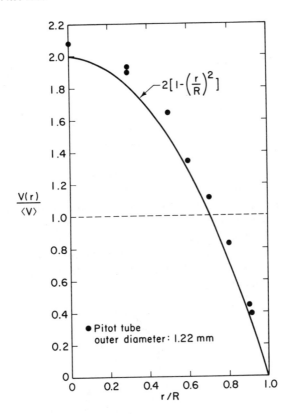

Figure 6-8. Pitot tube data, sugar solution in a 27.3-mm diameter pipe. $\langle V \rangle = 1.08$ m/s, Re = 800. Data of D. C. Bogue.

the correctness of Eq. (6.32). It was shown in Example 5.4 that $\alpha = 2$ for this velocity distribution, and it is readily established that $\beta = \frac{4}{3}$.

APPENDIX 6.6.A

Laser Doppler Flowmeter

This is an appropriate place for a brief discussion of other methods of measuring point velocities in a flow field. Most effort in recent years has gone into the development of instruments that respond rapidly and do not interfere with the flow field. Noninterference with the flow field is particularly important for non-Newtonian fluids, where pitot tubes often do not work.

The simplest method of flow measurement for liquids is to insert very small reflective particles (suspended air bubbles often suffice), to light a portion of the flow field with rapid continuous pulses, and to follow the movement of the particles by time-exposure photography. The short distance traveled by a particle over each timed pulse defines the local fluid velocity. This is an accurate, but extremely tedious method.

The laser Doppler flowmeter is an instrument that provides automatic measurement of point velocities by use of the Doppler shift principle. A laser beam of known frequency is focused on a small segment of the flow field, where it is reflected from small particles moving with the fluid. The reflected beam will experience a small shift in frequency that is proportional to the particle velocity; this is the Doppler principle. By one of several optical arrangements, the frequencies of the original beam and the reflected beam are compared electronically and the point velocity is recorded. The method is thus quite straightforward in concept, although there are difficult problems in optics and electronics that have to be solved for practical implementation. Commercial instruments are now available. Typical data for laminar velocity profiles obtained from such an instrument are shown in Fig. 6-9. Note how close to the wall it was possible to obtain velocity measurements. The instrument can respond quickly enough to obtain information about rapid velocity fluctuations in turbulent flow.

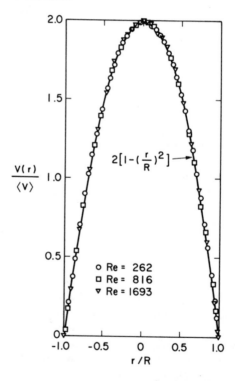

Figure 6-9. Velocity profile measured with a laser Doppler flowmeter. Water in a 13.6-mm glass tube. Data of A. V. Ramamurthy.

6.7 DIAMETER OF A FREE JET

The calculation of the diameter of a free jet illustrates some of the considerations in problems with a free (gas–liquid) interface. The process is shown schematically in Fig. 6-10. An incompressible Newtonian liquid is in laminar flow in a tube of diameter D. The liquid emerges into a horizontal jet. Because of the absence of resistance from the solid surface, the velocity profile rearranges in the jet and becomes uniform at some distance downstream of the exit from the tube. The downstream diameter will differ from the jet diameter.

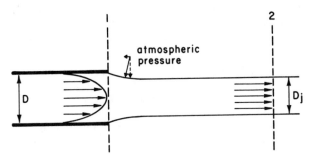

Figure 6-10. Schematic of a free jet.

We take plane 1 right at the tube exit and plane 2 sufficiently far downstream to assume that the final jet diameter has been reached. The continuity equation is then

$$\frac{\pi D^2}{4}\langle V\rangle_1 = \frac{\pi D_j^2}{4}\langle V\rangle_2 \tag{6.33}$$

The pressure at the liquid–air interface is atmospheric; hence, the assumption that pressure is uniform over the cross section means that $p_1 = p_2$ everywhere. The engineering Bernoulli equation, Eq. (5.34), is then

$$\frac{\alpha_2}{2}\langle V\rangle_2^2 = \frac{\alpha_1}{2}\langle V\rangle_1^2 - l_V \tag{6.34}$$

We expect the losses in this flow to be relatively small, and we shall set l_V to zero. If we further assume that the parabolic velocity profile for laminar flow in the tube, Eq. (6.32), persists right up to the exit, then we have $\alpha_1 = 2$. Since the downstream jet has a uniform profile, $\alpha_2 = 1$. Equation (6.34) then becomes

$$\tfrac{1}{2}\langle V\rangle_2^2 = \langle V_1\rangle^2 \tag{6.35}$$

The velocities can be eliminated between Eqs. (6.33) and (6.35), giving

$$D_j = (\tfrac{1}{2})^{1/4} D = 0.84 D \tag{6.36}$$

The jet is essentially a class III problem, and it can be solved using the momentum equation as well. Equation (5.46) can be written

$$w(\beta_1 \langle V \rangle_1 - \beta_2 \langle V \rangle_2) + p_1 \frac{\pi D^2}{4} - p_2 \frac{\pi D_j^2}{4} - F = 0 \qquad (6.37)$$

Now, $\beta_1 = \frac{4}{3}$ for laminar flow and $\beta_2 = 1$. $p_1 = p_2 = p_{atm}$, so the equation becomes

$$w(\tfrac{4}{3}\langle V \rangle_1 - \langle V \rangle_2) + p_{atm} \frac{\pi}{4}(D^2 - D_j^2) - F = 0 \qquad (6.38)$$

We now make an important observation, which is in fact the pedagogical reason for including this problem. There are two contributions to the force. One is the frictional drag between the air and the liquid, which can usually be neglected. The other is *the horizontal component of the force of the atmospheric air on the side surface of the jet*; see Fig. 6-10. This force equals p_{atm} times the projection of the change in surface area, or

$$F = P_{atm} \frac{\pi}{4}(D^2 - D_j^2) \qquad (6.39)$$

Thus, the atmospheric pressure term drops out and Eq. (6.38) simplifies to $\frac{4}{3}\langle V \rangle_1 = \langle V \rangle_2$, which combines with Eq. (6.33) to give

$$D_j = (\tfrac{3}{4})^{1/2} D = 0.866 D \qquad (6.40)$$

This is nearly the same as the result obtained from the engineering Bernoulli equation. The factor 0.87 is closer to experiment than 0.84, indicating that neglecting the air drag is a somewhat better approximation than neglecting the viscous losses.

The approximations in this problem are worth noting. The assumption that parabolic flow persists right up to the tube exit is valid for Reynolds numbers greater than 50 to 100. At very small Reynolds numbers, substantial velocity rearrangement takes place in a final length of tubing equal to about one diameter, and the solution calculated here does not apply; in fact, the jet diameter at very small Re is 10 to 15% *larger* than the tube diameter. This point is discussed further briefly in Chapter 17.

The assumption that the pressure in the jet at plane 1 is atmospheric neglects the fact that surface tension causes a pressure difference over a curved interface; see Sec. 2.6. Surface tension effects can be shown to be unimportant for this problem in the Reynolds number range required here, but they can be important for free jets of relatively inviscid liquids at much lower Reynolds numbers.

Finally, we have assumed that the jet reaches its final diameter over a sufficiently small length to neglect droop caused by gravity.

The behavior of jets of viscoelastic liquids, such as polymer solutions, is quite different. There is an additional axial stress in these fluids in pipe flow that is related to the presence of the additional material property, which we called a relaxation time in Sec. 2.5. The final jet diameter is larger than for

Newtonian fluids and may be larger than the tube diameter. The momentum balance then provides an experimental method of determining the relaxation time. This is one way that the small relaxation times cited for dilute polymer solutions (10^{-3} to 10^{-2} s) have been measured.

6.8 THE ROTAMETER

A *rotameter* is a device for measurement of volumetric flow rates. It consists of a vertical, slightly tapered tube with a bob or float, as shown in Fig. 6-11. Several different float geometries are commonly used, and the one shown in the figure is typical. Fluid flows upward, and the vertical position of the bob is determined by the flow rate. The relationship between the bob height and the flow rate can be determined by use of the macroscopic balances.

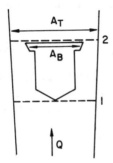

Figure 6-11. Schematic of a rotameter.

This problem does not fit nicely into any of our three categories, but the logical sequence that must be followed should now be clear. We take our two surfaces as shown in the figure, just below and just above the bob. We know nothing whatsoever about the pressure at either of these two points, so we shall have to eliminate the pressure difference between the energy and momentum equations. This will leave a single equation which, together with the continuity equation, expresses the velocity or flow rate in terms of geometric and fluid parameters and the losses and forces. We have no more equations available, so we shall have to make some assumptions about the nature of losses and forces in order to obtain the desired result.

Since the taper is gradual, the area A_T of the tube may be taken as constant between planes 1 and 2. The mean velocities are then

$$\langle V \rangle_1 = \langle V \rangle_2 = \frac{Q}{A_T} \qquad (6.41)$$

We will assume that the velocity is uniform over each surface where there is flow. Since the flow at plane 1 occurs over the entire cross section, we then have

$$\alpha_1 = \beta_1 = 1$$

At plane 2, on the other hand, the situation is like that in the jet ejector, Sec. 6.4, where the velocity is different over a center core and an annular space. Here, the velocity is zero in the center and flow occurs only through the annular area between the bob and the tube. The velocity in this area is $Q/(A_T - A_B)$, so we have

$$\beta_2 = \frac{\langle V^2 \rangle_2}{\langle V \rangle_2^2} = \frac{\left(\dfrac{Q}{A_T - A_B}\right)^2 \left(\dfrac{A_T - A_B}{A_T}\right)}{(Q/A_T)^2} = \frac{A_T}{A_T - A_B} \qquad (6.42)$$

Similarly,

$$\alpha_2 = \frac{\langle V^3 \rangle_2}{\langle V \rangle_2^3} = \left(\frac{A_T}{A_T - A_B}\right)^2 \qquad (6.43)$$

There is no shaft work, so the engineering Bernoulli equation, Eq. (5.34), for an incompressible fluid is

$$\tfrac{1}{2}[\alpha_1 \langle V \rangle_1^2 - \alpha_2 \langle V \rangle_2^2] + \frac{p_1 - p_2}{\rho} + g(h_1 - h_2) - l_V = 0 \qquad (6.44)$$

We shall need to do something about the losses. Based on our past experience, it seems likely that we can expect the losses to correlate with the maximum velocity, which occurs at the contraction; thus we assume a form

$$l_V = \tfrac{1}{2}K_R \left(\frac{Q}{A_T - A_B}\right)^2 \qquad (6.45)$$

The energy equation is then

$$\tfrac{1}{2}\rho \left(\frac{Q}{A_T}\right)^2 \left[1 - (1 + K_R)\left(\frac{A_T}{A_T - A_B}\right)^2\right] + (p_1 - p_2) + \rho g(h_1 - h_2) = 0 \qquad (6.46)$$

In writing the momentum equation, it is important to recall that the control volume consists of the entire tube cross section between 1 and 2. In particular, the control volume contains the float. The only forces acting on the control volume are those on the fluid from the side of the tube, and we shall neglect these for the small distance involved. The momentum equation, Eq. (5.46), is then

$$\rho Q[\beta_1 \langle V \rangle_1 - \beta_2 \langle V \rangle_2] + (p_1 - p_2)A_T$$
$$- \rho g[(h_2 - h_1)A_T - \mathcal{V}_B] - \rho_B g \mathcal{V}_B = 0 \qquad (6.47)$$

where ρ_B and $\mathcal{V}_B$ are the density and volume of the bob, respectively. $(h_2 - h_1)A_T - \mathcal{V}_B$ is the volume occupied by liquid, so $\rho g[(h_2 - h_1)A_T - \mathcal{V}_B]$ is the gravity term for the liquid in the control volume and $\rho_B g \mathcal{V}_B$ is the gravity term for the solid bob. The sign on each term is negative because gravity points down. Note that the "buoyancy" term, $\rho g \mathcal{V}_B$, appears in a perfectly natural manner. Substituting for β_2 and $\langle V \rangle$ and dividing by A_T

gives

$$p\left(\frac{Q}{A_T}\right)^2\left(1 - \frac{A_T}{A_T - A_B}\right) + (p_1 - p_2) + \rho g(h_1 - h_2) - \frac{g\mho_B}{A_T}(\rho_B - \rho) = 0$$

$$(6.48)$$

The pressure difference is now eliminated by subtracting Eq. (6.46) from (6.48); with a bit of rearrangement this gives

$$\frac{1}{2}\left(\frac{Q}{A_T}\right)^2\left(\frac{A_B}{A_T - A_B}\right)^2\left[1 + K_R\left(\frac{A_T}{A_B}\right)^2\right] - \frac{g\mho_B}{A_T}\left(\frac{\rho_B}{\rho} - 1\right) = 0$$

or

$$Q = (A_T - A_B)\sqrt{\frac{A_T/A_B}{1 + K_R(A_T/A_B)^2}}\sqrt{\frac{2g\mho_B}{A_B}\left(\frac{\rho_B}{\rho} - 1\right)} \qquad (6.49)$$

K_R needs to be determined experimentally. The dependence of A_T on vertical position determines the position of the float for a given flow rate; the ratio A_T/A_B will change with position as determined by the taper of the tube. Rotameters are usually calibrated for a particular fluid. According to Eq. (6.49), the calibration must be adjusted by a factor of $(\rho_B/\rho - 1)^{1/2}$ when the fluid is changed.

The first analysis of the rotameter was by Schoenborn and Colburn in 1939. They estimated the pressure drop by assuming that flow through the space between the bob and the tube was analogous to flow through an orifice, treated here in Sec. 6.5. They thus obtained the equation

$$Q = C_R(A_T - A_B)\sqrt{\frac{2g\mho_B}{A_B}\left(\frac{\rho_B}{\rho} - 1\right)} \qquad (6.50)$$

C_R is analogous to the orifice coefficient, which is shown as having a value of 0.62 in Eq. (6.27) for a sharp-edged orifice. The maximum value of C_R, which occurs in the absence of all losses, is unity. Equations (6.49) and (6.50) are equivalent if we assume that $A_T \simeq A_B$, in which case $C_R \simeq (1 + K_R)^{-1/2}$. Schoenborn and Colburn found that they could correlate C_R with the Reynolds number, using the velocity at the minimum spacing and the linear width of the gap. The latter is $2(\sqrt{A_T} - \sqrt{A_B})/\sqrt{\pi}$. The Reynolds number is thus

$$\text{Re} = \frac{2}{\sqrt{\pi}}(\sqrt{A_T} - \sqrt{A_B})\left(\frac{Q}{A_T - A_B}\right)\frac{\rho}{\eta} = \frac{2}{\sqrt{\pi}}\frac{\rho Q}{\eta(\sqrt{A_T} + \sqrt{A_B})} \qquad (6.51)$$

The correlation varied somewhat from rotameter to rotameter; Fig. 6-12 is typical. Note, however, that C_R exceeds unity at high Reynolds numbers, which is inconsistent with the orifice analogy.

The data used to construct Fig. 6-12 were used to calculate K_R. This is shown in Fig. 6-13, and it is clear that K_R does correlate well with Reynolds number over three decades. The correlation differs somewhat for bobs of different shape. Either Eq. (6.49) or (6.50) may be used for Re less than about 4000, but at higher Re the bob appears to be sufficiently high in the

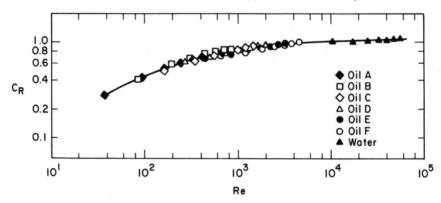

Figure 6-12. Rotameter coefficient ("Rotameter B") as a function of Reynolds number. Reproduced from Schoenborn and Colburn, *Trans. Am. Inst. Chem. Eng.*, **35**, 359 (1939), copyright by the American Institute of Chemical Engineers, by permission.

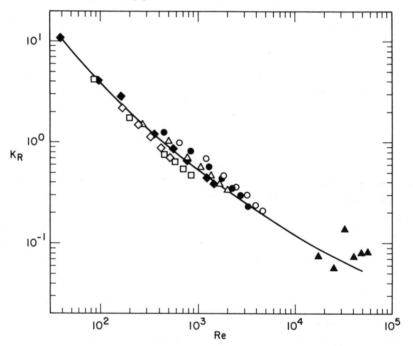

Figure 6-13. Loss coefficient as a function of Re for rotameter *B*.

tube to cause an error in the assumption that $A_T = A_B$, and Eq. (6.49) with K_R is preferable. There is considerable scatter in the data for water, but the water points (with one exception) do seem to follow the same correlation as the more viscous oils.

6.9 FLOW AND PRESSURE DISTRIBUTION IN A MANIFOLD

6.9 1 Description of a Manifold

A manifold is a device for distributing a liquid or gas, as shown in Fig. 6-14. The fluid is conveyed through a main tube and ejected through a series of side ports. The ports might simply be holes open to a surrounding uniform pressure; the burner on a kitchen gas range is the most familiar example of such a manifold, although gas distribution lines in liquid-phase reactors might have a similar configuration. The side ports might be long tubes, perhaps filled with catalyst, as in some packed bed reactors where highly exothermic reactions require the use of large numbers of small tubes to increase surface area and facilitate heat removal.

A schematic diagram of a return manifold would look the same, except that flow is from the side ports into the main tube. A U-shaped manifold consists of a distribution manifold and a return manifold, with the side ports consisting of tubes connecting the two. For simplicity, we will consider only distribution manifolds but the same principles apply more generally.

The basic design problem is to obtain a desired flow distribution through side ports. In most cases we would want approximately equal flows through all ports. This design problem can be solved using the macroscopic balances, although some of the required experiments are difficult to carry out reproducibly and research is still being done.

Figure 6-14. Schematic of a manifold.

6.9.2 Pressure Recovery

The first task in analyzing a manifold is to determine the change in the pressure distribution that results from fluid flow through the side ports. This is a class III problem that cannot be solved exactly, but by use of the momentum and energy balances considerable insight can be obtained, suggesting a single experiment which completes the description.

Flow past a single port is shown schematically in Fig. 6-15. The dashed region is the control volume for application of the momentum equation. We neglect the small frictional force on the tube walls from one side of the hole to the other. We also assume that the flow V_e through the side hole is perpen-

125

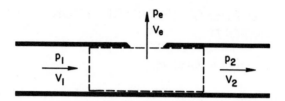

Figure 6-15. Flow past a single port.

dicular to the direction of flow, in which case the side flow contains no axial momentum. *This is a highly restrictive assumption* for a distribution manifold, although it is probably quite good for a return manifold. The momentum equation in the flow direction for a horizontal pipe then simplifies to

$$p_2 - p_1 = \rho(V_1^2 - V_2^2) \tag{6.52}$$

Since the flow rate decreases beyond the hole while the area stays the same, $V_2 < V_1$, so p_2 must be greater than p_1. That is, there is a *pressure increase* as the fluid moves past the hole.

The flow through the side hole is not, of course, exactly perpendicular to the flow direction, so the right-hand side of Eq. (6.52) should contain a negative term which accounts for the axial momentum loss through the hole. Thus, we should anticipate that Eq. (6.52) will overestimate the pressure recovery for a distribution manifold.

The energy equation can also be applied to this problem. It is convenient to visualize the flow as separating as shown schematically in Fig. 6-16. There

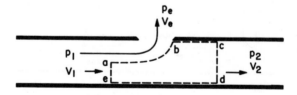

Figure 6-16. Assumed streamline pattern near a port.

is no flow across the line $\overline{ab}$; all fluid which flows across line $\overline{cd}$ enters the control volume across line $\overline{ae}$. This is, of course, a severe idealization of the real flow. Neglecting losses, the engineering Bernoulli equation applied to the control volume $\overline{abcdea}$ for a horizontal pipe then simplifies to

$$p_2 - p_1 = \frac{\rho}{2}(V_1^2 - V_2^2) \tag{6.53}$$

This equation also predicts a pressure increase, but it gives an estimate of the pressure recovery which is one-half that calculated from the momentum equation.

Each of the two estimates of the pressure recovery involves rather gross approximation, and it is not evident a priori which, if either, is the more accurate. Both suggest, however, that the pressure recovery should be approximately proportional to the difference of the squares of the upstream and downstream velocities, and we are motivated to use a relation

$$p_2 - p_1 = k\rho(V_1^2 - V_2^2) \tag{6.54}$$

Reported values of k range from 0.4 to 0.88 for distribution manifolds, but the best data appear to lie (with substantial scatter) close to $k = 0.45$, which is nearly the value given by the Bernoulli equation. (For return manifolds, k appears to be close to 1.0, the value given by the momentum equation.)

6.9.3 Side Flow

The flow out the side port can now be estimated. Let D denote the diameter of the manifold tube and d the diameter of the round side hole. The equation of conservation of mass for an incompressible fluid is then

$$D^2 V_1 = D^2 V_2 + d^2 V_e \tag{6.55}$$

If the side port is a sharp-edged orifice, with fluid discharging to a pressure p_e, it follows from Eq. (6.27) that the exit velocity is

$$V_e = 0.62 \sqrt{\frac{2}{\rho}\left(\frac{p_1 + p_2}{2} - p_e\right)} \tag{6.56}$$

Here, we have assumed that the pressure on the manifold side of the orifice is the arithmetic average of the upstream and downstream values. The velocity of approach term is not included, and the factor 0.62 is an average that might not apply to any particular orifice.

If the side port is a long tube of length l that exits to pressure p_e, then Eq. (3.6), which defines the friction factor, can be rearranged to the form

$$V_e = \sqrt{\frac{d}{4lf}} \sqrt{\frac{2}{\rho}\left(\frac{p_1 + p_2}{2} - p_e\right)} \tag{6.57}$$

Here we have ignored entry and exit losses in the side pipe. f will typically be of order 0.005, so if l/d is of order several hundred, the coefficient $\sqrt{d/4lf}$ will be of order 0.5 to 1.0.

In general, we can expect the velocity through the side port to be of the form

$$V_e = c\sqrt{\frac{2}{\rho}\left(\frac{p_1 + p_2}{2} - p_e\right)} \tag{6.58}$$

where c will be known for any given side port geometry and will be a number of order unity.

Equations (6.55), (6.56), and (6.58) are sufficient to solve for the three variables p_2, V_2, and V_e in terms of the upstream quantities p_1 and V_1. After some manipulation the pressure can be obtained explicitly as

$$p_2 = \frac{k - \gamma^2}{k + \gamma^2}p_1 + \frac{2\gamma^2}{k + \gamma^2}p_e + \frac{2\rho V_1^2 k^2 \gamma^2}{(k + \gamma^2)^2}\sqrt{1 + \frac{2(k + \gamma^2)}{\rho V_1^2 k \gamma^2}(p_1 - p_e)}$$

(6.59)

where the dimensionless parameter γ is defined

$$\gamma \equiv \frac{kcd^2}{D^2} \qquad (6.60)$$

This equation is a bit difficult to interpret, but a simplification is possible. The product kc will be less than unity, while d^2/D^2 will normally be quite small. Thus, we expect $\gamma \ll 1$, and we can expand the right-hand side of Eq. (6.59) about $\gamma = 0$ to obtain

$$p_2 = p_1 + \gamma V_1\sqrt{8\rho(p_1 - p_e)} + \text{terms of order } \gamma^2 \qquad (6.61)$$

Equation (6.58) can then be solved to the same approximation for V_e as

$$V_e = c\sqrt{\frac{2}{\rho}\{p_1 - p_e + \gamma V_1[8\rho(p_1 - p_e)]^{1/2}\}} \qquad (6.62)$$

V_2 is then obtained directly from Eq. (6.55) in the form

$$V_2 = V_1 - \frac{d^2}{D^2}V_e \qquad (6.63)$$

6.9.4 Overall Behavior

The calculations for a complete manifold can now be carried out in a sequential manner. The pressure drop from the entrance to the first port is computed from the equation for a straight length of pipe. The side flow at the first port and the downstream side pressure and velocity are computed from Eqs. (6.61) to (6.63). The pressure drop to the second port is then computed from the straight-pipe equations, and the process is repeated to the end of the manifold. It is observed experimentally that the friction factor for straight pipe in Fig. 3-1 can be used as long as the spacing between side ports is at least five pipe diameters. If the manifold is closed beyond the last port, there is no flow and $V_2 = 0$. In that case the initial pressure at the start of the manifold cannot be specified arbitrarily, but must be chosen in such a way that the condition $V_2 = 0$ beyond the last port is satisfied. This will generally require iterative calculation until the proper initial pressure is determined.

Figure 6-17 shows pressure measurements in a distribution manifold with four side ports, each a pipe with a diameter equal to one-fourth the distribution pipe diameter. The linear pressure drop between ports corresponds to normal pipe flow. Note that there is a net pressure *increase* from the beginning to the end of the manifold.

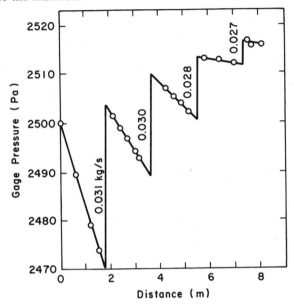

Figure 6-17. Pressure distribution for air flow along a 100-mm manifold with four 25-mm tubes as side ports, initial velocity 12.3 m/s. The numbers along each vertical line are the recorded mass flow rates at the ports. Data of Y.D. Miron.

6.10 CAVITATION

Cavitation is a phenomenon that needs to be mentioned briefly. The Bernoulli equation, Eq. (5.56), can be written for an incompressible liquid as

$$p_2 = p_1 + \tfrac{1}{2}\rho(V_1^2 - V_2^2) + g(h_1 - h_2) \qquad (6.64)$$

If V_2 is sufficiently greater than V_1, or if h_2 is sufficiently higher than h_1, we might come across circumstances in which the pressure p_2 is less than the vapor pressure of the liquid. In that case, vapor will form and calculations based on the assumption of a single liquid phase have no meaning. When vapor bubbles form within a liquid because of a local decrease in pressure, it is called *cavitation*. Cavitation often occurs near the tip of an impellor. Bubble formation can cause a surprising amount of damage to equipment when it occurs.

6.11 COMPRESSIBLE FLOW

6.11.1 General Comments

Compressibility of a gas is an important factor when the velocity becomes comparable to the velocity of sound. Sonic velocities are not usually reached in process equipment, but they can occur in nozzles and, of extreme importance from a safety point of view, in relief valves. When compressibility is important, there are important qualitative differences in the flow as compared to incompressible flow in the same geometry. Compressibility is usually dealt with in courses in thermodynamics and in more advanced treatises on fluid mechanics. We will present a brief introduction here in order to illustrate the phenomenon of *choking flow*.

The speed of sound is the speed at which small pressure waves travel through a fluid. For an isothermal ideal gas it is readily established in thermodynamics textbooks that the speed of sound is $(p/\rho)^{1/2}$ or, equivalently, $(R_g T/M_w)^{1/2}$, where R_g is the gas constant, T the absolute temperature, and M_w the molecular weight. At atmospheric pressure, approximately 10^5 Pa, and room temperature, the density of air is approximately $1.2 \, \text{kg/m}^3$, so the speed of sound is about 300 m/s.

6.11.2 Pipe Flow

The important features of compressible flow are illustrated by considering steady-state isothermal flow of an ideal gas in a horizontal smooth pipe. The more common case of adiabatic flow is carried out in an identical manner, with essentially the same results, but the analysis requires a bit more detail and manipulation.

We will be using the engineering Bernoulli equation, and to do so we need to make a few observations about the losses. It is found experimentally that the friction factor-Reynolds number function obtained for incompressible fluids applies to compressible fluids as well. The Reynolds number is written

$$\text{Re} = \frac{D\langle V\rangle\rho}{\eta} = \frac{4}{\pi D}\frac{(\pi D^2\langle V\rangle\rho/4)}{\eta} = \frac{4w}{\pi D\eta} \tag{6.65}$$

w is the mass flow rate, which is independent of axial position. For simplicity, we will assume that the viscosity does not change significantly over the length of the pipe despite pressure changes; in that case the Reynolds number is a constant over the entire length, and therefore so is the friction factor. The velocity does change continuously with position, however, so the losses must be written for a differential length, dz:

$$dl_V = 2\langle V\rangle^2 f\frac{dz}{D} \tag{6.66}$$

The engineering Bernoulli equation for a differential length, Eq. (5.33), is

$$\tfrac{1}{2}dV^2 + \frac{dp}{\rho} + 2V^2 f \frac{dz}{D} = 0 \qquad (6.67)$$

Here we have assumed that $\alpha = 1$. We now further assume that the gas is ideal,

$$\rho = \frac{M_w p}{R_g T} \qquad (6.68)$$

and we replace V by the mass flow rate, w, through the relation $w = \pi D^2 V \rho / 4$. Following some algebraic manipulation, and using the fact that $dV^2 = 2V\,dV$, Eq. (6.67) becomes

$$-\frac{dp}{p} + \frac{\pi^2 D^4 M_w}{16 w^2 R_g T} p\,dp + \frac{2f}{D}dz = 0 \qquad (6.69)$$

This can be integrated to give

$$-\ln \frac{p_2}{p_1} + \frac{\pi^2 D^4 M_w}{32 w^2 R_g T}(p_2^2 - p_1^2) + \frac{2fL}{D} = 0 \qquad (6.70)$$

or, following some rearrangement and the use of Eq. (6.68),

$$w^2 = \frac{\pi^2 D^4 \rho_1 p_1}{16}\left[\frac{1 - (p_2/p_1)^2}{(4fL/D) - \ln (p_2/p_1)^2}\right] \qquad (6.71)$$

Thus, we have the mass throughput in terms of conditions at the start of the pipe (p_1 and ρ_1) and the final pressure, p_2.

Let us take the conditions at the start of the pipe as fixed. Since the pressure must decrease, it follows that $0 < p_2 < p_1$. Now w^2 clearly goes to zero as $p_2 \to p_1$. The remarkable observation to be made from Eq. (6.71), however, is that w^2 *also goes to zero as* $p_2 \to 0$. Thus, there is a maximum throughput, and it occurs for an intermediate value of p_2.

The maximum is found by differentiating w^2 with respect to $(p_2/p_1)^2$ and setting the derivative to zero:

$$\frac{dw^2}{d(p_2/p_1)^2} = \frac{\pi^2 D^4 \rho_1 p_1}{16}\left\{-\frac{1}{\dfrac{4fL}{D} - \ln\left(\dfrac{p_2}{p_1}\right)^2} + \frac{1 - (p_2/p_1)^2}{\left(\dfrac{p_2}{p_1}\right)\left[\dfrac{4fL}{D} - \ln\left(\dfrac{p_2}{p_1}\right)^2\right]^2}\right\} = 0 \qquad (6.72)$$

This can be combined with Eq. (6.71) to eliminate the term involving fL/D and the logarithm, giving

$$w_{max}^2 = \left(\rho_2 \frac{\pi D^2}{4} V_{2,max}\right)^2 = \left(\frac{\pi D^2}{4}\right)^2 p_2^2 \frac{\rho_1}{p_1} \qquad (6.73)$$

Finally, using $p_1/\rho_1 = p_2/\rho_2$ for an isothermal ideal gas, we can solve for the exit velocity at the maximum throughput:

$$V_{2,max} = \sqrt{\frac{p_2}{\rho_2}} \qquad (6.74)$$

This is the speed of sound in an isothermal ideal gas, so we obtain the result that the *exit velocity cannot exceed sonic velocity*. If the pressure outside the pipe is lower than the pressure that gives the maximum throughput, the exit velocity will remain at the sonic velocity and there will be a standing expansion shock wave across which the pressure changes to the outside value at the exit. The existance of a maximum throughput is known as *choking*.

6.12 CONCLUDING REMARKS

The examples in this chapter have been selected to show the scope of applications of the macroscopic balances. These important tools are used frequently in process applications.

Application of the macroscopic balances may be the most difficult subject in a first course in fluid mechanics. This is because physically-based assumptions must always be made, and the student rarely has sufficient experience with the physical processes involved to feel comfortable with the assumptions. The examples in this chapter should serve as a guide, and confidence will come with repeated application.

We now leave macroscopic balances and turn to the analysis of detailed flow structure.

BIBLIOGRAPHICAL NOTES

This chapter is nicely complemented by several of the chapters in

LAPPLE, C. E., and COWORKERS, *Fluid and Particle Mechanics*, University of Delaware Press, Newark, Del., 1956.

For other examples, including applications to open channel flow and turbomachinary, see

STREETER, V. L., and WYLIE, E. B., *Fluid Mechanics*, 7th ed., McGraw-Hill Book Company, New York, 1979.

There is a good introduction to the problem of flow distribution in

ACRIVOS, A., BABCOCK, J., and PIGFORD, R. L., *Chem. Eng. Sci.*, **10**, 112 (1959).

See also the discussion in *Perry's Handbook*. A detailed discussion is

MILLER, D. S., "Internal Flow: A Guide to Losses in Pipe and Duct Systems," B.H.R.A. Report, Greenfield, England, 1971.

A recent survey and new data, including those in Fig. 6-17, may be found in

MIRON, Y., "Fluid Flow Through Piping Manifolds," M.Ch.E. thesis, University of Delaware, Newark, Del., 1977.

More detail on compressible flow may be found in Lapple and Streeter and Wylie and in many books on thermodynamics; see, for example, Chapter 8 of

BALZHISER, R. E., SAMUELS, M. R., and ELIASSEN, J. D., *Chemical Engineering Thermodynamics*, Prentice-Hall, Inc., Englewood Cliffs, N.J., 1972.

For an introduction to the use of the laser Doppler flowmeter, see

DURST, F., MELLING, A., and WHITELAW, J. H., *Laser Doppler Anemometry*, Academic Press, Inc., New York, 1978.

PROBLEMS

6.1. Relate Q to Δh for the venturi flowmeter shown in Fig. 6P1. The density of flowing fluid is ρ and the density of manometer fluid is ρ_m. (Losses may be almost entirely neglected in a well-designed venturi. Why? Typically, the actual flow rate is 98 to 99 percent of the flow rate computed by neglecting losses.)

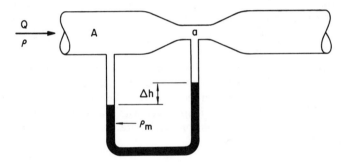

6.2. The siphon tube in Fig. 6P2 is 0.2 m in diameter. Estimate the flow rate if the tank contains water. What is the pressure at B?

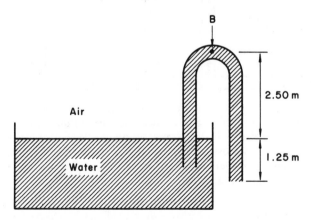

6.3. A major industrial disaster occurred at Flixborough, England when a pipeline containing a flammable hydrocarbon mixture burst. The break is believed to have occurred at a point where a reactor in a gravity-flow cascade was replaced by an *unsupported* double bend, as shown schematically in Fig. 6P3. Describe quantitatively why this is an unsafe configuration.

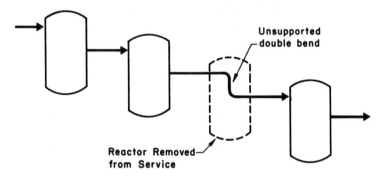

6.4. A flow distribution device is to deliver an equal volumetric flow to each of N exit ports. This is accomplished by changing the diameter of the distributor after each port. The initial diameter is D and the diameter of each port is d. How should the diameter vary?

6.5. A reactor consists of N parallel packed tubes, as shown in Fig. 6P5. Develop the design procedure to determine the flow in each packed section. All tubes have diameter D, and packed sections are of height h and equally spaced a distance l apart. The packing is spherical, with diameter d_p and void fraction ϵ. You may assume that fluid properties are constant, with density ρ and viscosity η.

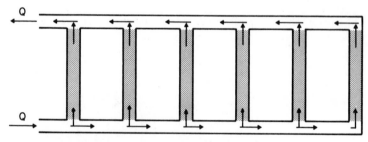

6.6. An open cylindrical tank of area A and height H empties to the atmosphere through a long horizontal pipe connected at the bottom. The pipe has length L and diameter d, $d \ll H$. If the tank is initially filled with a Newtonian liquid of density ρ and viscosity η, estimate the time for the tank to empty
a) if the pipe flow is laminar.
b) if the pipe flow is turbulent.

6.7. A plane jet of Newtonian fluid of density ρ and viscosity η impinges on a flat plate and splits into two streams, as shown in Fig. 6P7. Determine the force on the plate and the split in the flow. The jet approaches the plate with a flow rate Q_0 and thickness h_0; the width of the plane jet is W.

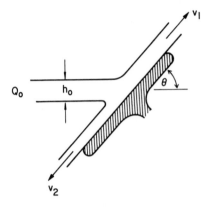

6.8. It is sometimes possible to induce a *hydraulic jump* in an open channel, where a rapidly flowing stream suddenly changes to a slowly flowing stream of increased depth; see Fig. 6P8. Given h_1 and v_1, show that a hydraulic jump is possible and compute h_2.

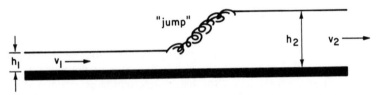

6.9. A high velocity water jet is used to cut rock. You may visualize the final stages of the cutting process as a jet impinging on a cylinder, as shown schematically in Fig. 6P9. The cylinder will fail in shear at a stress (force normal to the cylinder axis/cylinder cross-sectional area) of T_{fs}. Compute the velocity of the water jet required for cutting, if the nozzle diameter is d.

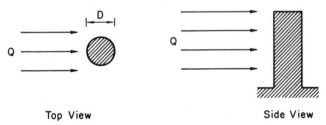

Top View   Side View

6.10. In the reaction injection molding ("RIM") process, stoichiometric amounts of a low viscosity reacting mixture are pumped through a mixing chamber into a mold, where a polymerization reaction occurs and a solid, shaped object is formed. RIM is a lower energy process than the more common injection molding process, where a high viscosity polymer melt is injected into the mold.

a) The flow system is shown schematically in Fig. 6P10. Estimate the flow rate of reactant A as a function of the pressure difference $p_A - p_0$.

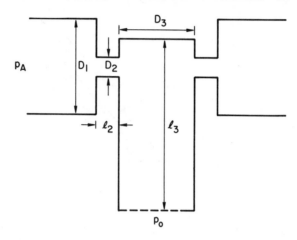

b) Lee and Macosko have described experiments on a system with the following dimensions:

$$D_1 = 9.5 \text{ mm} \qquad D_2 = 1 \text{ mm} \qquad D_3 = 3.2 \text{ mm}$$

$$l_2 = 1.5 \text{ mm} \qquad l_3 = 25.4 \text{ mm}$$

They studied water-glycerine systems with densities of order 1200 kg/m^3 and viscosities of order $0.1 \text{ Pa} \cdot \text{s}$. For flow rates in the range $0 - 3 \times 10^{-5}$ m^3/s, show that the only term that is important in determining the flow rate is the contraction loss from the reservoir.

6.11. a) Estimate the losses in a gradual expansion (Fig. 6P11) by assuming that the pressure variation through the expansion is approximately linear. (Note that the pressure difference is not known.)

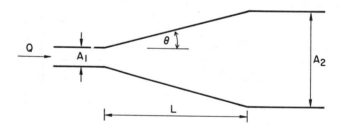

b) Estimate the losses in a gradual expansion by assuming that the flow at any axial position is approximately the same as turbulent flow in a pipe of the same diameter. (Hint: Compute the losses for a differential length from the results for a smooth pipe, and then integrate over the length of the contraction. The approach is analogous to the treatment of compressible flow in Sec. 6.11.)

c) Compare to the results of parts (a) and (b) and the value given by the Crane Co., $K = 2.6 \sin \theta (A_2/A_1 - 1)^2$, $0 \le \theta \le 22.5°$, and discuss any differences.

Part IV
Detailed Flow Structure

Microscopic Balances 7

7.1 INTRODUCTION

In many applications it is necessary to know the detailed flow structure. The macroscopic balances studied in Chapters 5 and 6 are averages over surfaces normal to the mean flow direction, and detailed information about structure is lost in that averaging. Thus, the macroscopic balances can never give us a description of fluid motion and forces on a fine scale, and we need to turn to an examination of the application of the conservation equations over a small spatial region.

The general approach is as shown in Fig. 7-1. We choose a small cube within the region of flow as a control volume and apply the conservation equations. We will obtain a set of equations describing the flow which will depend on position; in fact, they will be differential equations with coordinates x, y, and z, as well as time t, as independent variables. Integration of these differential equations will then provide a complete description of the flow at each point. The choice of a cube as the control volume means that the resulting equations will all be expressed in terms of a rectangular Cartesian coordinate system. This is for convenience only and we could derive the equations using some other shape as a control volume.

This chapter parallels Chapter 5, in that it is devoted entirely to derivation of the equations governing the flow. In the chapters that follow we will examine a variety of applications.

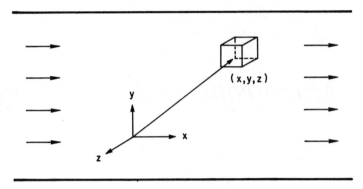

Figure 7-1. Microscopic control volume in a fluid.

7.2 CONSERVATION OF MASS

7.2.1 Continuity Equation

We first consider the equation of conservation of mass. The control volume is shown enlarged in Fig. 7-2. The small cube has faces of length Δx, Δy, and Δz, with one corner at position (x, y, z). The total volume is $\Delta x \, \Delta y \, \Delta z$. The velocity vector $\mathbf{v}$ at any position has components (v_x, v_y, v_z).

The principle of conservation of mass simply states that the rate of change of mass in the control volume equals the rate at which mass enters the control volume minus the rate at which mass leaves the control volume. The total mass is $\bar{\rho} \, \Delta x \, \Delta y \, \Delta z$, where $\bar{\rho}$ denotes the average density in the small cube.

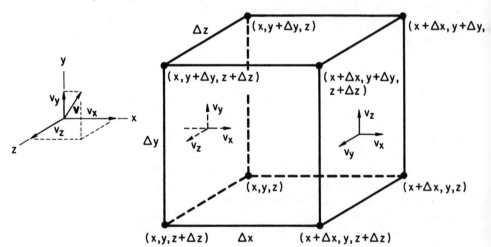

Figure 7-2. Cubic control volume.

Then

$$\text{rate of change of mass} = \frac{\partial}{\partial t} \bar{\rho} \, \Delta x \, \Delta y \, \Delta z \qquad (7.1)$$

The derivative is a *partial time derivative* because it denotes differentiation with respect to time at a particular position in space.

Fluid enters and leaves the control volume across faces normal to each of the coordinate directions. Let us focus first on the face at x, normal to the x direction, with area $\Delta y \, \Delta z$. We will refer to a face normal to the x direction as an *x face*. The normal component of velocity is v_x, and this is the only component involving flow across the face into the control volume, since v_y and v_z are parallel to the plane of the face. We will assume that flow is in the positive coordinate direction, as shown in the figure; if not, velocities will simply be negative numbers and the result will be unchanged. Flow at the x face located at x is then into the control volume, with a mass flow rate

$$\text{mass flow rate} = \langle \rho v_x \rangle \, \Delta y \, \Delta z |_x \qquad (7.2a)$$

where $\langle \rho v_x \rangle$ is the area average over the face. The symbol $|_x$ means "evaluated at x." Similarly, mass flow rates in over the y and z faces are, respectively,

$$\text{mass flow rate in, } y\text{-face} = \langle \rho v_y \rangle \, \Delta x \, \Delta z |_y \qquad (7.2b)$$

$$\text{mass flow rate in, } z\text{-face} = \langle \rho v_z \rangle \, \Delta x \, \Delta y |_z \qquad (7.2c)$$

Mass flows out in the x direction through the face at $x + \Delta x$, with a flow rate

$$\text{mass flow rate out, } x\text{-face} = \langle \rho v_x \rangle \, \Delta y \, \Delta z |_{x+\Delta x} \qquad (7.3a)$$

Similarly,

$$\text{mass flow rate out, } y\text{-face} = \langle \rho v_y \rangle \, \Delta x \, \Delta z |_{y+\Delta y} \qquad (7.3b)$$

$$\text{mass flow rate out, } z\text{-face} = \langle \rho v_z \rangle \, \Delta x \, \Delta y |_{z+\Delta z} \qquad (7.3c)$$

The conservation equation is then

$$\frac{\partial \bar{\rho}}{\partial t} \Delta x \, \Delta y \, \Delta z = \langle \rho v_x \rangle \, \Delta y \, \Delta z |_x + \langle \rho v_y \rangle \, \Delta x \, \Delta z |_y + \langle \rho v_z \rangle \, \Delta x \, \Delta y |_z$$
$$- \langle \rho v_x \rangle \, \Delta y \, \Delta z |_{x+\Delta x} - \langle \rho v_y \rangle \, \Delta x \, \Delta z |_{y+\Delta y} - \langle \rho v_z \rangle \, \Delta x \, \Delta y |_{z+\Delta z}$$
$$(7.4)$$

Dividing by the volume, $\Delta x \, \Delta y \, \Delta z$, Eq. (7.4) can be rewritten

$$\frac{\partial \bar{\rho}}{\partial t} = - \frac{\langle \rho v_x \rangle |_{x+\Delta x} - \langle \rho v_x \rangle |_x}{\Delta x} - \frac{\langle \rho v_y \rangle |_{y+\Delta y} - \langle \rho v_y \rangle |_y}{\Delta y}$$
$$- \frac{\langle \rho v_z \rangle |_{z+\Delta z} - \langle \rho v_z \rangle |_z}{\Delta z} \qquad (7.5)$$

Each term on the right-hand side of Eq. (7.5) is a difference quotient. We now let the volume $\Delta x \, \Delta y \, \Delta z$ shrink to zero. In that case,

$$\lim_{\substack{\Delta x \to 0 \\ \Delta y \to 0 \\ \Delta z \to 0}} \frac{\langle \rho v_x \rangle |_{x+\Delta x} - \langle \rho v_x \rangle |_x}{\Delta x} = \frac{\partial \rho v_x}{\partial x} \qquad (7.6a)$$

That is, as the area of the face shrinks to zero, the area average approaches the point value, and as the distance between faces shrinks to zero the difference quotient goes to the derivative. *We have a partial derivative because it represents the rate of change with respect to x at a particular y and z at a particular time, t.* Similarly,

$$\lim_{\substack{\Delta x \to 0 \\ \Delta y \to 0 \\ \Delta z \to 0}} \frac{\langle \rho v_y \rangle|_{y+\Delta y} - \langle \rho v_y \rangle|_y}{\Delta y} = \frac{\partial \rho v_y}{\partial y} \qquad (7.6b)$$

$$\lim_{\substack{\Delta x \to 0 \\ \Delta y \to 0 \\ \Delta z \to 0}} \frac{\langle \rho v_z \rangle|_{z+\Delta z} - \langle \rho v_z \rangle|_z}{\Delta z} = \frac{\partial \rho v_z}{\partial z} \qquad (7.6c)$$

As the volume shrinks to a point the density on each face and the average density, $\bar{\rho}$, simply approach the point density, ρ. Thus, we obtain, from Eq. (7.5),

$$\frac{\partial \rho}{\partial t} = -\frac{\partial \rho v_x}{\partial x} - \frac{\partial \rho v_y}{\partial y} - \frac{\partial \rho v_z}{\partial z} \qquad (7.7)$$

Equation (7.7) is known as the *continuity equation.* It is a partial differential equation whose solution establishes the relation between density and velocity at each point in the flow. Note that we have made explicit use of the *continuum hypothesis,* discussed in Sec. 2.3, by assuming that the concept of density remains valid as the control volume shrinks to a point.

The continuity equation is sometimes written in a slightly different form by expanding the derivative of each product to two terms:

$$\frac{\partial \rho v_x}{\partial x} = \rho \frac{\partial v_x}{\partial x} + v_x \frac{\partial \rho}{\partial x} \qquad (7.8)$$

and similarly for y and z. Equation (7.7) then becomes

$$\frac{\partial \rho}{\partial t} + v_x \frac{\partial \rho}{\partial x} + v_y \frac{\partial \rho}{\partial y} + v_z \frac{\partial \rho}{\partial z} = -\rho \left(\frac{\partial v_x}{\partial x} + \frac{\partial v_y}{\partial y} + \frac{\partial v_z}{\partial z} \right) \qquad (7.9a)$$

or, defining the symbol $D\rho/Dt$ as the left-hand side of Eq. (7.9a),

$$\frac{D\rho}{Dt} = -\rho \left(\frac{\partial v_x}{\partial x} + \frac{\partial v_y}{\partial y} + \frac{\partial v_z}{\partial z} \right) \qquad (7.9b)$$

$D(\ )/Dt$ is known as the *substantial derivative.* The physical meaning of this derivative is discussed below.

7.2.2 Substantial Derivative

The physical meaning of the substantial derivative is most easily understood by focusing attention on a small particle of fluid, as shown in Fig. 7-3. At time t, the position of the particle is $x(t)$, $y(t)$, $z(t)$. At a later time, $t + \Delta t$, the particle has moved to position $x(t + \Delta t)$, $y(t + \Delta t)$, $z(t + \Delta t)$. For suffi-

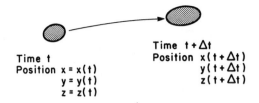

Figure 7-3. A particle of fluid changes spatial position with time.

ciently small Δt, we can write

$$x(t + \Delta t) = x(t) + \frac{dx}{dt} \Delta t = x(t) + v_x \Delta t \qquad (7.10a)$$

The substitution of v_x for dx/dt is possible since v_x is simply the rate of change of the x-coordinate position of the particle. Similarly,

$$y(t + \Delta t) = y(t) + v_y \Delta t \qquad (7.10b)$$

$$z(t + \Delta t) = z(t) + v_z \Delta t \qquad (7.10c)$$

Now consider any property ξ of the fluid particle that can change with time and position, such as its density. $\xi(t + \Delta t)$ is different from $\xi(t)$ because it is evaluated at a later time, but also because the particle is at a new position in space. Thus,

$$\xi(t + \Delta t) = \xi(t) + \frac{\partial \xi}{\partial t} \Delta t + \frac{\partial \xi}{\partial x} \Delta x + \frac{\partial \xi}{\partial y} \Delta y + \frac{\partial \xi}{\partial z} \Delta z \qquad (7.11a)$$

or, replacing Δx with $v_x \Delta t$ and similarly for y and z,

$$\xi(t + \Delta t) = \xi(t) + \frac{\partial \xi}{\partial t} \Delta t + \frac{\partial \xi}{\partial x} v_x \Delta t + \frac{\partial \xi}{\partial y} v_y \Delta t + \frac{\partial \xi}{\partial z} v_z \Delta t \qquad (7.11b)$$

If we divide by Δt and take the limit as Δt goes to zero, we thus obtain

$$\frac{D\xi}{Dt} = \lim_{\Delta t \to 0} \frac{\xi(t + \Delta t) - \xi(t)}{\Delta t} = \frac{\partial \xi}{\partial t} + v_x \frac{\partial \xi}{\partial x} + v_y \frac{\partial \xi}{\partial y} + v_z \frac{\partial \xi}{\partial z} \qquad (7.12)$$

Thus, we see that *the derivative $D\xi/Dt$ can be interpreted as the rate of change with time as recorded by an observer moving with a fluid particle.*

7.2.3 Vector Notation

A vector notation is often a useful shorthand in representing the continuity equation. In a rectangular Cartesian system the x, y, and z unit vectors are denoted $\mathbf{i}$, $\mathbf{j}$, and $\mathbf{k}$, respectively. Then the velocity vector is

$$\mathbf{v} = v_x \mathbf{i} + v_y \mathbf{j} + v_z \mathbf{k} \qquad (7.13)$$

The *gradient* of a scalar is the vector whose components are the partial derivatives; for example,

$$\text{grad } \xi = \nabla \xi = \frac{\partial \xi}{\partial x}\mathbf{i} + \frac{\partial \xi}{\partial y}\mathbf{j} + \frac{\partial \xi}{\partial z}\mathbf{k}$$

Both "grad" and "∇" are used in the literature to denote the gradient; the inverted triangle is called "nabla" or "del."

The "dot," or inner product of two vectors is a scalar formed by summing the products of x, y, and z components; this follows from the relations

$$\mathbf{i} \cdot \mathbf{i} = \mathbf{j} \cdot \mathbf{j} = \mathbf{k} \cdot \mathbf{k} = 1$$

$$\mathbf{i} \cdot \mathbf{j} = \mathbf{i} \cdot \mathbf{k} = \mathbf{j} \cdot \mathbf{k} = 0$$

Thus,

$$\mathbf{v} \cdot \nabla \xi = v_x \frac{\partial \xi}{\partial x} + v_y \frac{\partial \xi}{\partial y} + v_z \frac{\partial \xi}{\partial z} \tag{7.14}$$

and, from Eq. (7.12),

$$\frac{D\xi}{Dt} = \frac{\partial \xi}{\partial t} + \mathbf{v} \cdot \nabla \xi \tag{7.15}$$

It is useful to think of the gradient operator as a vector,

$$\nabla(\quad) = \frac{\partial(\quad)}{\partial x}\mathbf{i} + \frac{\partial(\quad)}{\partial y}\mathbf{j} + \frac{\partial(\quad)}{\partial z}\mathbf{k}$$

Then the inner product between ∇ and a vector $\mathbf{v}$ will be

$$\nabla \cdot \mathbf{v} = \frac{\partial v_x}{\partial x} + \frac{\partial v_y}{\partial y} + \frac{\partial v_z}{\partial z} \tag{7.16}$$

This is called the *divergence* of $\mathbf{v}$, and the symbol div $\mathbf{v}$ is sometimes used. Note that the order is important here, and that while $\nabla \cdot \mathbf{v}$ is a scalar, $\mathbf{v} \cdot \nabla$ is a differential operator:

$$\mathbf{v} \cdot \nabla(\quad) = v_x \frac{\partial(\quad)}{\partial x} + v_y \frac{\partial(\quad)}{\partial y} + v_z \frac{\partial(\quad)}{\partial z} \tag{7.17}$$

The continuity equation, Eq. (7.7), can thus be written in vector notation as

$$\frac{\partial \rho}{\partial t} = -\nabla \cdot \rho\mathbf{v} \tag{7.18}$$

The expanded version, Eq. (7.9), is written in vector notation as

$$\frac{D\rho}{Dt} = -\rho\nabla \cdot \mathbf{v} \tag{7.19}$$

with $D\rho/Dt$ given by Eq. (7.15) with ξ replaced by ρ.

7.2.4 Incompressible Fluid

If the density is constant in time and space, all its derivatives vanish. In that case the continuity equation simplifies to the form

$$\text{incompressible:} \quad \nabla \cdot \mathbf{v} = 0 \tag{7.20}$$

7.3 CONSERVATION OF MOMENTUM

7.3.1 Momentum Flow

We now apply the principle of conservation of linear momentum to the small control volume contained in the fluid. Momentum is a vector, mass times velocity; momentum per unit volume is density times velocity. The principle of conservation of linear momentum states that for each coordinate direction the rate of change of linear momentum in the control volume equals the rate at which momentum flows into the control volume, minus the rate at which momentum flows out of the control volume, plus the sum of all forces acting on the control volume.

We will carry out the derivation for the x component of linear momentum; the y- and z-component equations follow by analogy and can be obtained from the x equation by permutation of x, y, and z. x momentum per unit volume is $\overline{\rho v_x}$; thus, total x momentum is $\overline{\rho v_x} \, \Delta x \, \Delta y \, \Delta z$, and

$$\text{rate of change of } x \text{ momentum} = \frac{\partial}{\partial t} \overline{\rho v_x} \, \Delta x \, \Delta y \, \Delta z \qquad (7.21)$$

The rate at which x momentum flows into the control volume across a face is the product of x momentum per unit volume, ρv_x, and the volumetric flow rate across the face. There is flow into the control volume across *three* faces, an x face, a y face, and a z face, and x momentum is carried into the control volume across each. The volumetric flow rate across an x face is $v_x \, \Delta y \, \Delta z$. Across a y face the volumetric flow rate is $v_y \, \Delta x \, \Delta z$, while the volumetric flow rate across a z face is $v_z \, \Delta x \, \Delta y$. Thus,

$$\begin{array}{l}\text{rate at which } x \\ \text{momentum flows} \\ \text{into control volume}\end{array} = (\rho v_x)(v_x \, \Delta y \, \Delta z)|_x + (\rho v_x)(v_y \, \Delta x \, \Delta z)|_y$$
$$+ (\rho v_x)(v_z \, \Delta x \, \Delta y)|_z \qquad (7.22a)$$

Similarly,

$$\begin{array}{l}\text{rate at which} \\ x \text{ momentum flows out} \\ \text{of control volume}\end{array} = (\rho v_x)(v_x \, \Delta y \, \Delta z)|_{x+\Delta x} + (\rho v_x)(v_y \, \Delta x \, \Delta z)|_{y+\Delta y}$$
$$+ (\rho v_x)(v_z \, \Delta x \, \Delta y)|_{x+\Delta z} \qquad (7.22b)$$

(We have not included the triangular brackets $\langle \rangle$ to denote the area average over each face in order to facilitate identification of the individual momentum and flow rate terms. For a sufficiently small control volume, the area average and the point value on the face are the same, and all averages disappear in the limiting process in any event.) Thus far, then, the x component of the momentum equation can be written in a mixture of words and symbols as

$$\frac{\partial}{\partial t}\overline{\rho v_x}\,\Delta x\,\Delta y\,\Delta z = +\rho v_x v_x\,\Delta y\,\Delta z|_x - \rho v_x v_x\,\Delta y\,\Delta z|_{x+\Delta x}$$
$$+ \rho v_x v_y\,\Delta x\,\Delta z|_y - \rho v_x v_y\,\Delta x\,\Delta z|_{y+\Delta y}$$
$$+ \rho v_x v_z\,\Delta x\,\Delta y|_z - \rho v_x v_z\,\Delta x\,\Delta y|_{z+\Delta z}$$
$$+ \text{sum of all forces acting on control volume}$$

(7.23)

7.3.2 Stress

A force will be exerted on a face of the control volume because of the presence of fluid adjacent to that face. Let us focus on an x face. The force acting on the x face because of the surrounding fluid is denoted $\boldsymbol{\sigma}_x\,\Delta y\,\Delta z$, where $\boldsymbol{\sigma}_x$ is a stress, or force per unit area. $\boldsymbol{\sigma}_x$ can be resolved into its x, y, and z components, as shown in Fig. 7-4, and written

$$\boldsymbol{\sigma}_x = \sigma_{xx}\mathbf{i} + \tau_{xy}\mathbf{j} + \tau_{xz}\mathbf{k} \tag{7.24a}$$

where σ_{xx} denotes the stress component normal to the surface and represents a tension or compression. τ_{xy} and τ_{xz} denote stress components parallel to the face and represent shear. It is common to use the σ–τ notation introduced here; the reason will become clearer subsequently. The first subscript (x in this case) denotes the face; the second subscript denotes the coordinate direction in which the stress is acting. In a similar manner, the force acting on a y face is $\boldsymbol{\sigma}_y\,\Delta x\,\Delta z$, with components as shown in the figure,

$$\boldsymbol{\sigma}_y = \tau_{yx}\mathbf{i} + \sigma_{yy}\mathbf{j} + \tau_{yz}\mathbf{k} \tag{7.24b}$$

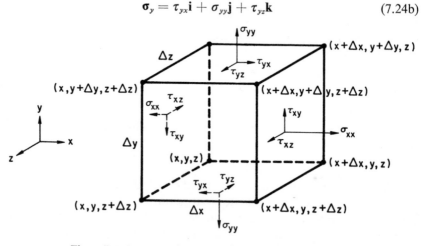

Figure 7-4. Stresses acting on faces of the control volume.

In the z direction,

$$\boldsymbol{\sigma}_z = \tau_{zx}\mathbf{i} + \tau_{zy}\mathbf{j} + \sigma_{zz}\mathbf{k} \qquad (7.24c)$$

The components of $\boldsymbol{\sigma}_z$ are not shown in the figure.

At this point it is necessary to adopt a convention on signs. We will use the convention that the fluid on the side of the face with the greater (algebraic) value of the coordinate exerts positive stresses on the fluid with the smaller coordinate value. This is illustrated schematically for x and y faces in Fig. 7-5. σ_{xx} and τ_{xy} are shown as acting in the positive coordinate directions on

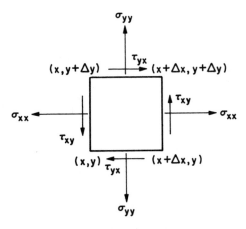

Figure 7-5. Convention for positive and negative stresses on a face.

the x face at $x + \Delta x$, since the fluid outside the control volume is the farther from the origin here. For the x face at x, the fluid inside the control volume is farther from the origin and, by the convention, would exert stresses acting in the positive direction on the fluid on the other side of the face, outside the control volume. But the fluid outside the control volume must exert an equal and opposite stress on the fluid in the control volume, so the stresses exerted by the fluid outside the control volume on the x face at x must act in the negative coordinate direction, as shown. Similar results follow for the y and z faces.

It is evident from the figure that this convention is equivalent to stating that *the normal stresses exerted by the surroundings put the control volume in tension when they are positive.* We could have equally well adopted the convention that the surrounding fluid puts the control volume in compression when the stresses are positive. This latter approach has a logical basis in considering analogies among molecular transport of mass, momentum, and energy, but it is contrary to the convention usually used in mechanics and is rarely employed.

We can now go on with the development of the x-momentum equation. The x components of stress from $\boldsymbol{\sigma}_x$, $\boldsymbol{\sigma}_y$, and $\boldsymbol{\sigma}_z$, respectively, are σ_{xx}, τ_{yx}, and τ_{zx}. With our convention on signs, we therefore have

$$
\begin{aligned}
\text{x-direction forces} \\
\text{exerted by surrounding} &= \sigma_{xx}\,\Delta y\,\Delta z|_{x+\Delta x} - \sigma_{xx}\,\Delta y\,\Delta z|_x \\
\text{fluid on control volume} & \\
&+ \tau_{yx}\,\Delta x\,\Delta z|_{y+\Delta y} - \tau_{yx}\,\Delta x\,\Delta z|_y \\
&+ \tau_{zx}\,\Delta x\,\Delta y|_{z+\Delta z} - \tau_{zx}\,\Delta x\,\Delta y|_z
\end{aligned}
\tag{7.25}
$$

7.3.3 Body Force

Finally, we must consider forces that act on the body of the fluid rather than on the surface. The only body force that we will consider will be the force of gravity, which has a value

$$
\text{body force in x direction} = \bar{\rho} g_x\,\Delta x\,\Delta y\,\Delta z
\tag{7.26}
$$

where g_x is the x component of the gravitational acceleration vector, $\mathbf{g}$. The only other important body force that arises in applications is an electrical force that occurs when a conducting fluid moves through a magnetic field.

7.3.4 Cauchy Momentum Equation

Equations (7.23), (7.25), and (7.26) combine to form the complete expression for conservation of x momentum,

$$
\begin{aligned}
\frac{\partial}{\partial t}\overline{\rho v_x}\,\Delta x\,\Delta y\,\Delta z = &+\rho v_x v_x\,\Delta y\,\Delta z|_x - \rho v_x v_x\,\Delta y\,\Delta z|_{x+\Delta x} \\
&+ \rho v_x v_y\,\Delta x\,\Delta z|_y - \rho v_x v_y\,\Delta x\,\Delta z|_{y+\Delta y} \\
&+ \rho v_x v_z\,\Delta x\,\Delta y|_z - \rho v_x v_z\,\Delta x\,\Delta y|_{z+\Delta z} \\
&+ \sigma_{xx}\,\Delta y\,\Delta z|_{x+\Delta x} - \sigma_{xx}\,\Delta y\,\Delta z|_x \\
&+ \tau_{yx}\,\Delta x\,\Delta z|_{y+\Delta y} - \tau_{yx}\,\Delta x\,\Delta z|_y \\
&+ \tau_{zx}\,\Delta x\,\Delta y|_{z+\Delta z} - \tau_{zx}\,\Delta x\,\Delta y|_z + \bar{\rho} g_x\,\Delta x\,\Delta y\,\Delta z
\end{aligned}
\tag{7.27}
$$

Dividing by the volume, $\Delta x\,\Delta y\,\Delta z$, and rearranging slightly, this can be written

$$
\begin{aligned}
\frac{\partial \overline{\rho v_x}}{\partial t} &+ \frac{\rho v_x v_x|_{x+\Delta x} - \rho v_x v_x|_x}{\Delta x} + \frac{\rho v_y v_x|_{y+\Delta y} - \rho v_y v_x|_y}{\Delta y} \\
&+ \frac{\rho v_z v_x|_{z+\Delta z} - \rho v_z v_x|_z}{\Delta z} = \frac{\sigma_{xx}|_{x+\Delta x} - \sigma_{xx}|_x}{\Delta x} \\
&+ \frac{\tau_{yx}|_{y+\Delta y} - \tau_{yx}|_y}{\Delta y} + \frac{\tau_{zx}|_{z+\Delta z} - \tau_{zx}|_z}{\Delta z} + \bar{\rho} g_x
\end{aligned}
\tag{7.28}
$$

In the limit as $\Delta x\,\Delta y\,\Delta z \to 0$ each difference quotient becomes a partial derivative, and we obtain the partial differential equation

$$\frac{\partial \rho v_x}{\partial t} + \frac{\partial \rho v_x v_x}{\partial x} + \frac{\partial \rho v_y v_x}{\partial y} + \frac{\partial \rho v_z v_x}{\partial z} = \frac{\partial \sigma_{xx}}{\partial x} + \frac{\partial \tau_{yx}}{\partial y} + \frac{\partial \tau_{zx}}{\partial z} + \rho g_x \quad (7.29)$$

One further simplification can be obtained. The derivatives of the products on the left side of Eq. (7.29) can be expanded to two terms and written

$$\text{left-hand side} = \begin{cases} +\rho \dfrac{\partial v_x}{\partial t} + \rho v_x \dfrac{\partial v_x}{\partial x} + \rho v_y \dfrac{\partial v_x}{\partial y} + \rho v_z \dfrac{\partial v_x}{\partial z} \\[2mm] +v_x \dfrac{\partial \rho}{\partial t} + v_x \dfrac{\partial \rho v_x}{\partial x} + v_x \dfrac{\partial \rho v_y}{\partial y} + v_x \dfrac{\partial \rho v_z}{\partial z} \end{cases} \quad (7.30)$$

The second row sums to zero from continuity, Eq. (7.7), and the final form of the x-momentum equation is

$$\rho \frac{Dv_x}{Dt} = \rho \frac{\partial v_x}{\partial t} + \rho v_x \frac{\partial v_x}{\partial x} + \rho v_y \frac{\partial v_x}{\partial y} + \rho v_z \frac{\partial v_x}{\partial z}$$

$$= \frac{\partial \sigma_{xx}}{\partial x} + \frac{\partial \tau_{yx}}{\partial y} + \frac{\partial \tau_{zx}}{\partial z} + \rho g_x \quad (7.31a)$$

The y- and z-momentum equations are, respectively,

$$\rho \frac{Dv_y}{Dt} = \rho \frac{\partial v_y}{\partial t} + \rho v_x \frac{\partial v_y}{\partial x} + \rho v_y \frac{\partial v_y}{\partial y} + \rho v_z \frac{\partial v_y}{\partial z}$$

$$= \frac{\partial \tau_{xy}}{\partial x} + \frac{\partial \sigma_{yy}}{\partial y} + \frac{\partial \tau_{zy}}{\partial z} + \rho g_y \quad (7.31b)$$

$$\rho \frac{Dv_z}{Dt} = \rho \frac{\partial v_z}{\partial t} + \rho v_x \frac{\partial v_z}{\partial x} + \rho v_y \frac{\partial v_z}{\partial y} + \rho v_z \frac{\partial v_z}{\partial z}$$

$$= \frac{\partial \tau_{xz}}{\partial x} + \frac{\partial \tau_{yz}}{\partial y} + \frac{\partial \sigma_{zz}}{\partial z} + \rho g_z \quad (7.31c)$$

The latter two equations follow from the permutation $xyz \longrightarrow yzx \longrightarrow zxy$. Equations (7.31) are components of what is sometimes called the *Cauchy momentum equation.*

7.3.5 Stress Symmetry

The three components of the Cauchy momentum equation contain the three velocity components, but they also contain the nine components of the stress: $\sigma_{xx}, \sigma_{yy}, \sigma_{zz}, \tau_{xy}, \tau_{xz}, \tau_{yx}, \tau_{yz}, \tau_{zx}, \tau_{zy}$. In order to have a closed system of equations, we will require an additional nine relationships between the variables. Three of these follow from the fact that the stress can generally be taken as symmetric; that is,

$$\tau_{xy} = \tau_{yx} \quad (7.32a)$$

$$\tau_{xz} = \tau_{zx} \quad (7.32b)$$

$$\tau_{yz} = \tau_{zy} \quad (7.32c)$$

The symmetry of the stress is usually deduced by application of the principle of conservation of angular momentum to the control volume in Fig. 7-5. The small cubic region is taken as a rigid body, in which case the principle of conservation of angular momentum states that *the rate of change of moment of momentum equals the sum of the imposed torques.* Consider rotation about the z axis. The force on the x face at $x + \Delta x$ causing rotation is $\tau_{xy} \Delta y \Delta z$, and it acts through a moment arm $\Delta x/2$ about the center, causing a torque $\tau_{xy} \Delta x \Delta y \Delta z/2$. By the right-hand rule, this is a positive torque, directed in the positive z direction. Similarly, there is a positive torque $\tau_{xy} \Delta x \Delta y \Delta z/2$ exerted by the force on the face at x. The torques exerted on the y faces are negative and equal to $\tau_{yx} \Delta x \Delta y \Delta z/2$, giving

$$\text{net torque} = \tfrac{1}{2}(\tau_{xy}|_x + \tau_{xy}|_{x+\Delta x} - \tau_{yx}|_y - \tau_{yx}|_{y+\Delta y}) \Delta x \Delta y \Delta z \quad (7.33)$$

The moment of momentum equals $\rho \Delta x \Delta y \Delta z\, r_g^2\, \Omega$, where Ω is the angular velocity and r_g is the *radius of gyration*, which equals $(\Delta x \Delta y/6)^{1/2}$ for the square face. Thus, dividing both sides by the volume, $\Delta x \Delta y \Delta z$, we obtain

$$\frac{\rho}{6} \frac{d\Omega}{dt} \Delta x \Delta y = \tfrac{1}{2}(\tau_{xy}|_x + \tau_{xy}|_{x+\Delta x} - \tau_{yx}|_y - \tau_{yx}|_{y+\Delta y}) \quad (7.34)$$

In the limit as Δx and Δy go to zero, $\tau_{xy}|_x$ and $\tau_{xy}|_{x+\Delta x}$ approach the same value, which we simply denote as τ_{xy}; similarly, $\tau_{yx}|_y$ and $\tau_{yx}|_{y+\Delta y}$ approach a common value of τ_{yx}. We thus have

$$\lim_{\substack{\Delta x \to 0 \\ \Delta y \to 0}} \frac{\rho}{6} \frac{d\Omega}{dt} \Delta x \Delta y = \tau_{xy} - \tau_{yx} \quad (7.35)$$

The angular acceleration $d\Omega/dt$ must remain finite, so the left-hand side of Eq. (7.35) vanishes in the limit, giving the symmetry relation (7.32a). The other two symmetry relations follows in an identical manner.

It should be pointed out that the derivation of stress symmetry uses not only the continuum hypothesis, but a second and stronger assumption as well. We have assumed that the only torque acting on the control volume comes from the action of the surrounding fluid, and that there are no couples caused by the action of body forces. This is usually a good assumption, but we can imagine cases in which it might fail. Consider a highly polar material, for example. There is a dipole force which exerts a local couple; as the control volume shrinks to a sufficiently small size, the dipole forces will exert a body force couple that is not included in the balance in Eq. (7.35). Thus, for such a material the stress may not be symmetric. The proof of stress symmetry is thus equivalent to a specific assumption about the nature of the fluid—that *there are no internal couples arising from fluid structure.* This assumption seems to be a good one for all fluids commonly encountered in processing, although anisotropic materials such as *liquid crystals* might fail to satisfy the symmetry property.

7.4 NEWTONIAN FLUID

7.4.1 Intuitive Development

The three symmetry relations reduce the number of unknown stress components in the momentum equation to six. To proceed further, we need to establish a *constitutive relation* between the stress and the velocity field. We will restrict attention here to *Newtonian fluids*, continuing the development begun in Sec. 2.4.2.

We defined a Newtonian fluid as one in which the shear stress is directly proportional to the shear rate. In Sec. 2.4.2 we focused on motion between two planes (Fig. 2-1) and defined the shear rate as the relative velocity divided by the spacing. We now consider two fluid planes that are only differentially apart, with all motion in the x direction, as shown in Fig. 7-6. The relative

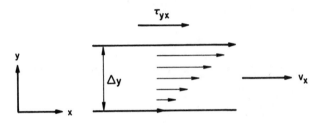

Figure 7-6. Shearing between two planes of fluid.

velocity divided by the spacing is then $\Delta v_x/\Delta y$ or, in the limit, dv_x/dy. In place of τ_s for the shear stress we write τ_{yx}, and the analogous definition of a Newtonian fluid is

$$x\text{-direction motion only:} \quad \tau_{yx} = \eta \frac{dv_x}{dy} = \tau_{xy} \qquad (7.36a)$$

The second equality follows from the symmetry relation. η is a constant.

Suppose that we now rotate the system by 90° so that all motion is in the y direction. In that case we can interchange x and y in Fig. 7-6 and write

$$y\text{-direction motion only:} \quad \tau_{xy} = \eta \frac{dv_y}{dx} = \tau_{yx} \qquad (7.36b)$$

The motion will not usually be restricted to one coordinate direction only. A generalization of Eqs. (7.36) that reduces properly to the two special cases is

$$\tau_{xy} = \tau_{yx} = \eta \left(\frac{\partial v_x}{\partial y} + \frac{\partial v_y}{\partial x} \right) \qquad (7.37a)$$

Here we have written partial derivatives, because v_x may depend on coordinate directions other than y, and v_y on other directions than x. If the only velocity component is v_x, then Eq. (7.36a) is recovered, while Eq. (7.36b) is recovered if the only velocity component is v_y. The expressions for τ_{xz} and τ_{yz} follow by permutation of x, y, and z, which is equivalent to rotating the system successively by 90°:

$$\tau_{xz} = \tau_{zx} = \eta\left(\frac{\partial v_x}{\partial z} + \frac{\partial v_z}{\partial x}\right) \tag{7.37b}$$

$$\tau_{yz} = \tau_{zy} = \eta\left(\frac{\partial v_y}{\partial z} + \frac{\partial v_z}{\partial y}\right) \tag{7.37c}$$

This intuitive generalization of the discussion in Sec. 2.4.2 of the Newtonian fluid establishes the proper format for the relation between stress and rate of deformation: the shearing stresses are linear, homogeneous functions of the derivatives of the velocity.

The relations for the normal stresses, σ_{xx}, σ_{yy}, and σ_{zz}, require a more rigorous treatment. This is because the fluid can sustain a uniform pressure even in the absence of motion and because the normal stresses must account for compressibility effects.

7.4.2 Constitutive Equation

A Newtonian fluid is defined as a fluid with the following properties:

· The stress is symmetric.
· The stress at a point in the fluid depends only on the instantaneous value of the velocity gradients at the point.
· The stress is a linear function of the velocity gradients.
· The stress is isotropic when there is no motion.

Linearity is required for consistency with the definition in Sec. 2.4.2. Dependence on the instantaneous value of the velocity gradient is required to rule out the viscoelastic effects discussed in Sec. 2.5 and shown in Fig. 2-13. *Isotropic* means "the same in all directions"; isotropy at rest is required for consistency with the common understanding of pressure and to rule out Bingham plastic behavior such as that discussed in Sec. 2.4.3 and shown in Fig. 2-10.

The symmetry arguments used in the preceding section can be made rigorous and combined with the four properties defining a Newtonian fluid to establish the form of the relation between the stress and deformation gradients. We shall not present the detailed analysis, which is done most easily using standard theorems in matrix algebra, but only present the results. The stress components τ_{xy}, τ_{yx}, τ_{xz}, τ_{zx}, τ_{yz}, and τ_{zy} are given by Eqs. (7.37). The normal stresses are given as follows:

$$\sigma_{xx} = -p + 2\eta\frac{\partial v_x}{\partial x} + (\kappa - \tfrac{2}{3}\eta)\nabla \cdot \mathbf{v} \tag{7.38a}$$

$$\sigma_{yy} = -p + 2\eta\frac{\partial v_y}{\partial y} + (\kappa - \tfrac{2}{3}\eta)\nabla \cdot \mathbf{v} \tag{7.38b}$$

$$\sigma_{zz} = -p + 2\eta \frac{\partial v_z}{\partial z} + (\kappa - \tfrac{2}{3}\eta)\nabla \cdot \mathbf{v} \qquad (7.38c)$$

$\nabla \cdot \mathbf{v}$ is the divergence of the velocity, given by Eq. (7.16). There are two material constants,* κ and η, and one function, p. κ is known as the bulk viscosity; it can be shown to be zero for monotonic gases. Very large expansions would be necessary to measure nonzero values of κ for polyatomic gases and compressible liquids, and there seems to be no experimentally measurable error involved in setting κ equal to zero. We shall do so here, in which case the constitutive equation for a Newtonian fluid has a single material constant, the viscosity, η. The equations for the normal stress are then

$$\sigma_{xx} = -p + 2\eta \frac{\partial v_x}{\partial x} - \tfrac{2}{3}\eta \nabla \cdot \mathbf{v} \qquad (7.39a)$$

$$\sigma_{yy} = -p + 2\eta \frac{\partial v_y}{\partial y} - \tfrac{2}{3}\eta \nabla \cdot \mathbf{v} \qquad (7.39b)$$

$$\sigma_{zz} = -p + 2\eta \frac{\partial v_z}{\partial z} - \tfrac{2}{3}\eta \nabla \cdot \mathbf{v} \qquad (7.39c)$$

Equations (7.39) can be added together to give

$$\sigma_{xx} + \sigma_{yy} + \sigma_{zz} = -3p + 2\eta\left(\frac{\partial v_x}{\partial x} + \frac{\partial v_y}{\partial y} + \frac{\partial v_z}{\partial z}\right) - \left(\frac{2}{3} + \frac{2}{3} + \frac{2}{3}\right)\eta \nabla \cdot \mathbf{v}$$

The quantity in parentheses in the second term on the right is simply $\nabla \cdot \mathbf{v}$, so the last two terms are equal and we have

$$p = -\tfrac{1}{3}(\sigma_{xx} + \sigma_{yy} + \sigma_{zz}) \qquad (7.40)$$

That is, *p is a compressive stress that is equal to the mean normal stress* on the control volume. This is consistent with our notion of *pressure*, and p is, in fact, the thermodynamic pressure.

It is conventional to introduce an additional bit of nomenclature. The *deviatoric stress* is defined as the stress in excess of the isotropic pressure and is defined as

$$\tau_{xx} = \sigma_{xx} + p = 2\eta \frac{\partial v_x}{\partial x} - \tfrac{2}{3}\eta \nabla \cdot \mathbf{v} \qquad (7.41a)$$

$$\tau_{yy} = \sigma_{yy} + p = 2\eta \frac{\partial v_y}{\partial y} - \tfrac{2}{3}\eta \nabla \cdot \mathbf{v} \qquad (7.41b)$$

$$\tau_{zz} = \sigma_{zz} + p = 2\eta \frac{\partial v_z}{\partial z} - \tfrac{2}{3}\eta \nabla \cdot \mathbf{v} \qquad (7.41c)$$

Note that $\tau_{xx} + \tau_{yy} + \tau_{zz} = 0$.

*Students who have previously studied the theory of elasticity will recognize that two material constants is a characteristic of a linear stress equation. In linear elasticity the two constants are Young's modulus and Poisson's ratio.

7.4.3 Aside on Pressure

Readers planning to go on to the study of the behavior of non-Newtonian fluids should be aware of a certain arbitrariness in the definition of pressure. If the fluid is incompressible, the thermodynamic pressure is undefined. Thus, instead of defining τ_{xx}, τ_{yy}, and τ_{zz} as in Eqs. (7.41), we could have defined them as $\sigma_{xx} + \hat{p}$, $\sigma_{yy} + \hat{p}$, $\sigma_{zz} + \hat{p}$, where $\hat{p}$ is *any* isotropic function. For the Newtonian fluid, even when incompressible, it is best to identify $\hat{p}$ with p as defined by Eq. (7.40). For some incompressible viscoelastic liquids, however, a different definition can be more appropriate. This point has caused a great deal of confusion in the non-Newtonian fluid mechanics literature.

7.4.4 Momentum Equation

Using the definition $\tau_{xx} = \sigma_{xx} + p$, the x component of the Cauchy momentum equation, Eq. (7.31a), can be written

$$\rho \frac{Dv_x}{Dt} = -\frac{\partial p}{\partial x} + \frac{\partial \tau_{xx}}{\partial x} + \frac{\partial \tau_{yx}}{\partial y} + \frac{\partial \tau_{zx}}{\partial z} + \rho g_x \qquad (7.42)$$

Substitution of Eqs. (7.37) and (7.41) then gives

$$\rho \frac{Dv_x}{Dt} = -\frac{\partial p}{\partial x} + \frac{\partial}{\partial x}\left[2\eta \frac{\partial v_x}{\partial x} - \frac{2}{3}\eta \nabla \cdot \mathbf{v}\right] + \frac{\partial}{\partial y}\left[\eta\left(\frac{\partial v_x}{\partial y} + \frac{\partial v_y}{\partial x}\right)\right]$$
$$+ \frac{\partial}{\partial z}\left[\eta\left(\frac{\partial v_x}{\partial z} + \frac{\partial v_z}{\partial x}\right)\right] + \rho g_x \qquad (7.43a)$$

Similar equations are obtained for y and z by the permutation $xyz \longrightarrow yzx \longrightarrow zxy$:

$$\rho \frac{Dv_y}{Dt} = -\frac{\partial p}{\partial y} + \frac{\partial}{\partial y}\left[2\eta \frac{\partial v_y}{\partial y} - \frac{2}{3}\eta \nabla \cdot \mathbf{v}\right] + \frac{\partial}{\partial z}\left[\eta\left(\frac{\partial v_y}{\partial z} + \frac{\partial v_z}{\partial y}\right)\right]$$
$$+ \frac{\partial}{\partial x}\left[\eta\left(\frac{\partial v_y}{\partial x} + \frac{\partial v_x}{\partial y}\right)\right] + \rho g_y \qquad (7.43b)$$

$$\rho \frac{Dv_z}{Dt} = -\frac{\partial p}{\partial z} + \frac{\partial}{\partial z}\left[2\eta \frac{\partial v_z}{\partial z} - \frac{2}{3}\eta \nabla \cdot \mathbf{v}\right] + \frac{\partial}{\partial x}\left[\eta\left(\frac{\partial v_z}{\partial x} + \frac{\partial v_x}{\partial z}\right)\right]$$
$$+ \frac{\partial}{\partial y}\left[\eta\left(\frac{\partial v_z}{\partial y} + \frac{\partial v_y}{\partial z}\right)\right] + \rho g_z \qquad (7.43c)$$

Equations (7.43) define a set of three equations for the four variables v_x, v_y, v_z, and p. The fourth equation is the continuity equation, Eq. (7.9). If the fluid is compressible, the density is also a variable, and the required fifth equation is the constitutive equation relating density to pressure for the particular material being studied.

7.4.5 Navier–Stokes Equations

In many applications the viscosity can be taken as a constant, independent of spatial position. We can then take the viscosity outside the differentiation and write Eq. (7.43a) as

$$\rho \frac{Dv_x}{Dt} = -\frac{\partial p}{\partial x} + \eta \frac{\partial^2 v_x}{\partial x^2} + \eta \frac{\partial^2 v_x}{\partial x^2} - \frac{2}{3}\eta \frac{\partial}{\partial x}(\nabla \cdot \mathbf{v}) + \eta \frac{\partial^2 v_x}{\partial y^2}$$
$$+ \eta \frac{\partial^2 v_y}{\partial x\, \partial y} + \eta \frac{\partial^2 v_x}{\partial z^2} + \eta \frac{\partial^2 v_z}{\partial z\, \partial x} + \rho g_x \qquad (7.44)$$

We have written the term $2\eta \partial^2 v_x/\partial x^2$ as the sum of two terms. The three underlined terms sum to $\eta\, \partial/\partial x(\partial v_x/\partial x + \partial v_y/\partial y + \partial v_z/dz)$, or $\eta\, \partial(\nabla \cdot \mathbf{v})/\partial x$, so Eq. (7.44) simplifies to

$$\rho \frac{Dv_x}{Dt} = -\frac{\partial p}{\partial x} + \eta\left(\frac{\partial^2 v_x}{\partial x^2} + \frac{\partial^2 v_x}{\partial y^2} + \frac{\partial^2 v_x}{\partial z^2}\right) + \frac{1}{3}\eta \frac{\partial}{\partial x}(\nabla \cdot \mathbf{v}) + \rho g_x \qquad (7.45a)$$

The corresponding equations for the y and z components are

$$\rho \frac{Dv_y}{Dt} = -\frac{\partial p}{\partial y} + \eta\left(\frac{\partial^2 v_y}{\partial x^2} + \frac{\partial^2 v_y}{\partial y^2} + \frac{\partial^2 v_y}{\partial z^2}\right) + \frac{1}{3}\eta \frac{\partial}{\partial y}(\nabla \cdot \mathbf{v}) + \rho g_y \qquad (7.45b)$$

$$\rho \frac{Dv_z}{Dt} = -\frac{\partial p}{\partial z} + \eta\left(\frac{\partial^2 v_z}{\partial x^2} + \frac{\partial^2 v_z}{\partial y^2} + \frac{\partial^2 v_z}{\partial z^2}\right) + \frac{1}{3}\eta \frac{\partial}{\partial z}(\nabla \cdot \mathbf{v}) + \rho g_z \qquad (7.45c)$$

Equations (7.45) are known as the *Navier–Stokes equations*. They were first derived for incompressible fluids using molecular arguments by Navier in France in 1822, and in complete generality by Stokes in England in 1845. Equations (7.45) are the basic equations that we will apply.

The operator $\partial^2/\partial x^2 + \partial^2/\partial y^2 + \partial^2/\partial z^2$ can be interpreted as $\nabla \cdot \nabla$, which is usually written ∇^2 and is known as the *Laplacian*. We can then write Eqs. (7.45) in the vector shorthand

$$\rho \frac{D\mathbf{v}}{Dt} = -\nabla p + \eta \nabla^2 \mathbf{v} + \tfrac{1}{3}\eta \nabla(\nabla \cdot \mathbf{v}) + \rho \mathbf{g} \qquad (7.46)$$

with

$$\frac{D\mathbf{v}}{Dt} = \frac{\partial \mathbf{v}}{\partial t} + (\mathbf{v} \cdot \nabla)\mathbf{v} \qquad (7.47)$$

7.4.6 Equivalent Pressure

We defined the sum $p + \rho gh$ as the equivalent pressure in Sec. 5.4.2. Recall that

$$g_x = -g\frac{\partial h}{\partial x} \qquad g_y = -g\frac{\partial h}{\partial y} \qquad g_z = -g\frac{\partial h}{\partial z} \qquad (7.48)$$

where the negative sign occurs because h is measured up and the gravitational force points down. It is convenient to generalize the definition slightly and *define the equivalent pressure* $\mathcal{P}$ as

$$\mathcal{P} = p + \rho g h - \tfrac{1}{3}\eta \nabla \cdot \mathbf{v} \tag{7.49}$$

Equation (7.46) can then be written in the compact form

$$\rho \frac{D\mathbf{v}}{Dt} = -\nabla \mathcal{P} + \eta \nabla^2 \mathbf{v} \tag{7.50}$$

This is the form that is most convenient for confined flows in an enclosed conduit.

7.5 CURVILINEAR COORDINATES

Although we have derived the Navier–Stokes and continuity equations in a rectangular control volume, it will be more convenient in some applications to have the equations available in other coordinate systems. This coordinate transformation is carried out by expressing the variables and coordinates in one system in terms of those in the other. In a cylindrical r, θ, z system, for example, we can write $r = (x^2 + y^2)^{1/2}$, $\theta = \arctan y/x$, $z = z$. The resulting equations are shown in Tables 7-1 through 7-10. The Navier–Stokes equations are written in terms of the equivalent pressure, $\mathcal{P}$, and apply to both incompressible and compressible Newtonian fluids with a constant viscosity. The individual pressure, gravity, and compressibility terms can be recovered by use of Eq. (7.49).

TABLE 7-1

CONTINUITY EQUATION IN RECTANGULAR, CYLINDRICAL, AND SPHERICAL COORDINATES

$$\frac{\partial \rho}{\partial t} = -\nabla \cdot (\rho \mathbf{v})$$

Rectangular (x, y, z) coordinates:

$$\nabla \cdot (\rho \mathbf{v}) = \frac{\partial}{\partial x}(\rho v_x) + \frac{\partial}{\partial y}(\rho v_y) + \frac{\partial}{\partial z}(\rho v_z)$$

Cylindrical (r, θ, z) coordinates:

$$\nabla \cdot (\rho \mathbf{v}) = \frac{1}{r}\frac{\partial}{\partial r}(\rho r v_r) + \frac{1}{r}\frac{\partial}{\partial \theta}(\rho v_\theta) + \frac{\partial}{\partial z}(\rho v_z)$$

Spherical (r, θ, ϕ) coordinates:

$$\nabla \cdot (\rho \mathbf{v}) = \frac{1}{r^2}\frac{\partial}{\partial r}(\rho r^2 v_r) + \frac{1}{r \sin \theta}\frac{\partial}{\partial \theta}(\rho v_\theta \sin \theta)$$
$$+ \frac{1}{r \sin \theta}\frac{\partial}{\partial \phi}(\rho v_\phi)$$

TABLE 7-2
Cauchy Momentum Equation
in Rectangular Cartesian (x, y, z) Coordinates

x component: $\quad \rho\left(\dfrac{\partial v_x}{\partial t} + v_x \dfrac{\partial v_x}{\partial x} + v_y \dfrac{\partial v_x}{\partial y} + v_z \dfrac{\partial v_x}{\partial z}\right) = -\dfrac{\partial p}{\partial x} + \dfrac{\partial \tau_{xx}}{\partial x} + \dfrac{\partial \tau_{yx}}{\partial y} + \dfrac{\partial \tau_{zx}}{\partial z} + \rho g_x$

y component: $\quad \rho\left(\dfrac{\partial v_y}{\partial t} + v_x \dfrac{\partial v_y}{\partial x} + v_y \dfrac{\partial v_y}{\partial y} + v_z \dfrac{\partial v_y}{\partial z}\right) = -\dfrac{\partial p}{\partial y} + \dfrac{\partial \tau_{xy}}{\partial x} + \dfrac{\partial \tau_{yy}}{\partial y} + \dfrac{\partial \tau_{zy}}{\partial z} + \rho g_y$

z component: $\quad \rho\left(\dfrac{\partial v_z}{\partial t} + v_x \dfrac{\partial v_z}{\partial x} + v_y \dfrac{\partial v_z}{\partial y} + v_z \dfrac{\partial v_z}{\partial z}\right) = -\dfrac{\partial p}{\partial z} + \dfrac{\partial \tau_{xz}}{\partial x} + \dfrac{\partial \tau_{yz}}{\partial y} + \dfrac{\partial \tau_{zz}}{\partial z} + \rho g_z$

TABLE 7-3
Stress Constitutive Equation
for a Newtonian Fluid
in Rectangular Cartesian
(x, y, z) Coordinates

$$\tau_{xx} = \eta\left[2\frac{\partial v_x}{\partial x} - \tfrac{2}{3}(\nabla \cdot \mathbf{v})\right]$$

$$\tau_{yy} = \eta\left[2\frac{\partial v_y}{\partial y} - \tfrac{2}{3}(\nabla \cdot \mathbf{v})\right]$$

$$\tau_{zz} = \eta\left[2\frac{\partial v_z}{\partial z} - \tfrac{2}{3}(\nabla \cdot \mathbf{v})\right]$$

$$\tau_{xy} = \tau_{yx} = \eta\left[\frac{\partial v_x}{\partial y} + \frac{\partial v_y}{\partial x}\right]$$

$$\tau_{yz} = \tau_{zy} = \eta\left[\frac{\partial v_y}{\partial z} + \frac{\partial v_z}{\partial y}\right]$$

$$\tau_{zx} = \tau_{xz} = \eta\left[\frac{\partial v_z}{\partial x} + \frac{\partial v_x}{\partial z}\right]$$

$$(\nabla \cdot v) = \frac{\partial v_x}{\partial x} + \frac{\partial v_y}{\partial y} + \frac{\partial v_z}{\partial z}$$

TABLE 7-4
Navier–Stokes Equations for a Newtonian Fluid
with a Constant Viscosity in Rectangular Cartesian
(x, y, z) Coordinates[a]

x component: $\quad \rho\left(\dfrac{\partial v_x}{\partial t} + v_x \dfrac{\partial v_x}{\partial x} + v_y \dfrac{\partial v_x}{\partial y} + v_z \dfrac{\partial v_x}{\partial z}\right) = -\dfrac{\partial \mathcal{P}}{\partial x} + \eta\left(\dfrac{\partial^2 v_x}{\partial x^2} + \dfrac{\partial^2 v_x}{\partial y^2} + \dfrac{\partial^2 v_x}{\partial z^2}\right)$

y component: $\quad \rho\left(\dfrac{\partial v_y}{\partial t} + v_x \dfrac{\partial v_y}{\partial x} + v_y \dfrac{\partial v_y}{\partial y} + v_z \dfrac{\partial v_y}{\partial z}\right) = -\dfrac{\partial \mathcal{P}}{\partial y} + \eta\left(\dfrac{\partial^2 v_y}{\partial x^2} + \dfrac{\partial^2 v_y}{\partial y^2} + \dfrac{\partial^2 v_y}{\partial z^2}\right)$

z component: $\quad \rho\left(\dfrac{\partial v_z}{\partial t} + v_x \dfrac{\partial v_z}{\partial x} + v_y \dfrac{\partial v_z}{\partial y} + v_z \dfrac{\partial v_z}{\partial z}\right) = -\dfrac{\partial \mathcal{P}}{\partial z} + \eta\left(\dfrac{\partial^2 v_z}{\partial x^2} + \dfrac{\partial^2 v_z}{\partial y^2} + \dfrac{\partial^2 v_z}{\partial z^2}\right)$

[a]The equations are written in terms of the equivalent pressure, $\mathcal{P}$.

TABLE 7-5
CAUCHY MOMENTUM EQUATION
IN CYLINDRICAL (r, θ, z) COORDINATES

r component: $\quad \rho\left(\dfrac{\partial v_r}{\partial t} + v_r \dfrac{\partial v_r}{\partial r} + \dfrac{v_\theta}{r}\dfrac{\partial v_r}{\partial \theta} - \dfrac{v_\theta^2}{r} + v_z \dfrac{\partial v_r}{\partial z}\right) = -\dfrac{\partial p}{\partial r}$

$$+ \frac{1}{r}\frac{\partial}{\partial r}(r\tau_{rr}) + \frac{1}{r}\frac{\partial \tau_{r\theta}}{\partial \theta} - \frac{\tau_{\theta\theta}}{r} + \frac{\partial \tau_{rz}}{\partial z} + \rho g_r$$

θ component: $\quad \rho\left(\dfrac{\partial v_\theta}{\partial t} + v_r \dfrac{\partial v_\theta}{\partial r} + \dfrac{v_\theta}{r}\dfrac{\partial v_\theta}{\partial \theta} + \dfrac{v_r v_\theta}{r} + v_z \dfrac{\partial v_\theta}{\partial z}\right) = -\dfrac{1}{r}\dfrac{\partial p}{\partial \theta}$

$$+ \frac{1}{r^2}\frac{\partial}{\partial r}(r^2\tau_{r\theta}) + \frac{1}{r}\frac{\partial \tau_{\theta\theta}}{\partial \theta} + \frac{\partial \tau_{\theta z}}{\partial z} + \rho g_\theta$$

z component: $\quad \rho\left(\dfrac{\partial v_z}{\partial t} + v_r \dfrac{\partial v_z}{\partial r} + \dfrac{v_\theta}{r}\dfrac{\partial v_z}{\partial \theta} + v_z \dfrac{\partial v_z}{\partial z}\right) = -\dfrac{\partial p}{\partial z}$

$$+ \frac{1}{r}\frac{\partial}{\partial r}(r\tau_{rz}) + \frac{1}{r}\frac{\partial \tau_{\theta z}}{\partial \theta} + \frac{\partial \tau_{zz}}{\partial z} + \rho g_z$$

TABLE 7-6
STRESS CONSTITUTIVE EQUATION FOR A NEWTONIAN
FLUID IN CYLINDRICAL (r, θ, z) COORDINATES

$$\tau_{rr} = \eta\left[2\frac{\partial v_r}{\partial r} - \tfrac{2}{3}(\nabla \cdot \mathbf{v})\right]$$

$$\tau_{\theta\theta} = \eta\left[2\left(\frac{1}{r}\frac{\partial v_\theta}{\partial \theta} + \frac{v_r}{r}\right) - \tfrac{2}{3}(\nabla \cdot \mathbf{v})\right]$$

$$\tau_{zz} = \eta\left[2\frac{\partial v_z}{\partial z} - \tfrac{2}{3}(\nabla \cdot \mathbf{v})\right]$$

$$\tau_{r\theta} = \tau_{\theta r} = \eta\left[r\frac{\partial}{\partial r}\left(\frac{v_\theta}{r}\right) + \frac{1}{r}\frac{\partial v_r}{\partial \theta}\right]$$

$$\tau_{\theta z} = \tau_{z\theta} = \eta\left[\frac{\partial v_\theta}{\partial z} + \frac{1}{r}\frac{\partial v_z}{\partial \theta}\right]$$

$$\tau_{zr} = \tau_{rz} = \eta\left[\frac{\partial v_z}{\partial r} + \frac{\partial v_r}{\partial z}\right]$$

$$(\nabla \cdot \mathbf{v}) = \frac{1}{r}\frac{\partial}{\partial r}(rv_r) + \frac{1}{r}\frac{\partial v_\theta}{\partial \theta} + \frac{\partial v_z}{\partial z}$$

TABLE 7-7
NAVIER–STOKES EQUATIONS FOR A NEWTONIAN FLUID
WITH A CONSTANT VISCOSITY IN CYLINDRICAL (r, θ, z) COORDINATES[a]

r component:
$$\rho\left(\frac{\partial v_r}{\partial t} + v_r\frac{\partial v_r}{\partial r} + \frac{v_\theta}{r}\frac{\partial v_r}{\partial \theta} - \frac{v_\theta^2}{r} + v_z\frac{\partial v_r}{\partial z}\right) = -\frac{\partial \mathcal{P}}{\partial r}$$
$$+ \eta\left[\frac{\partial}{\partial r}\left(\frac{1}{r}\frac{\partial}{\partial r}(rv_r)\right) + \frac{1}{r^2}\frac{\partial^2 v_r}{\partial \theta^2} - \frac{2}{r^2}\frac{\partial v_\theta}{\partial \theta} + \frac{\partial^2 v_r}{\partial z^2}\right]$$

θ component:
$$\rho\left(\frac{\partial v_\theta}{\partial t} + v_r\frac{\partial v_\theta}{\partial r} + \frac{v_\theta}{r}\frac{\partial v_\theta}{\partial \theta} + \frac{v_r v_\theta}{r} + v_z\frac{\partial v_\theta}{\partial z}\right) = -\frac{1}{r}\frac{\partial \mathcal{P}}{\partial \theta}$$
$$+ \eta\left[\frac{\partial}{\partial r}\left(\frac{1}{r}\frac{\partial}{\partial r}(rv_\theta)\right) + \frac{1}{r^2}\frac{\partial^2 v_\theta}{\partial \theta^2} + \frac{2}{r^2}\frac{\partial v_r}{\partial \theta} + \frac{\partial^2 v_\theta}{\partial z^2}\right]$$

z component:
$$\rho\left(\frac{\partial v_z}{\partial t} + v_r\frac{\partial v_z}{\partial r} + \frac{v_\theta}{r}\frac{\partial v_z}{\partial \theta} + v_z\frac{\partial v_z}{\partial z}\right) = -\frac{\partial \mathcal{P}}{\partial z}$$
$$+ \eta\left[\frac{1}{r}\frac{\partial}{\partial r}\left(r\frac{\partial v_z}{\partial r}\right) + \frac{1}{r^2}\frac{\partial^2 v_z}{\partial \theta^2} + \frac{\partial^2 v_z}{\partial z^2}\right]$$

[a] The equations are written in terms of the equivalent pressure, $\mathcal{P}$.

TABLE 7-8
CAUCHY MOMENTUM EQUATION
IN SPHERICAL (r, θ, ϕ) COORDINATES

r component:
$$\rho\left(\frac{\partial v_r}{\partial t} + v_r\frac{\partial v_r}{\partial r} + \frac{v_\theta}{r}\frac{\partial v_r}{\partial \theta} + \frac{v_\phi}{r\sin\theta}\frac{\partial v_r}{\partial \phi} - \frac{v_\theta^2 + v_\phi^2}{r}\right)$$
$$= -\frac{\partial p}{\partial r} + \frac{1}{r^2}\frac{\partial}{\partial r}(r^2\tau_{rr}) + \frac{1}{r\sin\theta}\frac{\partial}{\partial \theta}(\tau_{r\theta}\sin\theta) + \frac{1}{r\sin\theta}\frac{\partial \tau_{r\phi}}{\partial \phi} - \frac{\tau_{\theta\theta} + \tau_{\phi\phi}}{r} + \rho g_r$$

θ component:
$$\rho\left(\frac{\partial v_\theta}{\partial t} + v_r\frac{\partial v_\theta}{\partial r} + \frac{v_\theta}{r}\frac{\partial v_\theta}{\partial \theta} + \frac{v_\phi}{r\sin\theta}\frac{\partial v_\theta}{\partial \phi} + \frac{v_r v_\theta}{r} - \frac{v_\phi^2\cot\theta}{r}\right)$$
$$= -\frac{1}{r}\frac{\partial p}{\partial \theta} + \frac{1}{r^2}\frac{\partial}{\partial r}(r^2\tau_{r\theta}) + \frac{1}{r\sin\theta}\frac{\partial}{\partial \theta}(\tau_{\theta\theta}\sin\theta) + \frac{1}{r\sin\theta}\frac{\partial \tau_{\theta\phi}}{\partial \phi} + \frac{\tau_{r\theta}}{r} - \frac{\cot\theta}{r}\tau_{\phi\phi} + \rho g_\theta$$

ϕ component:
$$\rho\left(\frac{\partial v_\phi}{\partial t} + v_r\frac{\partial v_\phi}{\partial r} + \frac{v_\theta}{r}\frac{\partial v_\phi}{\partial \theta} + \frac{v_\phi}{r\sin\theta}\frac{\partial v_\phi}{\partial \phi} + \frac{v_\phi v_r}{r} + \frac{v_\theta v_\phi}{r}\cot\theta\right)$$
$$= -\frac{1}{r\sin\theta}\frac{\partial p}{\partial \phi} + \frac{1}{r^2}\frac{\partial}{\partial r}(r^2\tau_{r\phi}) + \frac{1}{r}\frac{\partial \tau_{\theta\phi}}{\partial \theta} + \frac{1}{r\sin\theta}\frac{\partial \tau_{\phi\phi}}{\partial \phi} + \frac{\tau_{r\phi}}{r} + \frac{2\cot\theta}{r}\tau_{\theta\phi} + \rho g_\phi$$

TABLE 7-9
STRESS CONSTITUTIVE EQUATION FOR A NEWTONIAN FLUID IN SPHERICAL (r, θ, ϕ) COORDINATES

$$\tau_{rr} = \eta\left[2\frac{\partial v_r}{\partial r} - \tfrac{2}{3}(\nabla \cdot \mathbf{v})\right]$$

$$\tau_{\theta\theta} = \eta\left[2\left(\frac{1}{r}\frac{\partial v_\theta}{\partial \theta} + \frac{v_r}{r}\right) - \tfrac{2}{3}(\nabla \cdot \mathbf{v})\right]$$

$$\tau_{\phi\phi} = \eta\left[2\left(\frac{1}{r\sin\theta}\frac{\partial v_\phi}{\partial \phi} + \frac{v_r}{r} + \frac{v_\theta\cot\theta}{r}\right) - \tfrac{2}{3}(\nabla \cdot \mathbf{v})\right]$$

$$\tau_{r\theta} = \tau_{\theta r} = \eta\left[r\frac{\partial}{\partial r}\left(\frac{v_\theta}{r}\right) + \frac{1}{r}\frac{\partial v_r}{\partial \theta}\right]$$

$$\tau_{\theta\phi} = \tau_{\phi\theta} = \eta\left[\frac{\sin\theta}{r}\frac{\partial}{\partial\theta}\left(\frac{v_\phi}{\sin\theta}\right) + \frac{1}{r\sin\theta}\frac{\partial v_\theta}{\partial \phi}\right]$$

$$\tau_{\phi r} = \tau_{r\phi} = \eta\left[\frac{1}{r\sin\theta}\frac{\partial v_r}{\partial \phi} + r\frac{\partial}{\partial r}\left(\frac{v_\phi}{r}\right)\right]$$

$$(\nabla \cdot \mathbf{v}) = \frac{1}{r^2}\frac{\partial}{\partial r}(r^2 v_r) + \frac{1}{r\sin\theta}\frac{\partial}{\partial\theta}(v_\theta\sin\theta) + \frac{1}{r\sin\theta}\frac{\partial v_\phi}{\partial\phi}$$

TABLE 7-10
NAVIER–STOKES EQUATIONS FOR A NEWTONIAN FLUID WITH A CONSTANT VISCOSITY IN SPHERICAL (r, θ, ϕ) COORDINATES[a]

r component:
$$\rho\left(\frac{\partial v_r}{\partial t} + v_r\frac{\partial v_r}{\partial r} + \frac{v_\theta}{r}\frac{\partial v_r}{\partial\theta} + \frac{v_\phi}{r\sin\theta}\frac{\partial v_r}{\partial\phi} - \frac{v_\theta^2 + v_\phi^2}{r}\right)$$
$$= -\frac{\partial\mathcal{P}}{\partial r} + \eta\left[\frac{1}{r^2}\frac{\partial}{\partial r}\left(r^2\frac{\partial v_r}{\partial r}\right) + \frac{1}{r^2\sin\theta}\frac{\partial}{\partial\theta}\left(\sin\theta\frac{\partial v_r}{\partial\theta}\right) + \frac{1}{r^2\sin^2\theta}\frac{\partial^2 v_r}{\partial\phi^2}\right.$$
$$\left. - \frac{2}{r^2}v_r - \frac{2}{r^2}\frac{\partial v_\theta}{\partial\theta} - \frac{2}{r^2}v_\theta\cot\theta - \frac{2}{r^2\sin\theta}\frac{\partial v_\phi}{\partial\phi}\right]$$

θ component:
$$\rho\left(\frac{\partial v_\theta}{\partial t} + v_r\frac{\partial v_\theta}{\partial r} + \frac{v_\theta}{r}\frac{\partial v_\theta}{\partial\theta} + \frac{v_\phi}{r\sin\theta}\frac{\partial v_\theta}{\partial\phi} + \frac{v_r v_\theta}{r} - \frac{v_\phi^2\cot\theta}{r}\right)$$
$$= -\frac{1}{r}\frac{\partial\mathcal{P}}{\partial\theta} + \eta\left[\frac{1}{r^2}\frac{\partial}{\partial r}\left(r^2\frac{\partial v_\theta}{\partial r}\right) + \frac{1}{r^2\sin\theta}\frac{\partial}{\partial\theta}\left(\sin\theta\frac{\partial v_\theta}{\partial\theta}\right) + \frac{1}{r^2\sin^2\theta}\frac{\partial^2 v_\theta}{\partial\phi^2}\right.$$
$$\left. + \frac{2}{r^2}\frac{\partial v_r}{\partial\theta} - \frac{v_\theta}{r^2\sin^2\theta} - \frac{2\cos\theta}{r^2\sin^2\theta}\frac{\partial v_\phi}{\partial\phi}\right]$$

ϕ component:
$$\rho\left(\frac{\partial v_\phi}{\partial t} + v_r\frac{\partial v_\phi}{\partial r} + \frac{v_\theta}{r}\frac{\partial v_\phi}{\partial\theta} + \frac{v_\phi}{r\sin\theta}\frac{\partial v_\phi}{\partial\phi} + \frac{v_\phi v_r}{r} + \frac{v_\theta v_\phi}{r}\cot\theta\right)$$
$$= -\frac{1}{r\sin\theta}\frac{\partial\mathcal{P}}{\partial\phi} + \eta\left[\frac{1}{r^2}\frac{\partial}{\partial r}\left(r^2\frac{\partial v_\phi}{\partial r}\right) + \frac{1}{r^2\sin\theta}\frac{\partial}{\partial\theta}\left(\sin\theta\frac{\partial v_\phi}{\partial\theta}\right) + \frac{1}{r^2\sin^2\theta}\frac{\partial^2 v_\phi}{\partial\phi^2}\right.$$
$$\left. - \frac{v_\phi}{r^2\sin^2\theta} + \frac{2}{r^2\sin\theta}\frac{\partial v_r}{\partial\phi} + \frac{2\cos\theta}{r^2\sin^2\theta}\frac{\partial v_\theta}{\partial\phi}\right]$$

[a]The equations are written in terms of the equivalent pressure, $\mathcal{P}$.

7.6 BOUNDARY CONDITIONS

The momentum and continuity equations are four differential equations for the three components of the velocity and for the pressure. Integration of these equations will lead to constants of integration, and these must be evaluated using known information about the fluid behavior at boundaries.

The most common boundary condition is the *no-slip* condition, which states that fluid adheres to a solid surface and moves with the velocity of the surface. Thus, the fluid velocity at the interface with a stationary wall will be zero. The no-slip condition is not intuitively obvious to everyone, but it does appear to be satisfied by all Newtonian fluids in conduits that are large relative to molecular dimensions. The data in Figs. 6-7, 6-8, and particularly 6-9 demonstrate the apparent vanishing of the velocity at the wall. The no-slip condition also appears to be satisfied by non-Newtonian fluids except perhaps in the case of certain polymer melts under extreme processing conditions.

At a fluid–fluid interface, both the velocity and tangential stress must be continuous across the interface. The stress normal to the interface is continuous except for an amount equal to the normal stress induced by surface tension. We will not consider surface tension effects in this book, but simply refer the reader to the more specialized literature.

In some cases the only boundary condition that can be stated is finiteness of a velocity or stress, or symmetry resulting from a geometrical symmetry.* We shall see examples of these in Chapter 8.

7.7 MACROSCOPIC EQUATIONS

The macroscopic balance equations derived in Chapter 5 can be obtained directly by integration of the microscopic equations over the finite control volume. This integration requires the use of *Green's theorem*, which is a theorem usually proved in courses in advanced calculus relating volume integrals of a divergence to integrals over the bounding surface. We can demonstrate the general approach and avoid the use of Green's theorem by restricting ourselves to the very special case of a conduit without bends and having a constant rectangular cross section, as shown in Fig. 7.7.

We will derive the engineering Bernoulli equation for illustration, subject to the following assumptions, all of which are made for convenience:

• *Assumption No. 1:* The conduit has a constant rectangular cross section and has no bends.

---

*Symmetry arguments must always be applied with caution. There are some counterexamples illustrating nonsymmetric solutions to problems with apparent geometrical symmetry.

- *Assumption No. 2:* The macroscopic system is at steady state.
- *Assumption No. 3:* The fluid is incompressible and Newtonian.
- *Assumption No. 4:* The system is isothermal, so there are no physical property changes.
- *Assumption No. 5:* There is no shaft work.

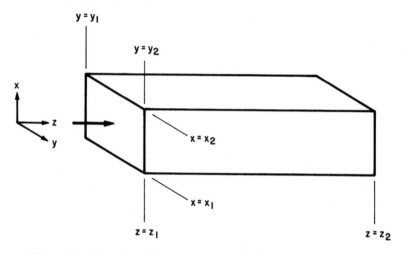

Figure 7-7. Schematic of flow in a straight conduit with constant rectangular cross section.

The approach parallels that used in Sec. 5.5.3. We first take the inner ("dot") product of the velocity vector **v** with each term in Eq. (7.50):

$$\rho \mathbf{v} \cdot \frac{\partial \mathbf{v}}{\partial t} + \rho \mathbf{v} \cdot [(\mathbf{v} \cdot \nabla)\mathbf{v}] = -\mathbf{v} \cdot \nabla(p + \rho gh) + \eta \mathbf{v} \cdot \nabla^2 \mathbf{v} \qquad (7.51)$$

By writing out the components in Cartesian coordinates we can easily establish the following identities, where in some cases incompressibility ($\nabla \cdot \mathbf{v} = 0$) is used:

$$\mathbf{v} \cdot \frac{\partial \mathbf{v}}{\partial t} = \frac{1}{2}\frac{\partial v^2}{\partial t}$$

$$\mathbf{v} \cdot [(\mathbf{v} \cdot \nabla)\mathbf{v}] = \tfrac{1}{2}\nabla \cdot (\mathbf{v}v^2)$$

$$\mathbf{v} \cdot \nabla(p + \rho gh) = \nabla \cdot \mathbf{v}(p + \rho gh)$$

v is the magnitude of **v**. The identities for the last term in Eq. (7.51) do not have a convenient compact form, and we shall not use them. Equation (7.51) can then be written

$$\frac{\partial}{\partial t}\tfrac{1}{2}\rho v^2 + \nabla \cdot \mathbf{v}(\tfrac{1}{2}\rho v^2 + p + \rho gh) = \eta \mathbf{v} \cdot \nabla^2 \mathbf{v} \qquad (7.52)$$

Note that the second term on the left already has the appearance of the Bernoulli equation.

We now integrate the equation term by term over the entire control volume:

$$\int_{\text{volume}} \frac{\partial}{\partial t} \tfrac{1}{2}\rho v^2 \, d\mho + \int_{\text{volume}} \nabla \cdot \mathbf{v}(\tfrac{1}{2}\rho v^2 + p + \rho gh) \, d\mho = \eta \int_{\text{volume}} \mathbf{v} \cdot \nabla^2 \mathbf{v} \, d\mho$$

(7.53)

The time derivative on the first term can be taken outside the integral over the fixed volume and we have

$$\int_{\text{volume}} \frac{\partial}{\partial t} \tfrac{1}{2}\rho v^2 \, d\mho = \frac{d}{dt} \int_{\text{volume}} \tfrac{1}{2}\rho v^2 \, d\mho = 0 \qquad (7.54)$$

The integral is the total kinetic energy, and its derivative is equal to zero because at steady state the total kinetic energy does not change.

The second term can be written

$$\int_{\text{volume}} \nabla \cdot \mathbf{v}(\tfrac{1}{2}\rho v^2 + p + \rho gh) \, d\mho$$

$$= \int_{x=x_1}^{x=x_2} \int_{y=y_1}^{y=y_2} \int_{z=z_1}^{z=z_2} \left\{ \frac{\partial}{\partial x}[v_x(\tfrac{1}{2}\rho v^2 + p + \rho gh)] + \frac{\partial}{\partial y}[v_y(\tfrac{1}{2}\rho v^2 + p + \rho gh)] \right.$$
$$\left. + \frac{\partial}{\partial z}[v_z(\tfrac{1}{2}\rho v^2 + p + \rho gh)] \right\} dx \, dy \, dz$$

$$= \int_{y=y_1}^{y=y_2} \int_{z=z_1}^{z=z_2} \{[v_x(\tfrac{1}{2}\rho v^2 + p + \rho gh)]|_{x=x_2}$$
$$- [v_x(\tfrac{1}{2}\rho v^2 + p + \rho gh)]|_{x=x_1}\} dy \, dz$$

$$+ \int_{x=x_1}^{x=x_2} \int_{z=z_1}^{z=z_2} \{[v_y(\tfrac{1}{2}\rho v^2 + p + \rho gh)]|_{y=y_2}$$
$$- [v_y(\tfrac{1}{2}\rho v^2 + p + \rho gh)]|_{y=y_1}\} dx \, dz$$

$$+ \int_{x=x_1}^{x=x_2} \int_{y=y_1}^{y=y_2} \{[v_z(\tfrac{1}{2}\rho v^2 + p + \rho gh)]|_{z=z_2}$$
$$- [v_z(\tfrac{1}{2}\rho v^2 + p + \rho gh)]|_{z=z_1}\} dx \, dy \qquad (7.55)$$

From the no-slip boundary condition we can set $v_x = v_y = 0$ at the walls, $x = x_1, x = x_2, y = y_1, y = y_2$. Thus, the first two integrals vanish. The third is the difference between area integrals at the two ends, and we can write

$$\int_{\text{volume}} \nabla \cdot \mathbf{v}(\tfrac{1}{2}\rho v^2 + p + \rho gh) \, d\mho = \Delta\{\langle\tfrac{1}{2}\rho v^2 V\rangle A + \langle pV\rangle A + \langle\rho ghV\rangle A\}$$

Here, we have replaced v_z with V to facilitate comparison with the earlier development.

Integration of the viscous stress term requires more extensive manipulation, and we shall simply present the result, which is[*]

$$\eta \int_{\text{volume}} \mathbf{v} \cdot \nabla^2 \mathbf{v} \, d\mathcal{V} = -\eta \int_{\text{volume}} \Phi \, d\mathcal{V} \tag{7.55a}$$

$$\Phi = 2\left[\left(\frac{\partial v_x}{\partial x}\right)^2 + \left(\frac{\partial v_y}{\partial y}\right)^2 + \left(\frac{\partial v_z}{\partial z}\right)^2\right] + \left[\frac{\partial v_y}{\partial x} + \frac{\partial v_x}{\partial y}\right]^2$$
$$+ \left[\frac{\partial v_z}{\partial y} + \frac{\partial v_y}{\partial z}\right]^2 + \left[\frac{\partial v_x}{\partial z} + \frac{\partial v_z}{\partial x}\right]^2 \tag{7.55b}$$

We thus obtain

$$\Delta\{\langle\tfrac{1}{2}\rho v^2 V\rangle A + \langle pV\rangle A + \langle \rho gh V\rangle A\} + \eta \int_{\text{volume}} \Phi \, d\mathcal{V} = 0 \tag{7.56}$$

Using the assumptions in Sec. 5.4.2, we immediately obtain the engineering Bernoulli equation, Eq. (5.34), with $\delta W_s = 0$ and the explicit expression for the losses,

$$l_V = \eta \int_{\text{volume}} \Phi \, d\mathcal{V} \bigg/ w \tag{7.57}$$

Φ is often called the *Rayleigh dissipation function*, for it represents the rate per unit volume at which heat is generated by viscous dissipation; compare Sec. 3.4.2. Note that $\Phi > 0$ and the losses can vanish identically only for $\eta \to 0$.

Equation (7.57) suggests why the losses may be expected to correlate with V^2, with a Reynolds number-dependent coefficient. If there is one characteristic velocity, V, and one characteristic spacing, D, over which the velocity changes rapidly (e.g., the minimum spacing in a rotameter), we expect

$$\Phi = \text{order of magnitude of } V^2/D^2 \tag{7.58a}$$

$$\int_{\text{volume}} \Phi \, d\mathcal{V} = \text{order of magnitude of } (V^2/D^2)\text{volume} \tag{7.58b}$$

$$w = \text{order of magnitude of } \rho V D^2 \tag{7.58c}$$

$$l_V = \text{order of magnitude of } \frac{\eta(V^2/D^2)\text{volume}}{\rho V D^2}$$

$$= \text{order of magnitude of } \left(\frac{\eta}{\rho VD}\right)\left(\frac{\text{volume}}{D^3}\right)V^2 \tag{7.58d}$$

Thus, in terms of order of magnitude, l_V separates into a product of reciprocal Reynolds number, a geometric term (volume/D^3), and a velocity head term (V^2).

---

[*]Φ is the same as the function $\tfrac{1}{2}\Pi$ defined in Sec. 8.8 in a different context. The function is given in cylindrical and spherical coordinates in Table 8-1.

7.8 CONCLUDING REMARKS

The remainder of the text is mostly devoted to the solution of the Navier–Stokes and continuity equations for flows of processing interest. It is important to commit these basic equations to memory, at least in the form given in Tables 7-1 and 7-4 for Cartesian coordinates. The stress equations for a Newtonian fluid, given in Table 7-3 for Cartesian coordinates, also need to be memorized because of their frequent application, as should the general structure (but not the detailed form) of the dissipation function, Eq. (7.55).

Many steps in the derivation of the Navier–Stokes equations involve explicit assumptions about the nature of the fluid (Newtonian) and the flow (sufficiently small variation in temperature and pressure to neglect changes in viscosity). It is important to understand the precise point at which each assumption is made to avoid misapplication of an equation. Attempting to use the Navier–Stokes equations in situations in which they do not apply is one of the most common errors in courses in fluid mechanics (and, one fears, in real process applications).

BIBLIOGRAPHICAL NOTES

The best rigorous treatment of the material in this chapter is in

SERRIN, J., "Mathematical Principles of Classical Fluid Mechanics," *Encyclopedia of Physics* (*Handbuch der Physik*), Vol. 8, No. 1, Springer-Verlag, Berlin, 1959.

This is an advanced treatment, but it should be examined at some time by every serious student of fluid mechanics. The basic assumptions are carefully defined, and the important results, including the stress equation for a Newtonian fluid, are derived with elegance and rigor. A similar treatment, in advanced textbook form, is

ARIS, R., *Vectors, Tensors, and the Basic Equations of Fluid Mechanics*, Prentice-Hall, Inc., Englewood Cliffs, N.J., 1962.

The bulk viscosity is estimated from molecular theory in

CHAPMAN, S., and COWLING, T. G., *The Mathematical Theory of Non-uniform Gases*, Cambridge University Press, New York, 1961.

HIRSCHFELDER, J. O., CURTISS, C. F., and BIRD, R. B., *The Molecular Theory of Gases and Liquids*, John Wiley & Sons, Inc., New York, 1954.

Experimental measurements are in

KARIM, S., and ROSENHEAD, L., *Rev. Mod. Phys.*, **24**, 108 (1952).

Further references are given by Serrin.

The arbitrariness of the isotropic portion of the normal stress for incompressible fluids alluded to in Sec. 7.4.3 is discussed by Serrin and in Chapter 1 of

ASTARITA, G., and MARRUCCI, G., *Principles of Non-Newtonian Fluid Mechanics*, McGraw-Hill Book Company, New York, 1974.

The role of surface tension is treated in depth in

LEVICH, V. G., *Physico-Chemical Hydrodynamics*, Prentice-Hall, Inc., Englewood Cliffs, N.J., 1962.

A general derivation of the engineering Bernoulli equation by integration of the Navier–Stokes equations may be found in

BIRD, R. B., *Chem. Eng. Sci.*, **6**, 123 (1957).

We will use the vector formulations of the equations of motion only as a convenient shorthand. The vector formulation, in fact, provides the potential for considerable analytical power, but to exploit this potential requires a background in vector analysis that has probably not yet been acquired by most readers of this book. The book by Aris, cited above, is a good introduction for advanced students. Perhaps the best basic introduction to vector analysis is still to be found in the lectures of J. Willard Gibbs, who is the founder of vector analysis as well as modern chemical thermodynamics:

WILSON, E. B., *Vector Analysis, Founded Upon the Lectures of J. Willard Gibbs*, Dover Publications, New York, 1960.

One-Dimensional Flows 8

8.1 INTRODUCTION

We now turn to the solution of the equations derived in Chapter 7. All but one of the examples in this chapter will deal with incompressible Newtonian fluids, so we will be solving the Navier–Stokes and continuity equations.

On first examination the Navier–Stokes and continuity equations seem rather foreboding. They are four coupled, nonlinear equations with derivatives in time and in each of three spatial directions. This is an extremely difficult system, and its behavior and analysis has commanded the attention of mathematicians for more than one hundred years. In a small number of cases the Navier–Stokes equations can be solved exactly. In most situations of practical interest, however, a good deal of approximation is required, often followed by machine computation. We will study some of the more common approximations in subsequent chapters.

Students are often disheartened by the apparent genius required to obtain solutions to complex equations like the ones being considered here. In reality, it is not mathematical brilliance that normally leads to solution. It is a simple fact that solutions to complex situations are obtained only by understanding the physics sufficiently to anticipate the form of the solution. The equations are then used to provide the details and to verify the assumed form. This sometimes looks like "guessing" the solution, but the mental processes involved are quite different.

The procedure for the solution of flow problems is as follows:

1. Utilize understanding of the process to determine upon which independent variables each of the velocity components depends. This step is a hypothesis that must be checked for possible contradiction at each subsequent step in the solution.
2. Substitute into the continuity equation to ensure that the dependence assumed in step 1 is consistent with continuity. In some cases the continuity equation simply guarantees (or shows lack of) consistency; in others, the continuity equation can be solved to obtain additional information about the velocity field.
3. Substitute into the momentum or Navier–Stokes equations, as appropriate, and solve. It is usually helpful to focus first on the pressure terms in these equations.

We will illustrate this solution process in this chapter with some examples of one-dimensional flows. It is important to emphasize that these solutions represent a very special situation, in that they are exact and free of approximation of any kind. Only a small number of such solutions is known to exist.

8.2 PLANE POISEUILLE FLOW

8.2.1 Problem Description

An incompressible Newtonian fluid flows at a steady rate through a rectangular channel of very large aspect ratio, as shown in Fig. 8-1. Flow is in the x direction. H/W and H/L are both very small numbers. The flow is caused by a pressure and/or elevation difference between $x = 0$ and $x = L$.

We will approximate the large aspect ratio as flow between plates of infinite width, $H/W \rightarrow 0$. The side walls should affect the flow only a small distance into the fluid, so for a large aspect ratio the "infinite" assumption should not cause substantial error. We will also assume that the flow development region near the entrance and exit is a negligible portion of the total distance L, so that we may take the flow as *fully developed;* this follows from the statement that H/L is small. Finally, we will assume that the flow is not turbulent.

8.2.2 Direct Solution

The problem description allows us to make some immediate hypotheses about the structure of the flow. First, we will assume steady state. If the flow is *fully developed*, observations of the flow field at all x positions should be

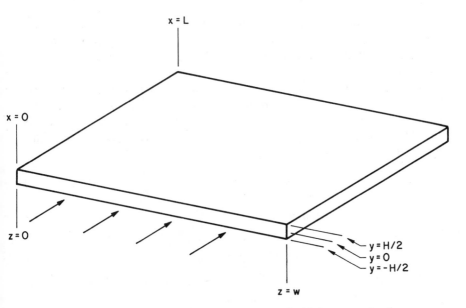

Figure 8-1. Schematic of flow in a plane channel with large aspect ratio.

the same:

$$\frac{\partial v_x}{\partial x} = \frac{\partial v_y}{\partial x} = \frac{\partial v_z}{\partial x} = 0 \tag{8.1}$$

Furthermore, if the flow field is infinite in expanse in the z direction, we can assume that there is no distinction between one z location and another; thus, the velocity field should not change in the z direction, or

$$\frac{\partial v_x}{\partial z} = \frac{\partial v_y}{\partial z} = \frac{\partial v_z}{\partial z} = 0 \tag{8.2}$$

From these two equations it follows that the components of the velocity vector will depend only on y. We can further anticipate that there will be no motion normal to the direction of mean flow ($v_z = 0$) and that all flow lines are parallel to the walls ($v_y = 0$); these last two assumptions will be examined more carefully in the next section. We thus hypothesize that the velocity field has a general form

$$v_x = v_x(y) \qquad v_y = v_z = 0 \tag{8.3}$$

We know that v_x must depend on y, since v_x will vanish at the walls ($y = \pm H/2$) because of the no-slip boundary condition, but there is a net flow in the x direction, so v_x cannot be zero for all values of y.

The continuity equation in Cartesian coordinates for an incompressible fluid is obtained from Table 7-1 as

$$\frac{\partial v_x}{\partial x} + \frac{\partial v_y}{\partial y} + \frac{\partial v_z}{\partial z} = 0$$

The first term vanishes because v_x does not depend on x, while the last two vanish because v_y and v_z are zero. Thus, the form assumed in Eq. (8.3) is consistent with the continuity equation, but no additional information is obtained.

The Navier–Stokes equations in Cartesian coordinates are given in Table 7-4. Most terms vanish upon substitution of Eq. (8.3); for example, the term $pv_x(\partial v_x/\partial x)$ vanishes because v_x does not depend on x, while $pv_y(\partial v_x/\partial y)$ vanishes because $v_y = 0$. When the zero terms are removed, the resulting forms of the equations are

$$x \text{ component: } \quad 0 = -\frac{\partial \mathcal{P}}{\partial x} + \eta \frac{d^2 v_x}{dy^2} \tag{8.4a}$$

$$y \text{ component: } \quad 0 = -\frac{\partial \mathcal{P}}{\partial y} \tag{8.4b}$$

$$z \text{ component: } \quad 0 = -\frac{\partial \mathcal{P}}{\partial z} \tag{8.4c}$$

We have written $d^2 v_x/dy^2$ instead of $\partial^2 v_x/\partial y^2$ because v_x is a function of only y, so all derivatives are ordinary derivatives.

We examine the equivalent pressure terms first. From Eqs. (8.4b) and (8.4c) it follows that $\mathcal{P}$ does not depend on y or z; thus, $\mathcal{P}$ is a function only of x,

$$\mathcal{P} = \mathcal{P}(x) \tag{8.5}$$

We thus replace $\partial \mathcal{P}/\partial x$ by $d\mathcal{P}/dx$ and write Eq. (8.4a) as

$$\eta \frac{d^2 v_x}{dy^2} = \frac{d\mathcal{P}}{dx} \tag{8.6}$$

We can now carry out an interesting logical exercise on Eq. (8.6). The left-hand side depends only on the y coordinate, while the right-hand side depends only on the x coordinate. x and y are independent variables, so one can be changed while the other is held constant. *If the right-hand side of Eq. (8.6) is independent of y, so must be the left-hand side.* Similarly, if the left-hand side is independent of x, so must be the right-hand side. Thus, we obtain the further relation that both sides of Eq. (8.6) are independent of x and y (as well as z), and are thus constants:

$$\eta \frac{d^2 v_x}{dy^2} = \frac{d\mathcal{P}}{dx} = \text{constant} \equiv \frac{\Delta \mathcal{P}}{L} \tag{8.7}$$

Here, $\Delta \mathcal{P} = \mathcal{P}(x = L) - \mathcal{P}(x = 0)$. That is, *the equivalent pressure varies linearly in the flow direction, and the pressure gradient is a constant.*

Equation (8.7) states that the derivative of dv_x/dy equals a constant, so we may integrate once with respect to y to obtain

$$\frac{dv_x}{dy} = \frac{1}{\eta} \frac{\Delta \mathcal{P}}{L} y + C_1 \tag{8.8}$$

where C_1 is a constant of integration. A second integration yields

$$v_x = \frac{1}{2\eta}\frac{\Delta\mathcal{P}}{L}y^2 + C_1 y + C_2 \tag{8.9}$$

with C_2 a second constant of integration. The constants of integration can be evaluated from the no-slip boundary condition, discussed in Sec. 7.6, which requires that v_x vanish at the stationary solid surfaces, $y = \pm H/2$:

$$y = +\frac{H}{2}: \quad v_x = 0 = \frac{1}{2\eta}\frac{\Delta\mathcal{P}}{L}\left(\frac{H}{2}\right)^2 + C_1\left(\frac{H}{2}\right) + C_2 \tag{8.10a}$$

$$y = -\frac{H}{2}: \quad v_x = 0 = \frac{1}{2\eta}\frac{\Delta\mathcal{P}}{L}\left(-\frac{H}{2}\right)^2 + C_1\left(-\frac{H}{2}\right) + C_2 \tag{8.10b}$$

Solution of these two simultaneous equations for C_1 and C_2 gives

$$C_1 = 0 \tag{8.11a}$$

$$C_2 = -\frac{1}{2\eta}\frac{\Delta\mathcal{P}}{L}\frac{H^2}{4} \tag{8.11b}$$

and substitution back into Eq. (8.9) then gives the complete solution,

$$v_x = \frac{H^2}{8\eta}\left(-\frac{\Delta\mathcal{P}}{L}\right)\left[1 - \left(\frac{2y}{H}\right)^2\right] \tag{8.12}$$

$\Delta\mathcal{P}$ is negative (the flow is caused by a pressure *drop*), so Eq. (8.12) predicts a positive velocity. The velocity profile is parabolic, with a maximum at the centerline, as shown in Fig. 8-2.

The flow rate is computed from the relation

$$Q = WH\langle v_x\rangle = \int_{area} v_x\, dA = W\int_{y=-H/2}^{H/2} v_x\, dy \tag{8.13}$$

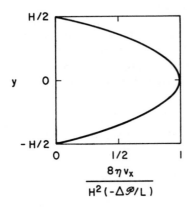

Figure 8-2. Velocity profile for laminar flow with a pressure gradient between infinite stationary planes.

Because of the assumption of infinite width, it is perhaps more consistent to work with the average velocity, which is then

$$\langle v_x \rangle = \frac{1}{H} \int_{-H/2}^{H/2} v_x \, dy = \frac{H^2}{8\eta H} \left(-\frac{\Delta \mathcal{P}}{L} \right) \int_{-H/2}^{H/2} \left[1 - \left(\frac{2y}{H} \right)^2 \right] dy$$

$$= \frac{H^2}{16\eta} \left(-\frac{\Delta \mathcal{P}}{L} \right) \int_{-1}^{1} (1 - \xi^2) \, d\xi = \frac{H^2}{12\eta} \left(-\frac{\Delta \mathcal{P}}{L} \right) \tag{8.14}$$

The calculation was simplified by use of the change of variable $\xi = 2y/H$. Equation (8.12) can then be written

$$v_x = \tfrac{3}{2} \langle v_x \rangle \left[1 - \left(\frac{2y}{H} \right)^2 \right] \tag{8.15}$$

Thus, we see that the maximum (centerline) velocity is equal to three-halves the mean velocity.

8.2.3 Symmetry Boundary Condition

The constant of integration C_1 could have been evaluated in an alternative way. The shear stress, τ_{yx}, must be continuous everywhere, and in particular across the center plane, $y = 0$. Thus, since τ_{yx} is proportional to dv_x/dy (since $\partial v_y / \partial x = 0$), dv_x/dy must be continuous at $y = 0$. We expect symmetry of the velocity about $y = 0$, so dv_x/dy must equal zero at $y = 0$; the only other way that symmetry could prevail would be with a discontinuous derivative and hence a discontinuous shear stress.

From Eq. (8.8), if we set $y = 0$ and require that dv_x/dy equal zero at $y = 0$, it follows immediately that $C_1 = 0$.

8.2.4 Relaxed Assumptions*

The assumption which we made about the flow field at the beginning of Sec. 8.2.1 were all physically reasonable, but it was not, in fact, necessary to make the last two. We need only assume that the flow is the same across any plane orthogonal to the x direction and any plane orthogonal to the z direction:

$$v_x = v_x(y) \qquad v_y = v_y(y) \qquad v_z = v_z(y) \tag{8.16}$$

which are equivalent to Eqs. (8.1) and (8.2). Specifically, we do not have to assume that $v_y = v_z = 0$. The continuity equation reduces to

$$\frac{\partial v_y}{\partial y} = 0 \qquad v_y = \text{constant} \tag{8.17}$$

Since v_y must vanish at $y = \pm H/2$ (there is no flow through the walls), it

---

*This section may be omitted on a first reading.

follows that the constant must be zero and v_y vanishes everywhere. The Navier–Stokes equations then reduce to

$$x \text{ component:} \quad 0 = -\frac{\partial \mathcal{P}}{\partial x} + \eta \frac{d^2 v_x}{dy^2} \tag{8.18a}$$

$$y \text{ component:} \quad 0 = -\frac{\partial \mathcal{P}}{\partial y} \tag{8.18b}$$

$$z \text{ component:} \quad 0 = -\frac{\partial \mathcal{P}}{\partial z} + \eta \frac{d^2 v_z}{dy^2} \tag{8.18c}$$

The y-component equation states that $\mathcal{P}$ cannot be a function of y. It is convenient to write the x and z components as

$$x \text{ component:} \quad \eta \frac{d^2 v_x}{dy^2} = \frac{\partial \mathcal{P}}{\partial x} \tag{8.19a}$$

$$z \text{ component:} \quad \eta \frac{d^2 v_z}{dy^2} = \frac{\partial \mathcal{P}}{\partial z} \tag{8.19b}$$

The same logic applied in Sec. 8.2.1 demonstrates that $\partial \mathcal{P}/\partial x$ and $\partial \mathcal{P}/\partial z$ cannot depend on y, so the two equations are uncoupled and can be solved separately.

Since $\partial \mathcal{P}/\partial z$ is independent of y, we can integrate Eq. (8.19b) twice to obtain

$$v_z = \frac{H^2}{8\eta}\left(-\frac{\partial \mathcal{P}}{\partial z}\right)\left[1 - \left(\frac{2y}{H}\right)^2\right] \tag{8.20}$$

We have already incorporated the boundary conditions $v_z = 0$ at $y = \pm H/2$. We can further write

$$\langle v_z \rangle = \frac{1}{H}\int_{-H/2}^{H/2} v_z \, dy = \frac{H^2}{12\eta}\left(-\frac{\partial \mathcal{P}}{\partial z}\right) \tag{8.21}$$

There is no net flow in the z direction, so $\langle v_z \rangle = 0$, in which case $\partial \mathcal{P}/\partial z = 0$ and we recover Eq. (8.5), $\mathcal{P} = \mathcal{P}(x)$. From Eq. (8.20) it then also follows that $v_z = 0$. Thus, we have *proved* that $v_y = v_z = 0$, and the solution for v_x then proceeds as in Sec. 8.2.1.

8.2.5 Solution Logic

It is important to reexamine the logic used in the solution of this problem, for it is the logic that we shall repeat in every case for which a solution to the Navier–Stokes equations is obtained. We made some assumptions initially about the nature of the velocity field, specifically that the only velocity changes possible are in the y direction. We then obtained a solution to the continuity and Navier–Stokes equations that followed from those assumptions. No results contradicting the assumptions developed in the course of the solution process, so we have an internally consistent exact solution to the equations.

The fact that we have found *a* solution to the set of nonlinear equations that we set out to solve does not guarantee that we have found *all* solutions. In the absence of a proof of uniqueness we must accept the possibility that another solution to the equations, which does not satisfy our starting assumptions, could also exist. Thus, our solutions must not only withstand the test of internal consistency, as this one does, but they must also be subjected to experiment. Equation (8.12) agrees with experimental data for large-aspect-ratio channels (W/H greater than 10) up to a Reynolds number ($\mathrm{Re} = H\langle v_z \rangle \rho / \eta$) of about 1000, after which the assumption of steady state breaks down and there is a transition to turbulence.

8.3 PLANE COUETTE FLOW

Plane Couette flow, or drag flow, is shown schematically in Fig. 8-3. Two large ("infinite") plates are separated by a distance H. The bottom plate is stationary, while the top moves in the x direction with a constant speed U. This is the flow used in Sec. 2.4.1 to define the viscosity, and we wish to show that definition is consistent with our development in Chapter 7. We assume that the fluid is incompressible and Newtonian.

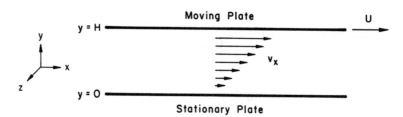

Figure 8-3. Schematic of plane Couette flow.

As in the preceding problem, we assume that the flow is entirely in the x direction, and that the only variation is in the y direction:

$$v_x = v_x(y) \qquad v_y = v_z = 0 \tag{8.22}$$

We have already seen that this form for the velocity satisfies continuity, but that no additional information is obtained. The Navier–Stokes equations are then

$$x \text{ component:} \quad 0 = -\frac{\partial \mathcal{P}}{\partial x} + \eta \frac{d^2 v_x}{dy^2} \tag{8.23a}$$

$$y \text{ component:} \quad 0 = -\frac{\partial \mathcal{P}}{\partial y} \tag{8.23b}$$

$$z \text{ component:} \quad 0 = -\frac{\partial \mathcal{P}}{\partial z} \tag{8.23c}$$

It thus immediately follows that $\mathcal{P} = \mathcal{P}(x)$.

In this problem we do not impose a pressure gradient by external means. We, therefore, have to ask whether one is induced by the flow itself. In this regard it is helpful to recall the engineering Bernoulli equation, Eq. (5.34). In the absence of work and losses, since the spacing is constant so that $\langle v_x \rangle$ is independent of x position, we will have $\mathcal{P}$ also independent of x. Work is being done to move the plates relative to one another, but this can be balanced by the losses. Thus, the Bernoulli equation does not require a pressure change. Since we have not imposed a pressure change, and one is not required to satisfy a macroscopic balance, we *hypothesize* that $\partial \mathcal{P}/\partial x = 0$, or $\mathcal{P} = $ constant. The test of this hypothesis will be the existence of a consistent solution that follows from it.

Equations (8.23) now reduce to

$$\frac{d^2 v_x}{dy^2} = 0 \tag{8.24}$$

This can be integrated twice to obtain

$$v_x = C_1 y + C_2 \tag{8.25}$$

where C_1 and C_2 are constants of integration. The no-slip boundary condition requires that the fluid adhere to the solid surfaces; thus, at the stationary plate, $y = 0$, we have

$$y = 0: \quad v_x = 0 = C_2 \tag{8.26a}$$

At the top plate the fluid moves with the speed U of the plate, so

$$y = H: \quad v_x = U = C_1 H + C_2 \tag{8.26b}$$

It thus follows that $C_1 = U/H$, $C_2 = 0$, and Eq. (8.24) becomes

$$v_x = \frac{Uy}{H} \tag{8.27}$$

It is of interest to calculate the shear stress. From Table 7-3 we have

$$\tau_{yx} = \eta \left(\frac{\partial v_x}{\partial y} + \frac{\partial v_y}{\partial x} \right) = \eta \frac{dv_x}{dy} = \eta \frac{U}{H} \tag{8.28}$$

That is, the shear stress is independent of position. We defined the shear rate, Γ_s, as U/H in Eq. (2.4), so Eq. (8.28) is identical to Eq. (2.7), which was used to define viscosity in Chapter 2.

As an aside, it is useful to note that if we assume that there are no x and z variations, it follows from the momentum equation, Table 7-2, that τ_{yx} is a constant at each point in the gap for plane Couette flow of *any* fluid, not just a Newtonian fluid.

8.4 POISEUILLE FLOW

8.4.1 Problem Description

We now consider steady flow of an incompressible Newtonian fluid in a long, smooth, round tube of radius R, as shown in Fig. 8-4. This is the pipe flow problem studied in detail in Chapter 3. We assume steady state, so any solution that we obtain will apply only in the laminar region, Re < 2100. We also assume that we are sufficiently far from the pipe entrance or exit that we may take the flow as fully developed, so that the velocity is independent of axial position. Finally, we assume that there is no variation in the circumferential direction (i.e., the pipe looks the same from all angles), so all velocity components are independent of θ.

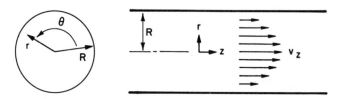

Figure 8-4. Schematic of laminar flow in a long, smooth, round tube.

8.4.2 Direct Solution

It is convenient to make one further assumption about the velocity. We assume that all flow lines are parallel to the pipe wall, so that there is no radial or circumferential flow. The velocity field then has the form

$$v_z = v_z(r) \qquad v_r = v_\theta = 0 \tag{8.29}$$

We will use cylindrical coordinates for this problem. From Table 7-1 the continuity equation is

$$\frac{1}{r}\frac{\partial}{\partial r}(rv_r) + \frac{1}{r}\frac{\partial v_\theta}{\partial \theta} + \frac{\partial v_z}{\partial z} = 0$$

Since v_r and v_θ are zero, while v_z is not a function of z, each term is zero and the continuity equation is satisfied. The equation does not contain any additional information, however.

The Navier–Stokes equations are given in cylindrical coordinates in Table 7-7. When Eqs. (8.29) are substituted into the Navier–Stokes equations, the nonlinear terms on the left all vanish, as do most of the terms on the right, and we obtain

$$r \text{ component:} \quad 0 = -\frac{\partial \mathcal{P}}{\partial r} \tag{8.30a}$$

178

θ component: $\quad 0 = -\dfrac{1}{r}\dfrac{\partial \mathcal{P}}{\partial \theta}$ (8.30b)

z component: $\quad 0 = -\dfrac{\partial \mathcal{P}}{\partial z} + \eta \dfrac{1}{r}\dfrac{d}{dr}\left(r\dfrac{dv_z}{dr}\right)$ (8.30c)

We have written dv_z/dr in place of $\partial v_z/\partial r$ because v_z is a function only of r.

The development from this point on essentially parallels the solution for plane Poiseuille flow in Sec. 8.2.2. It follows from the r- and θ-component equations that $\mathcal{P}$ is independent of r and θ,

$$\mathcal{P} = \mathcal{P}(z)$$ (8.31a)

We can then write Eq. (8.30c) as

$$\frac{1}{r}\frac{d}{dr}r\frac{dv_z}{dr} = \frac{1}{\eta}\frac{d\mathcal{P}}{dz}$$ (8.31b)

The left-hand side is a function only of r, and the right-hand side is a function only of z. Thus, each must be independent of r and z and equal to a constant,

$$\frac{1}{r}\frac{d}{dr}r\frac{dv_z}{dr} = \frac{1}{\eta}\frac{d\mathcal{P}}{dz} = \text{constant} \equiv \frac{1}{\eta}\frac{\Delta\mathcal{P}}{L}$$ (8.32)

Here, $\Delta\mathcal{P} = \mathcal{P}(z = L) - \mathcal{P}(z = 0)$. We have thus proved that the pressure gradient is linear in pipe length; this was assumed in Sec. 3.3.1.

We rewrite Eq. (8.32) as

$$\frac{d}{dr}r\frac{dv_z}{dr} = \frac{1}{\eta}\frac{\Delta\mathcal{P}}{L}r$$ (8.33)

A first integration gives

$$r\frac{dv_z}{dr} = \frac{1}{2\eta}\frac{\Delta\mathcal{P}}{L}r^2 + C_1$$ (8.34a)

or, dividing by r,

$$\frac{dv_z}{dr} = \frac{1}{2\eta}\frac{\Delta\mathcal{P}}{L}r + \frac{C_1}{r}$$ (8.34b)

C_1 is a constant of integration. A second integration gives

$$v_z = \frac{1}{4\eta}\frac{\Delta\mathcal{P}}{L}r^2 + C_1 \ln r + C_2$$ (8.35)

where C_2 is a second constant of integration.

The constant C_1 must equal zero. This follows most easily by noting that because of the logarithmic term, v_z will become infinite as r goes to zero unless C_1 equals zero. Alternatively, we could have allowed r to approach zero in Eq. (8.34a) or (8.34b); as long as dv_z/dr is required to remain finite, it then follows that $C_1 = 0$.

The no-slip boundary condition requires that v_z equal zero on the pipe wall, $r = R$:

$$r = R: \quad v_z = 0 = \frac{1}{4\eta} \frac{\Delta \mathcal{P}}{L} R^2 + C_2 \tag{8.36}$$

Substituting into Eq. (8.35) then gives the velocity profile,

$$v_z(r) = \frac{R^2}{4\eta} \left(-\frac{\Delta \mathcal{P}}{L} \right) \left[1 - \left(\frac{r}{R} \right)^2 \right] \tag{8.37}$$

$(-\Delta \mathcal{P})$ is a positive number.

We have already computed the average velocity for this parabolic profile in Example 5.1, Sec. 5.2, and found

$$\langle v_z \rangle = \frac{1}{\pi R^2} \int_{\text{area}} v_z \, dA = \frac{R^2}{8\eta} \left(-\frac{\Delta \mathcal{P}}{L} \right) \tag{8.38}$$

This is equivalent to the Hagen–Poiseuille equation, Eq. (3.9). Thus, we can write Eq. (8.37) in the alternative form

$$v_z(r) = 2\langle v_z \rangle \left[1 - \left(\frac{r}{R} \right)^2 \right] \tag{8.39}$$

Equation (8.39) was shown to agree very well with experimental data in Figs. 6-8 and 6-9. Note that the maximum velocity, which occurs at the centerline, is equal to twice the average velocity.

The friction factor was defined in Eq. (3.6) as

$$f = \frac{(-\Delta \mathcal{P}) R}{\rho \langle v_z \rangle^2 L} \tag{8.40}$$

Here, we have replaced $|\Delta p|$ with $(-\Delta \mathcal{P})$ and $D/2$ with R. Solving Eq. (8.36) for $(-\Delta \mathcal{P})/L$ and substituting into Eq. (8.40) then gives

$$f = \frac{8\eta}{\rho \langle v_z \rangle R} = \frac{16\eta}{\rho \langle v_z \rangle D} = \frac{16}{\text{Re}} \tag{8.41}$$

This is Eq. (3.8), which was found empirically to fit experimental pipe flow data, as shown in Fig. 3-1. We have now derived it here from first principles.

8.4.3 Relaxed Assumptions*

It is possible to obtain the solution in the preceding section with fewer assumptions. We need only assume fully developed flow and no θ variation in velocity, or

$$v_z = v_z(r) \qquad v_r = v_r(r) \qquad v_\theta = v_\theta(r) \tag{8.42}$$

The continuity equation then simplifies to

$$\frac{1}{r} \frac{d}{dr} (r v_r) = 0 \tag{8.43}$$

*This section may be omitted on a first reading.

This integrates immediately to $rv_r = $ constant, or

$$v_r = \frac{\text{constant}}{r} \tag{8.44}$$

The constant must be zero to ensure that v_r is finite at the centerline, $r = 0$; alternatively, the constant must be zero so that v_r vanishes at the pipe wall, or else the fluid would flow through the wall. Thus, $v_r = 0$.

Next, we turn to the Navier–Stokes equations. We need only write the r and θ components to establish the further results that $v_\theta = 0$ and $\mathcal{P} = \mathcal{P}(z)$. When Eqs. (8.42) and (8.44) are substituted into the equations in cylindrical coordinates in Table 7.7, we obtain

$$r \text{ component:} \quad -\rho \frac{v_\theta^2}{r} = -\frac{\partial \mathcal{P}}{\partial r} \tag{8.45a}$$

$$\theta \text{ component:} \quad 0 = -\frac{1}{r}\frac{\partial \mathcal{P}}{\partial \theta} + \eta \frac{d}{dr}\frac{1}{r}\frac{d}{dr}(rv_\theta) \tag{8.45b}$$

From Eq. (8.45b) it follows that $\partial \mathcal{P}/\partial \theta$ must be independent of θ and depends at most on r, since the other term in the equation depends only on r. Thus, we can write

$$\frac{\partial \mathcal{P}}{\partial \theta} = f_1(r) \tag{8.46}$$

and integrate once* to obtain

$$\mathcal{P} = \mathcal{P}_0(r, z) + f_1(r)\theta \tag{8.47}$$

Note that the "constant of integration" $\mathcal{P}_0$ can depend on r and z, since it is a constant with respect to θ changes. $\mathcal{P}$ must be periodic, with the same value at $\theta = 0$ and $\theta = 2\pi$, which correspond to the same position in the pipe; this can be true only if $f_1(r) = 0$, and we have therefore proved that $\partial \mathcal{P}/\partial \theta = 0$.

We can now write Eq. (8.45b) as

$$\frac{d}{dr}\frac{1}{r}\frac{d}{dr}(rv_\theta) = 0 \tag{8.48}$$

One integration gives

$$\frac{1}{r}\frac{d}{dr}(rv_\theta) = C_1 \tag{8.49}$$

where C_1 is a constant. A second integration then gives

$$rv_\theta = \tfrac{1}{2}C_1 r^2 + C_2 \tag{8.50a}$$

or, dividing by r,

$$v_\theta = \tfrac{1}{2}C_1 r + \frac{C_2}{r} \tag{8.50b}$$

---

*θ and r are independent variables, so $f_1(r)$ is a constant as far as θ integration and differentiation are concerned.

C_2 must be zero in order to keep v_θ finite at $r = 0$, and then C_1 must vanish in order to satisfy the no-slip condition ($v_\theta = 0$) at $r = R$. Thus, we have established that $v_\theta = 0$. It then follows at once from Eq. (8.45a) that $\partial \mathcal{P}/\partial r = 0$, and $\mathcal{P} = \mathcal{P}(z)$. We have therefore derived all the conditions leading to Eqs. (8.31), and the development then continues as in the preceding section.

8.4.4 Range of Solution

In our discussion of flow in a pipe in Chapter 3 we noted some restrictions on the application of the results derived here for laminar flow. First, the pipe must be long relative to an entry length of approximately $0.055 D$ Re [Eq. (3.14a)]. We will subsequently be able to derive an order-of-magnitude estimate of this entry length. Second, the Reynolds number must not exceed the critical value of approximately 2100, where the flow becomes turbulent. Note that the solution for the velocity, Eq. (8.37), does not involve the Reynolds number and applies in principle for all Re, $0 \leq \text{Re} < \infty$. Yet there is some physical mechanism that prevents the assumptions which we have made from being satisfied for Re > 2100. Poiseuille flow is a solution to the Navier–Stokes equations for large Reynolds numbers, but it is an *unstable solution* in that it cannot be maintained in practice. A chaotic, time-dependent solution which we call *turbulence* is the solution to the Navier–Stokes equations which is stable for Re > 2100.

The experimental observation that the parabolic velocity profile is unstable beyond a critical Reynolds number can be somewhat unsettling. When dealing with complex nonlinear equations we must be prepared for the possibility that a steady-state solution is not unique and might apply only in a limited range of some parameter, or perhaps not at all. Stability of steady-state solutions of the Navier–Stokes equations and other physical and physicochemical systems is treated in specialized texts. For those to whom this concept is new, the following simple example might provide some insight.

Consider the ordinary differential equation

$$\frac{dv}{dt} = (v - 1)(v - 2) \qquad v(0) = A \tag{8.51}$$

At steady-state $dv/dt = 0$, and we have the algebraic equation

$$(v - 1)(v - 2) = 0 \tag{8.52}$$

This has two solutions, $v = 1$ and $v = 2$. The differential equation is readily integrated to yield a solution

$$v(t) = \frac{2(A - 1) - (A - 2)e^t}{A - 1 - (A - 2)e^t} \tag{8.53}$$

For $A < 2$, $v(t)$ will always go to a value of 1 as $t \to \infty$. For $A > 2$, $v(t)$

will never go to a steady state and any information obtained by solving the steady-state algebraic equation is totally misleading. Compare this observation with the case of pipe flow for Re > 2100. For the mathematical example here, $v(t)$ increases in magnitude and becomes infinite for a value $t = \ln (A - 1)/(A - 2)$. The computed steady state $v = 2$ never occurs unless the system starts there with $A = 2$ and never changes. This is an unstable situation, since the slightest perturbation would move $v(t)$ to a value on one side or the other of $v = 2$, and it would then continue to move according to the differential equation to a value $v = 1$ or $v \rightarrow \infty$. Infinitesimal perturbations cannot be kept out of real physical systems, so an unstable steady state like $v = 2$ would never be observed in practice.

8.5 WIRE COATING

In the manufacture of coated wires the wire is pulled through a bath of the coating liquid and then through a die which "wipes" the liquid and leaves a coating of the desired thickness. It is a useful exercise to consider the situation shown in Fig. 8-5 in order to relate the ultimate radius of the coated wire, R_c, to the radius of the uncoated wire, R_w, and the die radius, R_d. The wire moves with velocity V_w and the die length L is assumed to be long enough to ignore entrance effects.

To compute R_c we will need to know the radial dependence of velocity within the die. This follows from a simple mass balance. Far downstream from the die, where there are no shearing forces on the coating, the entire coating moves as a plug with the wire. The problem is therefore analogous to the free jet in Sec. 6.7, where a change in velocity profile resulted in a change in jet diameter. At some control surface downstream, the volumetric flow rate of coating is

$$Q = V_w(\pi R_c^2 - \pi R_w^2) \tag{8.54}$$

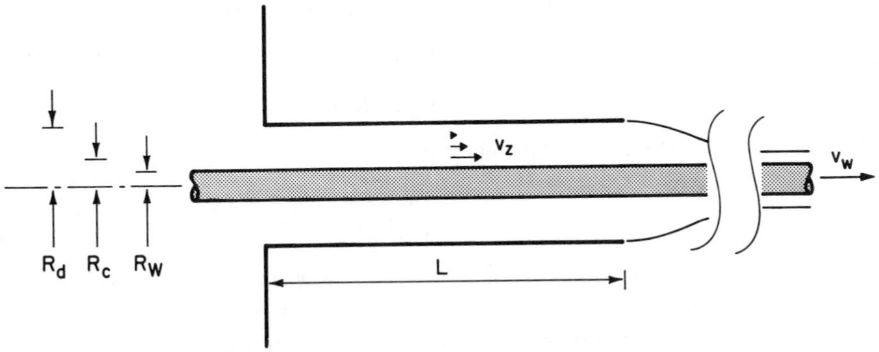

Figure 8-5. Schematic of wire coating.

On the other hand, at a cross section within the die, the volumetric flow rate is

$$Q = \int_{R_w}^{R_d} 2\pi r v_z(r)\, dr \tag{8.55}$$

where z is the axial and r the radial direction. Axial symmetry is assumed, so v_z depends only on r and not on θ. At steady state these two flow rates must be equal, so R_c is found to be

$$R_c = \left[R_w^2 + \frac{2}{V_w} \int_{R_w}^{R_d} r v_z(r)\, dr \right]^{1/2} \tag{8.56}$$

To obtain v_z we go to the Navier–Stokes and continuity equations in cylindrical coordinates. We assume axial streamlines, so v_z is a function only of r and $v_r = v_\theta = 0$. We have already seen that the continuity equation is satisfied by these assumptions. The flow in the die is a drag flow, with constant cross section and no imposed pressure gradient, analogous to the plane Couette flow in Sec. 8.3, so we assume that there is no axial variation in $\mathcal{P}$ in the die and $\partial \mathcal{P}/\partial z = 0$. With these assumptions the r and θ components of the Navier–Stokes equations are trivially satisfied with $\mathcal{P} = $ constant, and the z component simplifies to

$$0 = \eta \left[\frac{1}{r} \frac{d}{dr} \left(r \frac{dv_z}{dr} \right) \right] \tag{8.57}$$

A first integral of this equation is

$$r \frac{dv_z}{dr} = \text{constant} = C_1 \tag{8.58}$$

A second integration gives

$$v_z(r) = C_1 \ln r + C_2 \tag{8.59}$$

The constants C_1 and C_2 are evaluated from the boundary information. At $r = R_w$ the no-slip condition requires that the fluid move at the wire velocity, V_w. At $r = R_d$ the fluid adheres to the wall and has zero velocity. Thus,

$$r = R_w: \quad V_w = C_1 \ln R_w + C_2 \tag{8.60a}$$

$$r = R_d: \quad 0 = C_1 \ln R_d + C_2 \tag{8.60b}$$

Solving for C_1 and C_2 gives the velocity as

$$v_z(r) = V_w \frac{\ln r/R_d}{\ln R_w/R_d} \tag{8.61}$$

Note that the velocity is independent of viscosity, η. It depends only on the wire velocity and the geometry. Note also that the logarithmic term in the velocity is retained in this solution since it does not lead to a physically unacceptable infinite velocity.

The coating diameter is obtained by substituting Eq. (8.61) into Eq. (8.56) and carrying out the indicated integration, giving

$$R_c = \left(\frac{R_d^2 - R_w^2}{2 \ln R_d/R_w}\right)^{1/2} \tag{8.62}$$

The ultimate coating thickness depends only on the wire and die dimensions. It is independent of fluid properties and wire speed.

The force required to pull the wire through the die is computed by determining the shear stress at the wire surface. This is given by

$$\tau_{rz}\bigg|_{r=R_w} = \eta\left(\frac{\partial v_z}{\partial r} + \frac{\partial v_r}{\partial z}\right)\bigg|_{r=R_w} = \frac{\eta V_w}{r \ln R_w/R_d}\bigg|_{r=R_w} = \frac{\eta V_w}{R_w \ln R_w/R_d} \tag{8.63}$$

The force on the total wire surface in the die is

$$F_w = 2\pi R_w L \tau_{rz}\bigg|_{r=R_w} = \frac{2\pi\eta V_w L}{\ln R_w/R_d} \tag{8.64}$$

Note that $F_w < 0$, tending to hold the wire back. An equal and opposite force must be imposed on the wire to move it through the die at velocity V_w. This is not the total force, since there will also be resistance in the bath which is unaccounted for here. The linear dependence of F_w on length, L, shows that short dies should be used. There are also other reasons for using short dies which relate to the non-Newtonian behavior of most coating materials.

8.6 TORSIONAL FLOW

8.6.1 Problem Description

Many commercial viscometers are based on the measurement of the torque on a cylinder rotating in a liquid. The system is shown schematically in Fig. 8-6. In some cases only a thin layer of fluid is contained between an inner and an outer cylinder; this was the case discussed in Sec. 2.4.5. We consider here the other extreme, in which the outer wall is sufficiently removed from the rotating spindle that we may consider the cylinder to be in an infinite expanse of liquid. This flow is a prototype of other rotating systems, and it gives some insight as well into the problem of cavitation and conditions under which it might occur in operations such as mixing.

In the analysis we will assume that $L \gg R$, in which case we will ignore end effects at the cylinder bottom and at the free liquid interface. The basic hypothesis concerning the flow is that there is no axial or radial motion, so that the only nonzero velocity component in cylindrical coordinates is v_θ. We also assume angular symmetry, so that v_θ depends only on r. These assumptions about the velocity are consistent with the continuity equation.

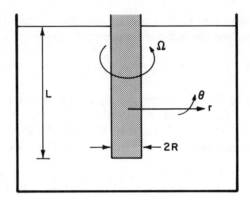

Figure 8-6. Schematic of torsional flow.

8.6.2 Velocity Field

This is a drag flow with no area change and no imposed pressure gradient in the flow direction. Thus, we assume that $\partial \mathcal{P}/\partial \theta = 0$. Together with the assumed form for the velocity,

$$v_\theta = v_\theta(r) \qquad v_r = v_z = 0 \tag{8.65}$$

the Navier–Stokes equations in cylindrical coordinates, Table 7-7, simplify to

$$r \text{ component:} \quad -\rho \frac{v_\theta^2}{r} = -\frac{\partial \mathcal{P}}{\partial r} \tag{8.66a}$$

$$\theta \text{ component:} \quad 0 = \eta \frac{d}{dr}\left[\frac{1}{r}\frac{d}{dr}(rv_\theta)\right] \tag{8.66b}$$

$$z \text{ component:} \quad 0 = -\frac{\partial \mathcal{P}}{\partial z} \tag{8.66c}$$

It follows from Eq. (8.66c) that $\mathcal{P}$ is independent of z. Note that Eq. (8.66a) contains a nonlinear term in velocity, and that the pressure does have a radial dependence. We will return to this point subsequently.

The velocity can be obtained directly from Eq. (8.66b). A first integration gives

$$\frac{1}{r}\frac{d}{dr}(rv_\theta) = \text{constant} = C_1 \tag{8.67a}$$

or, equivalently,

$$\frac{d}{dr}(rv_\theta) = C_1 r \tag{8.67b}$$

A second integration then gives

$$rv_\theta = \tfrac{1}{2}C_1 r^2 + C_2 \tag{8.68a}$$

or, dividing by r,

$$v_\theta = \tfrac{1}{2}C_1 r + \frac{C_2}{r} \tag{8.68b}$$

The fluid is assumed to be unbounded, so it is possible for r to increase without bound. C_1 must therefore be zero to ensure that v_θ will remain finite.

The no-slip condition at the rotating cylinder requires that the fluid move with the cylinder velocity. The cylinder moves with an angular velocity Ω and hence a linear velocity $R\Omega$; the no-slip condition is therefore

$$r = R: \quad v_\theta = R\Omega = \frac{C_2}{R} \qquad (8.69)$$

Thus, $C_2 = R^2\Omega$, and the velocity has the form

$$v_\theta = \frac{R^2\Omega}{r} \qquad (8.70)$$

Note that finiteness at $r = 0$ is not a problem, because the smallest radius encountered is $r = R$.

8.6.3 Torque

In order to find the torque we must first compute the shear stress on the cylinder. This is a stress on an r surface in the θ direction, so we require $\tau_{r\theta}$. From Table 7-6 we find that

$$\tau_{r\theta} = \eta\left[r\frac{\partial}{\partial r}\left(\frac{v_\theta}{r}\right) + \frac{1}{r}\frac{\partial v_r}{\partial \theta}\right] \qquad (8.71)$$

To obtain the shear stress at the cylinder surface we substitute Eq. (8.70) and evaluate Eq. (8.71) at $r = R$:

$$\tau_{r\theta}\Big|_{r=R} = \eta r\frac{d}{dr}\left(\frac{R^2\Omega}{r^2}\right)\Big|_{r=R} = -\frac{2\eta R^2\Omega r}{r^3}\Big|_{r=R} = -2\eta\Omega \qquad (8.72)$$

The calculation of the torque then follows in a manner identical to that in Sec. 2.4.5. Referring to Fig. 2-12, the differential force is $LR\tau_{r\theta}\,d\theta$, and the differential torque is $LR^2\tau_{r\theta}\,d\theta$. The total torque is therefore $2\pi LR^2\tau_{r\theta}$, or

$$G = -4\pi R^2 L\eta\Omega \qquad (8.73)$$

The negative sign arises because this is the torque exerted on the cylinder by the fluid; a positive torque of the same magnitude is required to rotate the cylinder. The viscosity can be determined from a single reading, but it is preferable to use the slope of a plot of $G/4\pi R^2 L$ versus Ω.

8.6.4 Pressure and Cavitation

The pressure is independent of θ and z, so we can write Eq. (8.66a) as

$$\frac{d\mathcal{P}}{dr} = \frac{\rho v_\theta^2}{r} = \frac{\rho(R^2\Omega)^2}{r^3} \qquad (8.74)$$

This pressure variation is required to balance the *centrifugal force*. Equation

(8.74) can be integrated to give

$$\mathcal{P} = \mathcal{P}_0 - \frac{\rho(R^2\Omega)^2}{2r^2} \qquad (8.75)$$

Here, $\mathcal{P}_0$ is a constant, equal to the pressure far from the rotating cylinder. $\mathcal{P}_0$ is the maximum pressure; the minimum pressure occurs at the cylinder surface.

The phenomenon of *cavitation* was introduced in Sec. 6.10. This is the formation of vapor bubbles when the pressure falls below the vapor pressure of the liquid. Conditions for cavitation can be computed from Eq. (8.75). We set $r = R$ to obtain the minimum pressure, and we set $\mathcal{P} = \mathcal{P}_{vp}$, the vapor pressure. Cavitation will then occur for

$$\text{cavitation:} \quad R\Omega \geq \sqrt{\frac{2(\mathcal{P}_0 - \mathcal{P}_{vp})}{\rho}} \qquad (8.76)$$

An illustrative calculation is helpful here. We take $\mathcal{P}_0$ as atmospheric pressure, approximately 10^5 Pa. The vapor pressure of water at room temperature is approximately 0.025×10^5 Pa, which can be neglected relative to $\mathcal{P}_0$. $\rho = 1000$ kg/m^3. Substituting these values into the right-hand side of Eq. (8.76) gives a critical value of $R\Omega$ of about 14 m/s. Such a speed is often obtainable on a rotating shaft. Cavitation is a violent phenomenon that can cause a surprising amount of damage.

8.7 RECTILINEAR FLOW AND HYDRAULIC DIAMETER*

We have obtained solutions for pressure-driven flow in slits and round tubes. Solutions can be obtained for other regular cross sections, although the forms are not as simple as the ones derived here. One useful result for an arbitrary cross section can be obtained without a complete solution.

The concept of hydraulic diameter was introduced in Sec. 3.6 to enable us to deal with flow in noncircular cross sections. In examining the data available, we found that in the laminar region the friction factor was inversely proportional to the Reynolds number, but the constant of proportionality depended on the shape. We demonstrate that analytically here.

The flow geometry is shown in Fig. 8-7. The conduit surface is described by a function $R(\theta)$ in a polar coordinate system. Flow is in the z direction, and the cross section is constant. There is a pressure difference $\Delta\mathcal{P}$ over a length L.

We assume that all flow is parallel to the conduit walls, so that the only nonzero velocity component is v_z. v_z will depend on both r and θ in this irregular cross section, but we assume that the flow is fully developed and indepen-

---

*This section may be omitted on a first reading.

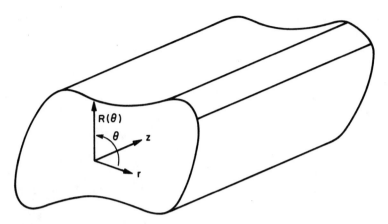

Figure 8-7. Schematic of flow in a straight conduit of arbitrary constant cross section.

dent of z. The continuity equation is then automatically satisfied, and the components of the Navier–Stokes equations in cylindrical coordinates become

$$r \text{ component:} \quad 0 = -\frac{\partial \mathcal{P}}{\partial r} \tag{8.77a}$$

$$\theta \text{ component:} \quad 0 = -\frac{1}{r}\frac{\partial \mathcal{P}}{\partial \theta} \tag{8.77b}$$

$$z \text{ component:} \quad 0 = -\frac{\partial \mathcal{P}}{\partial z} + \eta\left[\frac{1}{r}\frac{\partial}{\partial r}\left(r\frac{\partial v_z}{\partial r}\right) + \frac{1}{r^2}\frac{\partial^2 v_z}{\partial \theta^2}\right] \tag{8.77c}$$

It follows from Eqs. (8.77a) and (8.77b) that $\mathcal{P} = \mathcal{P}(z)$; since the second term in Eq. (8.77c) depends only on r and θ, we can thus replace $\partial \mathcal{P}/\partial z$ by $\Delta \mathcal{P}/L$ and write

$$\frac{\Delta \mathcal{P}}{L} = \eta\left[\frac{1}{r}\frac{\partial}{\partial r}\left(r\frac{\partial v_z}{\partial r}\right) + \frac{1}{r^2}\frac{\partial^2 v_z}{\partial \theta^2}\right] \tag{8.78}$$

We now multiply both sides of the equation by the differential area in cylindrical coordinates, $r\, dr\, d\theta$, integrate over the area, and divide by the total cross-sectional area, to obtain

$$\frac{\Delta \mathcal{P}}{L} = \frac{\eta}{A}\int_{\theta=0}^{2\pi}\int_{r=0}^{R(\theta)}\left[\frac{\partial}{\partial r}\left(r\frac{\partial v_z}{\partial r}\right) + \frac{1}{r}\frac{\partial^2 v_z}{\partial \theta^2}\right] dr\, d\theta \tag{8.79}$$

We now introduce a change of variables and define a dimensionless velocity, u, and a dimensionless distance, ξ. Let

$$v_z = u\langle v_z\rangle \qquad r = \frac{\xi D_H}{2} \tag{8.80a, b}$$

D_H is the hydraulic diameter. The dimensionless location of the perimeter is defined by

$$R(\theta) = \frac{\Xi(\theta)D_H}{2} \tag{8.80c}$$

Equation (8.79) can then be written

$$\frac{\Delta\mathcal{P}}{L} = \frac{\eta}{A}\langle v_z \rangle \int_{\theta=0}^{2\pi} \int_{\xi=0}^{\Xi(\theta)} \left[\frac{\partial}{\partial\xi}\left(\xi\frac{\partial u}{\partial\xi}\right) + \frac{1}{\xi}\frac{\partial^2 u}{\partial\theta^2}\right] d\xi \, d\theta \tag{8.81}$$

The double integral on the right depends on the *shape* of the cross section, but not on the length scale or flow rate.

The hydraulic diameter is defined in terms of the area, A, and perimeter, p, as

$$D_H = \frac{4A}{p} = \frac{4A}{\int_{\theta=0}^{2\pi} R(\theta)\,d\theta} = \frac{4A}{\frac{1}{2}D_H \int_{\theta=0}^{2\pi} \Xi(\theta)\,d\theta} \tag{8.82}$$

Thus, we can express the area in terms of D_H, and rewrite Eq. (8.82) as

$$\frac{\Delta\mathcal{P}}{2L} = \frac{16\eta\langle v_z \rangle}{D_H^2} F_s \tag{8.83}$$

where F_s is a *shape factor* that depends only on the shape of the conduit cross section but not the size:

$$F_s = \frac{\int_{\theta=0}^{2\pi} \int_{\xi=0}^{\Xi(\theta)} \left[\frac{\partial}{\partial\xi}\left(\xi\frac{\partial u}{\partial\xi}\right) + \frac{1}{\xi}\frac{\partial^2 u}{\partial\theta^2}\right] d\xi \, d\theta}{4\int_{\theta=0}^{2\pi} \Xi(\theta)\,d\theta} \tag{8.84}$$

Equation (8.83) can then be rearranged to the form

$$\frac{\Delta\mathcal{P}}{2\rho\langle v_z \rangle^2}\frac{D_H}{L} = \frac{16\eta}{\rho\langle v_z \rangle D_H} F_s \tag{8.85a}$$

or, equivalently,

$$f = \frac{16F_s}{Re} \tag{8.85b}$$

8.8 TUBE FLOW OF A POWER-LAW FLUID

All the examples considered thus far in this chapter have been for incompressible Newtonian fluids. It is useful for comparison to examine the tube flow of a non-Newtonian fluid in order to see the differences in behavior and the consequences of the Newtonian assumption.

The power-law fluid was defined in Eq. (2.10) for plane Couette flow. The general form of the equation is analogous to the three-dimensional form of the Newtonian fluid in Tables 7-3, 7-6 and 7-9 except that the viscosity is a function of the velocity gradients in the form

$$\eta = K |\tfrac{1}{2}\text{II}|^{(n-1)/2} \tag{8.86}$$

The function II is shown in Table 8-1 for three coordinate systems.

TABLE 8-1

THE FUNCTION $\tfrac{1}{2}$II IN RECTANGULAR, CYLINDRICAL, AND SPHERICAL COORDINATES

Rectangular: $\tfrac{1}{2}\text{II} = 2\left[\left(\dfrac{\partial v_x}{\partial x}\right)^2 + \left(\dfrac{\partial v_y}{\partial y}\right)^2 + \left(\dfrac{\partial v_z}{\partial z}\right)^2\right]$

$\qquad + \left[\dfrac{\partial v_y}{\partial x} + \dfrac{\partial v_x}{\partial y}\right]^2 + \left[\dfrac{\partial v_z}{\partial y} + \dfrac{\partial v_y}{\partial z}\right]^2 + \left[\dfrac{\partial v_x}{\partial z} + \dfrac{\partial v_z}{\partial x}\right]^2$

Cylindrical: $\tfrac{1}{2}\text{II} = 2\left[\left(\dfrac{\partial v_r}{\partial r}\right)^2 + \left(\dfrac{1}{r}\dfrac{\partial v_\theta}{\partial \theta} + \dfrac{v_r}{r}\right)^2 + \left(\dfrac{\partial v_z}{\partial z}\right)^2\right]$

$\qquad + \left[r\dfrac{\partial}{\partial r}\left(\dfrac{v_\theta}{r}\right) + \dfrac{1}{r}\dfrac{\partial v_r}{\partial \theta}\right]^2 + \left[\dfrac{1}{r}\dfrac{\partial v_z}{\partial \theta} + \dfrac{\partial v_\theta}{\partial z}\right]^2$

$\qquad + \left[\dfrac{\partial v_r}{\partial z} + \dfrac{\partial v_z}{\partial r}\right]^2$

Spherical: $\tfrac{1}{2}\text{II} = 2\left[\left(\dfrac{\partial v_r}{\partial r}\right)^2 + \left(\dfrac{1}{r}\dfrac{\partial v_\theta}{\partial \theta} + \dfrac{v_r}{r}\right)^2\right.$

$\qquad \left. + \left(\dfrac{1}{r\sin\theta}\dfrac{\partial v_\phi}{\partial \phi} + \dfrac{v_r}{r} + \dfrac{v_\theta \cot\theta}{r}\right)^2\right]$

$\qquad + \left[r\dfrac{\partial}{\partial r}\left(\dfrac{v_\theta}{r}\right) + \dfrac{1}{r}\dfrac{\partial v_r}{\partial \theta}\right]^2$

$\qquad + \left[\dfrac{\sin\theta}{r}\dfrac{\partial}{\partial \theta}\left(\dfrac{v_\phi}{\sin\theta}\right) + \dfrac{1}{r\sin\theta}\dfrac{\partial v_\theta}{\partial \phi}\right]^2$

$\qquad + \left[\dfrac{1}{r\sin\theta}\dfrac{\partial v_r}{\partial \phi} + r\dfrac{\partial}{\partial r}\left(\dfrac{v_\phi}{r}\right)\right]^2$

The development parallels that in Sec. 8.4. We assume that the velocity field has the form

$$v_z = v_z(r) \qquad v_r = v_\theta = 0 \tag{8.87}$$

This form satisfies continuity. We cannot use the Navier–Stokes equations because we do not have a Newtonian fluid with a constant viscosity. Thus, we must use the momentum equations in Table 7.5. It readily follows from Eq. 8.87 that the only nonzero components of the stress are τ_{rz} and τ_{zr}, which depend only on r. The momentum equations therefore reduce to

$$r \text{ component:} \quad 0 = -\frac{\partial \mathcal{P}}{\partial r} \tag{8.88a}$$

$$\theta \text{ component:} \quad 0 = -\frac{1}{r}\frac{\partial \mathcal{P}}{\partial \theta} \tag{8.88b}$$

$$z \text{ component:} \quad 0 = -\frac{\partial \mathcal{P}}{\partial z} + \frac{1}{r}\frac{d}{dr}(r\tau_{zr}) \tag{8.88c}$$

It immediately follows from the r and θ equations that $\mathcal{P} = \mathcal{P}(z)$, so that Eq. (8.88c) can be rewritten

$$\frac{1}{r}\frac{d}{dr}r\tau_{rz} = \frac{d\mathcal{P}}{dz} = \frac{\Delta\mathcal{P}}{L} \tag{8.89}$$

We can replace $d\mathcal{P}/dz$ with the constant $\Delta\mathcal{P}/L$ because the left-hand side of the equation depends only on r and the right-hand side depends only on z, and therefore both must equal a constant.

Equation (8.89) can be integrated once to obtain

$$r\tau_{rz} = \frac{\Delta\mathcal{P}}{2L}r^2 + C_1 \tag{8.90a}$$

or, dividing by r,

$$\tau_{rz} = \frac{\Delta\mathcal{P}}{2L}r + \frac{C_1}{r} \tag{8.90b}$$

The constant C_1 must equal zero in order for the shear stress to remain finite at the centerline, and we have

$$\tau_{rz} = \frac{\Delta\mathcal{P}}{2L}r \tag{8.91}$$

Equation (8.91) is, in fact, a general result for fully developed pipe flow, although we have derived it here for the special case of a power-law fluid; see the previous macroscopic derivation in a footnote in Sec. 3.4.2.

From Eq. (8.86) and Table 8-1 we can now express τ_{rz} in terms of dv_z/dr and write Eq. (8.91) as

$$K\left|\frac{dv_z}{dr}\right|^{n-1}\frac{dv_z}{dr} = \frac{\Delta\mathcal{P}}{2L}r \tag{8.92}$$

$\Delta\mathcal{P}$ is negative, and thus dv_z/dr must be negative. Equation (8.92) is thus equivalent to

$$\frac{dv_z}{dr} = -\left(-\frac{\Delta\mathcal{P}}{2KL}\right)^{1/n}r^{1/n} \tag{8.93}$$

This is integrated once to obtain

$$v_z = -\frac{n}{n+1}\left(-\frac{\Delta\mathcal{P}}{2KL}\right)^{1/n}r^{(n+1)/n} + C_2 \tag{8.94}$$

The constant of integration is evaluated from the no-slip boundary condition, $v_z = 0$ at $r = R$, and the final result is

$$v_z(r) = \frac{n}{n+1}\left[\frac{R^{n+1}}{2K}\left(-\frac{\Delta\mathcal{P}}{L}\right)\right]^{1/n}\left[1 - \left(\frac{r}{R}\right)^{(n+1)/n}\right] \tag{8.95}$$

This reduces to Eq. (8.37) for the special case $n = 1$, which is the Newtonian fluid.

The average velocity is obtained from area integration:

$$\langle v_z \rangle = \frac{1}{\pi R^2}\int_0^R 2\pi r v_z(r)\,dr = \frac{n}{(3n+1)}\left[\frac{R^{n+1}}{2K}\left(-\frac{\Delta\mathcal{P}}{L}\right)\right]^{1/n} \tag{8.96}$$

Thus, Eq. (8.95) can be written equivalently as

$$v_z(r) = \frac{3n+1}{n+1}\langle v_z \rangle \left[1 - \left(\frac{r}{R} \right)^{(n+1)/n} \right] \qquad (8.97)$$

Figure 8-8 shows velocity profile data obtained by streak photography for a polymer solution with a value of $n = 0.48$ obtained from a shear experiment in a viscometer, together with Eq. (8.97) for $n = 0.48$. It is to be noted that the profile is blunter than that for a Newtonian fluid.

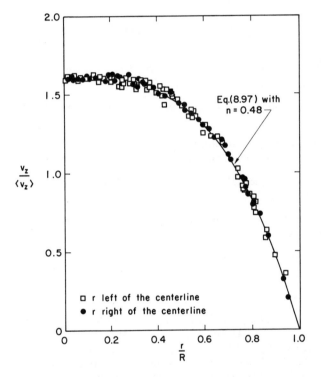

Figure 8-8. Velocity profile in a tube for a solution of polyacrylamide in water. The solution is approximately a power-law fluid with $n = 0.48$. The data were obtained by P.J. Cable using streak photography.

8.9 CONCLUDING REMARKS

The solutions to flow problems that were obtained in this chapter represent important, but very special cases. The anticipated velocity dependence on the spatial variables was such that the nonlinear inertial terms were identically zero and the pressure distribution had a particularly convenient form.

Thus, the mathematical problem reduced to the solution of elementary linear ordinary differential equations, for which exact solutions could be found. Only a small number of such solutions has ever been obtained.

The solutions in this chapter are also special in that the absence of any inertial contribution masks some of the important qualitative features of fluid flow problems. These features are brought out in the examples in the next two chapters.

BIBLIOGRAPHICAL NOTES

The development in Sec. 8.7 is adapted from a more general treatment that includes developing flow in Chapter 4 of

MIDDLEMAN, S., *Fundamentals of Polymer Processing*, McGraw-Hill Book Company, New York, 1977.

Chapter 5 of Middleman includes some of the flow problems in this chapter and some extensions, as well as additional flows for power-law fluids.

The concept of the stability of a solution, which is introduced in Sec. 8.4.4, is treated in detail in

DENN, M. M., *Stability of Reaction and Transport Processes*, Prentice-Hall, Englewood Cliffs, N.J., 1975.

PROBLEMS

8.1. An incompressible Newtonian fluid is contained between two long concentric cylinders of radii λR and R, $\lambda < 1$ (Fig. 8P1). The inner cylinder rotates with an angular velocity Ω. Compute the velocity distribution between the cylinders, and the torque required to hold the outer cylinder stationary. End effects caused by the finite length of the cylinders may be neglected.

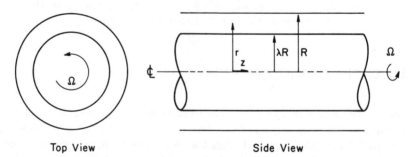

Top View   Side View

8.2. An incompressible Newtonian fluid flows through an annulus formed by two cylinders of radii λR and R, $\lambda < 1$ (Fig. 8P2). There is a pressure drop $|\Delta \mathcal{P}|$ over a length L.

a) Compute the velocity distribution and flow rate.

b) Compute the friction factor—Reynolds number relation in terms of the hydraulic diameter (Eq. 3.40). How good an approximation is $f = 16/\text{Re}$?

c) Show that the limiting solution for the velocity and flow rate as $\lambda \rightarrow 1$ is equivalent to Eqs. 8.12 and 8.14 for flow between parallel plates.

d) Compare the behavior as $\lambda \rightarrow 0$ with Eqs. 8.37 and 8.38 for flow in a pipe.

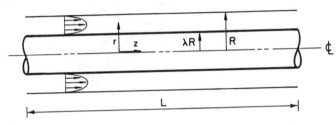

8.3. Many two-phase contacting devices require the flow of a film of liquid over a solid surface. An incompressible Newtonian liquid flows under the influence of gravity down an inclined plane at angle β to the horizontal (Fig. 8P3). The flow rate per unit width is q. Compute the velocity distribution in the film and the film thickness. The liquid film is in contact with a gas that may be taken as inviscid. (Hint: The key to this problem is the proper formulation of the boundary condition at the liquid-gas interface. What is the significance of the fact that the gas is inviscid?)

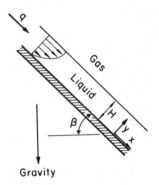

Gravity

8.4. A laminated coating process requires the co-current flow of two incompressible Newtonian fluids down an inclined plane at angle β to the horizontal (Fig. 8P4). Determine the velocity distributions and layer thicknesses for given flow rates per unit width q_I and q_II. (Hint: What are the boundary conditions?)

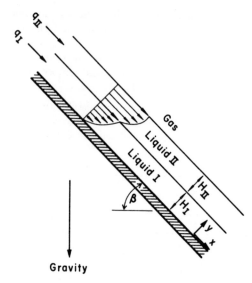

Gravity

8.5. An incompressible Newtonian fluid is contained between two disks separated by a spacing H. One disk rotates relative to the other with an angular velocity Ω.

a) Write the boundary conditions for this problem and use them to rationalize a flow field of the form $v_r = v_z = 0$, $v_\theta = rf(z)$.

b) Show that a solution to the Navier-Stokes equations does not exist for these assumed kinematics.

8.6. Show that the shear stress τ_{rz} is a linear function of radial position for fully-developed pipe flow of any fluid, Newtonian or non-Newtonian, as long as the extra-stress is a function only of velocity gradients.

8.7. The flow shown in Fig. 8P7 is an idealization of a gear pump. You may assume that $H \ll L$, so that the flow reversal near the teeth can be neglected and only the region where the flow is parallel to the walls needs to be analyzed. (This assumption is not a good one for a real gear pump.)

a) Compute the velocity profile and the pressure gradient. (Hint: Note that the net flow rate is zero. Why?)

b) Compute the stress on the moving surface.

c) Describe how this result could be used to compute the torque on a gear pump.

Surface Moving at Constant Speed

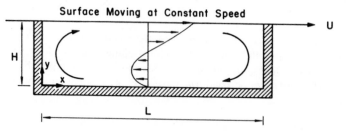

8.8. A single screw extruder is a device that is commonly used in plastics processing to build up pressure in a liquid (among other uses) so that the liquid can be extruded through a die of specified shape. (A meat grinder is a screw extruder.) Following certain geometric approximations, the screw extruder can be analyzed in terms of the flow process shown in Fig. 8P8. Fluid enters the channel at pressure $\mathcal{P}_0$ and leaves the extruder at pressure $\mathcal{P}_1$. U is related to the relative speed between screw shaft and barrel, H is the depth of the channel, and L is the distance that a fluid particle moves along the helical screw channel from entrance to exit.

a) For $H \ll L$, the flow rearrangement near the entrance and exit can be neglected, and only the region where the flow is parallel to the solid surfaces needs to be considered. Compute the velocity profile and flow rate in terms of U and $\mathcal{P}_1 - \mathcal{P}_0$ for a positive flow rate. Find equations.

b) Show that the maximum pressure increase occurs as the flow rate goes to zero.

c) Compute the power input in terms of U and $\mathcal{P}_1 - \mathcal{P}_0$.

d) Fluid is to be extruded at a flow rate per unit width q between parallel plates with a spacing H and length L. What is the required value of U?

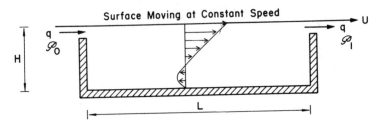

8.9. An incompressible Newtonian fluid is pumped isothermally through a long horizontal pipe at very high pressure, where the pressure dependence of viscosity must be taken into account. Compute the flow rate-pressure change equation when the viscosity-pressure dependence is of the form $\eta = \eta_0 \exp kp$. (Hint: Neglect the radial variation of pressure and use your solution to estimate the error involved in this approximation.)

8.10. An incompressible Newtonian fluid flows radially outward between two long porous cylinders, as shown in Fig. 8P10.

a) Compute the pressure change from the outside of the inner cylinder (pt. B) to the inside of the outer cylinder (pt. C).

b) What value would a pressure transducer placed at C record? (Consider your answer carefully!)

c) The flow rate per unit length through each porous cylinder is given by the equation

$$q = k|\Delta p|\bar{R}/\eta$$

where k is a constant and $\bar{R}$ is the mean radius of the cylinder. Compute the pressure change from the inside of the inner cylinder (pt. A) to the outside of the outer cylinder (pt. D).

d) Show that it is possible to move fluid from A to D with no net pressure change. Should this be patented as a design for a catalytic reactor that requires no net energy input? Why or (obviously) why not? Be specific; vague references to perpetual motion machines are not sufficient.

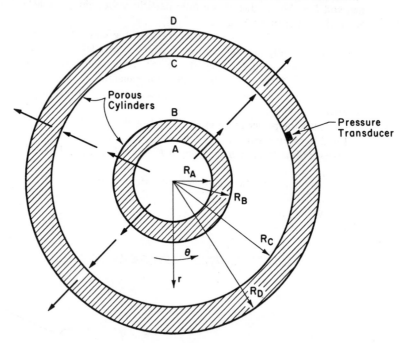

8.11. The viscosity of a certain Newtonian liquid depends on temperature according to the relation

$$\eta(T) = \frac{\eta_0}{1 + \beta(T - T_0)}$$

where T_0 is a reference temperature. The fluid flows under the influence of a pressure gradient between two flat plates, as shown in Fig. 8P11. The walls are maintained at temperatures T_0 and T_1, where T_0 is the reference temperature and $T_1 > T_0$. The temperature can be taken, to a first approximation,

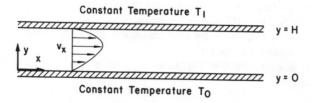

to be a linear function of position:

$$T \approx T_0 + (T_1 - T_0)y/H$$

Determine the flow rate-pressure drop relation.

8.12. Two immiscible incompressible Newtonian fluids flow cocurrently in a plane channel, as shown in Fig. 8P12.
a) Determine the velocity distribution.
b) Compute the flow rate of each phase, and compare the flow rate ratio to the *in situ* volumetric ratio (i.e., the volume ratio of fluid in the channel).

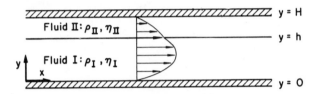

8.13. It has been suggested that the no-slip boundary condition should be replaced by the following:

The tangential velocity at a solid surface is proportional to the shear stress. Solve the problem of laminar flow in a pipe (Sec. 8.4) using this boundary condition. Show that the flow rate for a fixed pressure drop will depend on R^4 in large pipes and on R^3 in sufficiently small pipes, and that $f = 16/\mathrm{Re}$ only in the limit $\rho\langle v \rangle \longrightarrow 0$. Thus, the R^4 dependence of flow rate or the f-Re relation over a wide range is experimental support for the no-slip condition.

Accelerating Flow **9**

9.1 INTRODUCTION

In this chapter we consider the solution of a transient startup problem for an incompressible Newtonian fluid as a first introduction to the role of the inertial terms in the Navier–Stokes equations. A transient problem necessarily requires the solution of a partial differential equation, and the startup problem described in this chapter is one of the few transient cases for which an easily interpretable analytical solution exists.

The solution of even elementary partial differential equations is probably unfamiliar to many readers of this text. We shall therefore first state the solution without proof and examine the important physical consequences. We will then derive the solution by two different methods, one of them an approximate procedure that is based on a thorough physical understanding of the process.

9.2 PROBLEM DESCRIPTION

We consider the case in which a large ("infinite") plate is suddenly set in motion in an infinite expanse of fluid. The situation is shown schematically in Fig. 9-1. For all time $t \leq 0$ the plate and fluid are at rest. At $t = 0^+$ the velocity of the plate in the x direction changes suddenly to a value U. The fluid immediately adjacent to the plate must also take on the finite velocity of

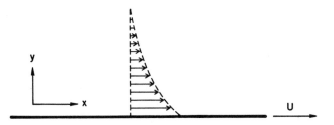

Figure 9-1. Schematic of a plate in an infinite extent of fluid. The plate is set in motion with constant velocity U at $t = 0$.

the plate for $t > 0$, because of the no-slip boundary condition. The inertia of the large mass of fluid above the plate prevents the remainder of the fluid from adjusting instantaneously to this new motion of the plate, however, so there will be a velocity variation in the direction normal to the plate. This velocity variation will change with time, for we know that increasing amounts of fluid will move with the plate as time goes on.

The fluid motion is assumed to be parallel to the plate; thus, motion takes place only in the x direction, and varies only in the y direction. The velocity then has the form

$$v_x = v_x(y, t) \qquad v_y = v_z = 0 \tag{9.1}$$

This form satisfies the continuity equation, but continuity provides no additional information.

This is a drag flow with no imposed pressure gradient and no change in cross section, so we may presume that $\partial \mathcal{P}/\partial x = 0$. In that case the Navier–Stokes equations in Table 7-4 simplify to

$$x \text{ component:} \qquad \rho \frac{\partial v_x}{\partial t} = \eta \frac{\partial^2 v_x}{\partial y^2} \tag{9.2a}$$

$$y \text{ component:} \qquad 0 = -\frac{\partial \mathcal{P}}{\partial y} \tag{9.2b}$$

$$z \text{ component:} \qquad 0 = -\frac{\partial \mathcal{P}}{\partial z} \tag{9.2c}$$

The y and z components simply state that $\mathcal{P}$ is a constant. Note that Eq. (9.2a) contains a nonzero inertial term, $\rho \, \partial v_x/\partial t$, but that the equation is linear in the unknown, v_x.

Equation (9.2a) must be integrated once with respect to time, so an initial condition is required to evaluate the resulting integration constant. The initial condition is no flow at $t = 0$:

$$v_x = 0 \text{ at } t = 0 \qquad y \geq 0 \tag{9.3a}$$

The equation must be integrated twice with respect to y, requiring two spatial boundary conditions. The first is no-slip at the surface of the plate:

$$v_x = U \text{ at } y = 0 \qquad t > 0 \qquad (9.3b)$$

The second condition is that v_x remain finite throughout the infinite spatial expanse. This is, in fact, equivalent to a stronger statement; an infinite time is needed to overcome the inertia of an infinite amount of fluid. Thus, the fluid at $y \rightarrow \infty$ will not move until $t \rightarrow \infty$:

$$v_x = 0 \text{ at } y = \infty \qquad t < \infty \qquad (9.3c)$$

Equation (9.2a) is commonly known as the *diffusion equation,* because it describes the diffusion of mass and heat as well as the situation described here. The solution is well known, and can be obtained in a number of ways. Students who have studied the Laplace transform would probably choose to transform the time dependence, solve the resulting ordinary differential equation in y, and then find the inverse transform from a table of transforms. We shall obtain the solution in different ways in later sections.

9.3 BOUNDARY LAYER

The solution of Eq. (9.2a) for the velocity, with boundary conditions (9.3), is

$$v_x(y, t) = U\left[1 - \text{erf}\left(\frac{y}{\sqrt{4\eta t/\rho}}\right)\right] \qquad (9.4)$$

The *error function* is a tabulated function, like the sine, cosine, or exponential, and can be found in any good set of tables of mathematical functions. It can also be found in statistical tables, because it is the integral of the statistical "normal distribution." The error function is defined as

$$\text{erf}(\zeta) = \sqrt{\frac{2}{\pi}} \int_0^\zeta e^{-\xi^2} \, d\xi \qquad (9.5)$$

Note that v_x can be expressed in terms of a *combination* of the independent variables y and t.

The velocity distribution defined by Eq. (9.4) is plotted in Fig. 9-2 as v_x/U. The important observation to be made here is that the velocity drops nearly to zero for $y \sim 2\sqrt{4\eta t/\rho}$. Thus, we find that at any time the influence of the wall is transmitted by viscous shearing only a finite distance into the fluid; this distance is proportional to the square root of the product of kinematic viscosity (η/ρ) and time. The wall region, where viscous forces are important, is known as the *boundary layer.*

The boundary layer is one of the most important concepts in fluid mechanics. If we may generalize from this single example, it is evident that there are two qualitatively different regions in the flow field when both viscous and inertial forces are important. There is the boundary layer region near the solid surface, proportional in thickness to $\sqrt{\eta t/\rho}$, where the presence of the solid surface

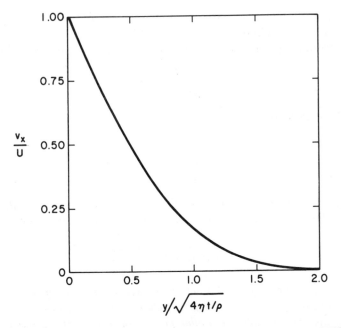

Figure 9-2. Velocity as a function of the combined variable $y/\sqrt{4\eta t/\rho}$ for a plate suddenly set in motion.

is a factor because of the transmission of the viscous stress. There is the remainder of the fluid, beyond the boundary layer, where the presence of the wall is not felt and the fact of viscosity is irrelevant. The precise demarcation between these two regions is ambiguous, but it clearly exists. The concept of the boundary layer is very important in obtaining approximate solutions to the Navier–Stokes equations.

9.4 DIMENSIONAL ANALYSIS (SIMILARITY) SOLUTION

The solution to Eqs. (9.2a) and (9.3) is most easily obtained by dimensional analysis. The procedure that we will use works only for problems in which there is no characteristic length, however.

Examination of the equations shows that there are six variables: v_x, U, t, y, ρ, and η. There are three dimensions, so there can be at most three independent dimensionless groups. Three such groups are readily found by inspection to be v_x/U, $y^2\rho/\eta t$, and $yU\rho/\eta$. Thus, we can write

$$\frac{v_x}{U} = \text{function of } \left(\frac{y^2\rho}{\eta t}, \frac{yU\rho}{\eta}\right) \tag{9.6}$$

It is easy to show, however, that v_x/U cannot depend on U, so there cannot be any dependence on $yU\rho/\eta$. This is done as follows:

We introduce a dimensionless velocity $u = v_x/U$. Equation (9.2a) then becomes

$$\rho \frac{\partial u}{\partial t} = \eta \frac{\partial^2 u}{\partial y^2} \tag{9.7}$$

and the boundary conditions (9.3) become

$$u = 0 \text{ at } t = 0 \qquad y \geq 0 \tag{9.8a}$$

$$u = 1 \text{ at } y = 0 \qquad t > 0 \tag{9.8b}$$

$$u = 0 \text{ at } y = \infty \qquad t < \infty \tag{9.8c}$$

Equations (9.7) and (9.8) contain the five variables $u = v_x/U$, t, y, ρ, and η, but not U. Thus, v_x/U cannot be a function of U. Equation (9.6) can therefore be written

$$\frac{v_x}{U} = \text{function of } \left(\frac{y^2 \rho}{\eta t} \right) \tag{9.9}$$

Note that $\sqrt{\eta t/\rho}$ represents a characteristic distance, and $y/\sqrt{\eta t/\rho}$ is a relative distance; a velocity profile in which the relative velocity is a function only of a relative distance is called a *similar profile*, or a *similarity solution* to the Navier–Stokes equations.

The essence of the solution is contained in Eq. (9.9), and the remainder is mechanical. We define the new variable

$$\zeta = \frac{y}{\sqrt{4\eta t/\rho}} \tag{9.10}$$

The square root and the factor of 4 are both for convenience and are introduced to make subsequent equations somewhat more compact. We now return to Eq. (9.2a) and write the derivatives as follows:

$$\frac{\partial v_x}{\partial t} = \frac{dv_x}{d\zeta} \frac{\partial \zeta}{\partial t} = -\frac{y}{\sqrt{16\eta t^3/\rho}} \frac{dv_x}{d\zeta} \tag{9.11a}$$

$$\frac{\partial v_x}{\partial y} = \frac{dv_x}{d\zeta} \frac{\partial \zeta}{\partial y} = \frac{1}{\sqrt{4\eta t/\rho}} \frac{dv_x}{d\zeta} \tag{9.11b}$$

$$\frac{\partial^2 v_x}{\partial y^2} = \frac{d}{d\zeta} \left(\frac{\partial v_x}{\partial y} \right) \frac{\partial \zeta}{\partial y} = \frac{1}{4\eta t/\rho} \frac{d^2 v_x}{d\zeta^2} \tag{9.11c}$$

Substituting into Eq. (9.2a), we then obtain the *ordinary differential equation*

$$-2\zeta \frac{dv_x}{d\zeta} = \frac{d^2 v_x}{d\zeta^2} \tag{9.12}$$

The initial and boundary conditions must also be written in terms of the new variable. At $y = 0$, $\zeta = 0$. Thus, Eq. (9.3b) can be written

$$v_x = U \text{ at } \zeta = 0 \tag{9.13a}$$

As $y \rightarrow \infty$ we have $\zeta \rightarrow \infty$, so Eq. (9.3c) can be written

$$v_x = 0 \text{ at } \zeta = \infty \tag{9.13b}$$

Two conditions are all that we are allowed for a second-order differential equation like Eq. (9.12), but we have not yet accounted for the initial condition, Eq. (9.3a). This states that $v_x = 0$ at $t = 0$. $t = 0$ corresponds to $\zeta = \infty$, however, so Eq. (9.3a) can also be written in the form of Eq. (9.13b). Thus, we have reduced the partial differential equation to an ordinary differential equation. (We have been able to carry out this reduction because of the absence of a characteristic length. For problems with a characteristic length it is often possible to introduce a change of variables that reduces a partial differential equation to an ordinary differential equation, but the boundary conditions can rarely be transformed satisfactorily.)

Equation (9.12) is a linear ordinary differential equation that is easily integrated. We note that $d^2v_x/d\zeta^2 = d(dv_x/d\zeta)/d\zeta$, so that Eq. (9.12) can be written equivalently as

$$\frac{d}{d\zeta}\left(\ln\frac{dv_x}{d\zeta}\right) = -2\zeta \tag{9.14}$$

This has a solution

$$\frac{dv_x}{d\zeta} = C_1 e^{-\zeta^2} \tag{9.15}$$

where C_1 is a constant of integration. A second integration then gives

$$v_x = C_1 \int_0^\zeta e^{-\zeta^2} \, d\zeta + C_2 \tag{9.16}$$

The constants C_1 and C_2 are evaluated using Eqs. (9.13) and noting that the definite integral $\int_0^\infty \exp(-\zeta^2) \, d\zeta$ equals $\sqrt{\pi}/2$, resulting in the solution given in Eq. (9.4).

9.5 INTEGRAL MOMENTUM APPROXIMATION

Solution of partial differential equations, even of the most elementary form, requires the development of certain analytical skills which most readers of this material have probably not yet mastered. The approach outlined above, using dimensional analysis, provides a nice rationalization for the introduction of the new independent variable and reduction of the mathematical problem to solution of an ordinary differential equation. Despite the logical simplicity, however, it is unlikely that someone unaccustomed to this way of solving problems could have devised the method.

With a firm understanding of the physical process it is often possible to devise approximate procedures for solving a problem. It is convenient to introduce one such method here, which is sometimes known as the *integral momentum* (or *von Kármán–Polhausen*) *approximation*. There are several approaches which are equivalent and lead to the same ultimate equation. The first and most physically grounded is to apply the principle of conservation of momentum to a macroscopic control volume of length (x direction) L and width (z direction) W, and extending far into the fluid ($y \to \infty$). The total x momentum in this control volume is

$$x \text{ momentum} = LW \int_0^\infty \rho v_x \, dy \tag{9.17}$$

The only x-direction force is that imposed by the moving flat plate:

$$\text{force} = -LW\tau_{yx}\big|_{y=0} = -LW\eta \frac{\partial v_x}{\partial y}\bigg|_{y=0} \tag{9.18}$$

The negative sign is needed because the sign convention requires that τ_{yx} be the stress exerted by the fluid on the plate. Equating the rate of change of momentum to the imposed force then gives

$$LW \frac{d}{dt} \int_0^\infty \rho v_x \, dy = -LW\eta \frac{\partial v_x}{\partial y}\bigg|_{y=0} \tag{9.19}$$

or, dividing by LW and noting that ρ is a constant,

$$\rho \frac{d}{dt} \int_0^\infty v_x \, dy = -\eta \frac{\partial v_x}{\partial y}\bigg|_{y=0} \tag{9.20}$$

Equation (9.20) is our starting point. It is useful to note that it can be derived directly from (9.2a), the diffusion equation. We simply integrate the entire equation from $y = 0$ to infinity to obtain*

$$\int_0^\infty \rho \frac{\partial v_x}{\partial t} \, dy = \rho \frac{d}{dt} \int_0^\infty v_x \, dy = \int_0^\infty \eta \frac{\partial^2 v_x}{\partial y^2} \, dy = -\eta \frac{\partial v_x}{\partial y}\bigg|_{y=0}$$

At this point we start to make use of our physical understanding of the problem. At *each time* there is a distance, $\delta(t)$, beyond which the effect of the wall is not felt and the velocity is essentially zero. Thus, we can *approximately* replace the infinite integral in Eq. (9.20) with integration from $y = 0$ to $y = \delta(t)$,

$$\rho \frac{d}{dt} \int_0^{\delta(t)} v_x \, dy = -\eta \frac{\partial v_x}{\partial y}\bigg|_{y=0} \tag{9.21}$$

Furthermore, we know that δ is the characteristic length for this problem,

---

*Note that there are some minor technical considerations here in the use of the calculus. Interchange of time differentiation and spatial integration in the first integral is possible only because of the fixed limits and the fact that there is no flow through these surfaces. The integration by parts assumes continuity of $\partial v_x/\partial y$ for all $0 < y < \infty$ and the vanishing of $\partial v_x/\partial y$ for $y \to \infty$.

despite the fact that it is not constant but increases with time. We thus *hypothesize* that the velocity is a function only of y/δ; that is, the shape of the velocity function relative to the characteristic penetration distance is always the same, although the value at any point changes with time as δ increases. We can therefore write

$$v_x = U\phi\left(\frac{y}{\delta}\right) \tag{9.22}$$

where ϕ is an as-yet-unspecified function. In order to satisfy the boundary conditions, it is necessary that ϕ satisfy two constraints:

$$\phi(0) = 1 \tag{9.23a}$$

$$\phi(1) = 0 \tag{9.23b}$$

The former condition is that of no-slip at the plate, while the latter is the requirement that $v_x \approx 0$ for $y \geq \delta$. We should also require $\phi'(1) = 0$ to satisfy the requirement of continuity of $\partial v_x/\partial y$ and hence the shear stress for all $0 < y < \infty$.

We can now perform some simple manipulations by combining Eqs. (9.21) and (9.22). First,

$$\eta \frac{\partial v_x}{\partial y}\bigg|_{y=0} = \eta U\phi'\left(\frac{y}{\delta}\right)\frac{1}{\delta}\bigg|_{y=0} = \eta \frac{1}{\delta} U\phi'(0) \tag{9.24}$$

Here, the prime denotes differentiation with respect to the argument, y/δ. Also,

$$\rho \int_0^\delta v_x \, dy = \rho U \int_0^\delta \phi\left(\frac{y}{\delta}\right) dy = \rho U\delta \int_0^\delta \phi\left(\frac{y}{\delta}\right)\frac{1}{\delta}\, dy = \rho U\delta \int_0^1 \phi(\xi)\, d\xi \tag{9.25}$$

The definite integral $\int_0^1 \phi(\xi)\, d\xi$ is simply a constant, so Eq. (9.21) becomes

$$\delta \frac{d\delta}{dt} = \frac{1}{2}\frac{d\delta^2}{dt} = \left[-\frac{\phi'(0)}{\int_0^1 \phi(\xi)\, d\xi}\right]\left(\frac{\eta}{\rho}\right) \tag{9.26}$$

This is a very elementary differential equation for $\delta(t)$. With the initial condition $\delta(0) = 0$ (no boundary layer at $t = 0$), it has the solution

$$\delta(t) = \left[-\frac{2\phi'(0)}{\int_0^1 \phi(\xi)\, d\xi}\right]^{1/2}\left(\frac{\eta t}{\rho}\right)^{1/2} \tag{9.27}$$

The coefficient of $(\eta t/\rho)^{1/2}$ is a constant. Thus, without making any explicit statement about the form of the function $\phi(y/\delta)$, we have established that the boundary layer thickness grows as $(\eta t/\rho)^{1/2}$. The actual value of the coefficient is relatively insensitive to the particular function ϕ. The quadratic satisfying $\phi(0) = 1$, $\phi(1) = \phi'(1) = 0$ is

$$\phi\left(\frac{y}{\delta}\right) = 1 - 2\frac{y}{\delta} + \left(\frac{y}{\delta}\right)^2 \tag{9.28}$$

Carrying out the indicated differentiation and integration gives

$$\delta(t) = 3.46\left(\frac{\eta t}{\rho}\right)^{1/2} \tag{9.29}$$

which agrees quite well with the exact boundary layer thickness shown in Fig. 9-2. The same result is obtained using a cubic function $\phi(y/\delta)$ with the additional smoothness requirement $\phi''(1) = 0$. Even the crudest possible approximation, $\phi(y/\delta) = 1 - (y/\delta)$, which ignores the requirement of continuity of $\partial v_x/\partial y$, gives a thickness $\delta(t) = 2.0 \, (\eta t/\rho)^{1/2}$. Once the function ϕ has been selected, the velocity follows directly from Eq. (9.22) and the stress at the wall from Eq. (9.24).

9.6 WEIGHTED RESIDUALS*

There is another way of deriving the differential equation (9.6) which is a bit more direct and general, although it does not allow the convenient physical interpretation of the integral momentum approach. We rewrite Eq. (9.2a) as

$$\rho\frac{\partial v_x}{\partial t} - \eta\frac{\partial^2 v_x}{\partial y^2} = 0 \tag{9.30}$$

If we substitute the approximation Eq. (9.22) into this equation, the two terms will not sum to zero, because the approximation is *not* a solution to the equation. There will be a residual which depends on y and t. Thus, we have

$$\text{residual} = \rho\frac{\partial}{\partial t}U\phi\left(\frac{y}{\delta}\right) - \eta\frac{\partial^2}{\partial y^2}U\phi\left(\frac{y}{\delta}\right) = -\frac{U}{\delta^2}\left(\rho y\phi'\frac{d\delta}{dt} + \eta\phi''\right) \neq 0 \tag{9.31}$$

Although the approximation does not satisfy the differential equation at each point, we can seek to satisfy the equation in some sense on the average. The most straightforward way of doing this, but not the only way, is to require that the average value of the residual over the flow field be zero:

$$\int_0^\delta (\text{residual}) \, dy = 0 \tag{9.32}$$

Integration is only from zero to δ because the velocity is assumed to vanish for $y \geq \delta$. We therefore obtain

$$\rho\frac{d\delta}{dt}\int_0^\delta y\phi' \, dy + \eta\int_0^\delta \phi'' \, dy = 0 \tag{9.33}$$

*This section can be omitted on a first reading.

With integration by parts of each integral, this simplifies to Eq. (9.26) for $\delta(t)$.

This last approach, which directly examines the error in satisfying the differential equation by use of an approximation, is a special case of a class of techniques known as *methods of weighted residuals*. They are of considerable use in problems in heat and mass transfer and reaction engineering as well as in fluid mechanics.

9.7 CONCLUDING REMARKS

The interaction of viscous and inertial stresses results in a boundary layer near a solid surface. It is only in this region that the viscous contribution is important. The boundary layer thickness is of the order of magnitude of $(\eta t/\rho)^{1/2}$, where t is the time that the bulk of the fluid has been in motion relative to the wall.

Exact solutions to the Navier–Stokes equations are rare, and it is usually necessary to make approximations to obtain solutions to flow problems. The notion of the boundary layer is the foundation of one of the most fruitful procedures for obtaining solutions to flows with large inertial terms. We shall devote a subsequent chapter to this topic. Since heat and mass transfer are often of most interest near a solid surface, it is evident that the boundary layer region will be of considerable importance in these transport processes.

BIBLIOGRAPHICAL NOTES

For an introduction to methods of weighted residuals, see

FINLAYSON, B. A., *The Method of Weighted Residuals and Variational Principles*, Academic Press, Inc., New York, 1972.

PROBLEMS

9.1. Consider an infinite expanse of an incompressible Newtonian fluid in which an infinite flat plate oscillates along the x-axis with displacement $L \sin \omega t$. Compute the velocity distribution in the fluid after initial transients have died out. Is there a boundary layer? (Hint: Look for a solution of the form $v_x = A(y) \sin \omega t + B(y) \cos \omega t$. Why?)

9.2. An incompressible Newtonian fluid is contained between two stationary infinite flat plates separated by a spacing H. At $t = 0$ the bottom plate is instantaneously accelerated to a velocity U relative to the upper plate.

   a) Show that the solution cannot be expressed solely in terms of the single variable ζ defined in Eq. (9.10).

b) If you have sufficient experience in solving linear partial differential equations, obtain the velocity as a function of position and time. Show that the velocity profile becomes linear as $t \longrightarrow \infty$. (Hint: the asymptotic result for $t \longrightarrow \infty$ can be obtained directly from a Laplace transform solution, without inversion, by application of the final value theorem.)

9.3. Repeat the integral momentum solution in Sec. 9.5 for a power law fluid (Eq. 8.86) and show that the boundary layer thickness grows as $t^{1/(n+1)}$.

Converging Flow **10**

10.1 INTRODUCTION

Exact solutions of the Navier–Stokes equations have been obtained in only a small number of cases, and most of these are flows like the ones in Chapters 8 and 9 for which the nonlinear inertial terms are identically zero because of the geometry. Fewer than 10 exact solutions have been obtained for flows of physical interest in which the nonlinear terms are nonzero. Thus, the primary skill that needs to be learned is not that of obtaining exact solutions; it is rather the reduction of the full Navier–Stokes equations through the use of physical experience to a problem that *can* be solved. Such approximation is the subject of the next portion of the text. This chapter is devoted to the analysis of one of the few existing exact solutions, because this solution provides insight into the kind of behavior that is to be expected and sought after in the development of physical approximations.

The flow geometry of practical interest that we would *like* to solve is shown in Fig. 10-1. There is a converging or diverging section between two regions of fully developed tube or channel flow. Such flows are commonly encountered in applications, including polymer processing operations. This problem is not amenable to analytical solution, however, and must be further simplified. We shall treat, instead, the idealized configuration shown in Fig. 10-2. Here, we consider the converging or diverging section by itself, and we ignore the entry and exit regions where the transition to Poiseuille flow must take place. The converging section is assumed to be continuous from the vertex

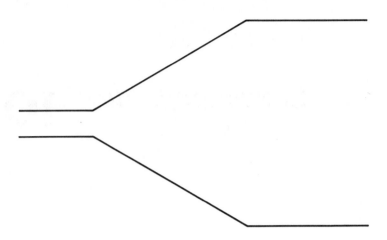

Figure 10-1. Schematic of a finite converging or diverging flow.

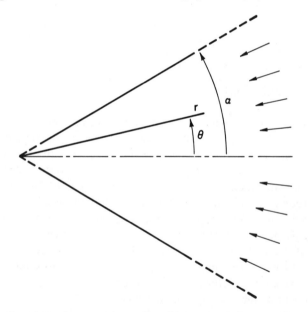

Figure 10-2. Schematic of an infinite converging or diverging flow.

to infinity. The presumption is that this idealization will represent the converging section in Fig. 10-1 except near the entrance and exit.

The Navier–Stokes equations can be solved exactly both for flow in a cone and for flow between converging flat plates. The planar flow involves slightly fewer manipulations and is adequate for our purposes, so we shall study only it. This flow problem was first solved by Hamel in 1916 and is known as *Hamel flow*.

10.2 SOLUTION

The flow is best described in a cylindrical (r, θ, z) coordinate system, as shown in Fig. 10-2. r is measured from the vertex of the plates. The total angle between the plates is 2α, with the walls at $\theta = \pm\alpha$. The fluid is assumed to be Newtonian and incompressible.

We will assume that there are no variations in the z direction, so that $v_z = 0$. We will also assume that the flow is entirely radial,* in that fluid particles move to or from the vertex along lines of constant θ, so that $v_\theta = 0$. The continuity equation in cylindrical coordinates from Table 7-1 is then

$$\frac{1}{r}\frac{\partial}{\partial r}(rv_r) = 0 \tag{10.1}$$

That is, rv_r is independent of r. We have also assumed that v_r is independent of z, so Eq. (10.1) integrates to

$$rv_r = \text{function only of } \theta = f(\theta) \tag{10.2a}$$

or

$$v_r = \frac{f(\theta)}{r} \tag{10.2b}$$

In this case, the continuity equation has given us important additional information about the flow. The fact that v_r becomes infinite at the vertex, $r = 0$, is not a cause for concern here, since we recognize that the flow which we are computing is not valid near the vertex of the plates.

The function $f(\theta)$ will be determined in the course of solution of the Navier–Stokes equations. Some restrictions on the function follow from boundary and flow conditions. For example, the velocity must vanish at the side walls, $\theta = \pm\alpha$, because of the no-slip condition, so the function $f(\theta)$ must vanish for $\theta = \pm\alpha$:

$$f(+\alpha) = f(-\alpha) = 0 \tag{10.3}$$

---

*As we shall see, this assumption leads to a self-consistent solution for Newtonian fluids, and it is also valid for power-law fluids. The assumption leads to a contradiction for some non-Newtonian fluid models, however, and there are experimental data for the converging flow of non-Newtonian polymeric liquids that do show a nonzero v_θ component.

The flow rate over a width W in the z direction is

$$Q = \int_{-\alpha}^{\alpha} v_r W r \, d\theta \tag{10.4}$$

It is more convenient to use the *flow rate per unit width*, $q = Q/W$; q is then

$$q = \int_{-\alpha}^{\alpha} v_r r \, d\theta = \int_{-\alpha}^{\alpha} f(\theta) \, d\theta \tag{10.5}$$

This relation partially specifies $f(\theta)$ for given q. Note that f is positive for diverging flow from the vertex and negative for converging flow toward the vertex. Thus, q defined in this way has an algebraic sign, with positive q representing outflow and negative q representing inflow.

The r and θ components of the Navier–Stokes equations, with $v_\theta = v_z = 0$ and $\partial/\partial t = \partial/\partial z = 0$, become

$$\rho v_r \frac{\partial v_r}{\partial r} = -\frac{\partial \mathcal{P}}{\partial r} + \eta\left[\frac{\partial}{\partial r}\frac{1}{r}\frac{\partial}{\partial r}(rv_r) + \frac{1}{r^2}\frac{\partial^2 v_r}{\partial \theta^2}\right] \tag{10.6a}$$

$$0 = -\frac{1}{r}\frac{\partial \mathcal{P}}{\partial \theta} + \eta\frac{2}{r^2}\frac{\partial v_r}{\partial \theta} \tag{10.6b}$$

Substituting Eq. (10.2) for v_r gives

$$-\rho\frac{f^2}{r^3} = -\frac{\partial \mathcal{P}}{\partial r} + \frac{\eta}{r^3}\frac{d^2 f}{d\theta^2} \tag{10.7a}$$

$$0 = -\frac{\partial \mathcal{P}}{\partial \theta} + \frac{2\eta}{r^2}\frac{df}{d\theta} \tag{10.7b}$$

Equation (10.7b) has been multiplied by r to simplify subsequent manipulations. The easiest way to proceed here is to eliminate the pressure. This is done by differentiating Eq. (10.7a) with respect to θ and (10.7b) with respect to r, giving

$$\frac{-2\rho}{r^3} f\frac{df}{d\theta} = -\frac{\partial^2 \mathcal{P}}{\partial\theta\,\partial r} + \frac{\eta}{r^3}\frac{d^3 f}{d\theta^3} \tag{10.8a}$$

$$0 = -\frac{\partial^2 \mathcal{P}}{\partial\theta\,\partial r} - \frac{4\eta}{r^3}\frac{df}{d\theta} \tag{10.8b}$$

Subtracting, we obtain

$$\frac{-2\rho}{r^3} f\frac{df}{d\theta} = \frac{\eta}{r^3}\frac{d^3 f}{d\theta^3} + \frac{4\eta}{r^3}\frac{df}{d\theta} \tag{10.9}$$

or, equivalently, multiplying each term by r^3/η,

$$\left(\frac{2\rho}{\eta}\right) f\frac{df}{d\theta} + \frac{d^3 f}{d\theta^3} + 4\frac{df}{d\theta} = 0 \tag{10.10}$$

Equation (10.10) does not involve r in any way, so our initial assumptions regarding the flow are consistent. It is a third-order, ordinary differential equation, requiring three conditions on $f(\theta)$. These are given by Eqs. (10.3) and (10.5).

It is useful for purposes of presentation to scale the dependent and independent variables. We define a normalized angle, ϕ, and a normalized flow variable, F, as follows:

$$\phi = \frac{\theta}{\alpha} \tag{10.11a}$$

$$F = \frac{\alpha f}{q} \tag{10.11b}$$

ϕ ranges between -1 and $+1$. Equation (10.10) and the boundary and flow conditions (10.3) and (10.5) then become

$$\Re F \frac{dF}{d\phi} + \frac{d^3 F}{d\phi^3} + 4\alpha^2 \frac{dF}{d\phi} = 0 \tag{10.12}$$

$$F(-1) = F(+1) = 0 \tag{10.13}$$

$$\int_{-1}^{1} F(\phi)\, d\phi = 1 \tag{10.14}$$

$$\Re = \frac{2pq\alpha}{\eta} \tag{10.15}$$

The parameter $\Re$ represents the ratio of inertial (order of pq^2) to viscous (order of ηq) stresses, so it plays a role analogous to the Reynolds number. There is no true Reynolds number as we have defined it previously in this problem because there is no characteristic length. $\Re$ can be positive or negative, depending on whether there is outflow or inflow, respectively; the range of $\Re$ is $-\infty < \Re < \infty$. Note that $\Re$ goes to zero both as $pq/\eta \to 0$ and as $\alpha \to 0$.

The fact that we have succeeded in reducing the set of partial differential equations to a single ordinary differential equation means that we can now obtain the solution with little further effort. Equations (10.12) through (10.14) can be solved analytically, although the solution is in terms of integrals that must be evaluated numerically. We shall not go through the development, which is similar to the one below for the limiting case $|\Re| \to \infty$. We rather show the result here of a numerical solution of Eq. (10.12) to obtain $(F(\phi)$, adjusting the unknown initial conditions $dF/d\phi$ and $d^2F/d\phi^2$ at $\phi = -1$ by trail and error until the conditions $F(+1) = 0$ and $\int_{-1}^{1} F(\phi)\, d\phi = 1$ are satisfied.

The function $F(\phi)$ is shown in Fig. 10-3 for $\alpha = \pi/4$ (45°). The curves for $\Re = +1$ and $\Re = -1$ cannot be distinguished from the curve for $\Re = 0$ on the scale of the figure. It is important to note the very different behavior for inflow and outflow. When $\Re$ is large and negative (convergent flow) the radial velocity is nearly constant over most of the included angle, approaching a value of $F(\phi) \approx \frac{1}{2}$, and all the velocity variation is in a small boundary layer near the wall. When $\Re$ is positive (divergent flow), on the other hand, $F(\phi)$

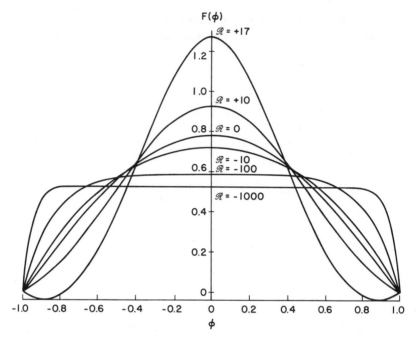

Figure 10-3. Dimensionless velocity function $F(\phi)$, $\alpha = \pi/4$ (45°). Curves for $\mathfrak{R} = +1$ and -1 do not differ significantly from the curve for $\mathfrak{R} = 0$.

becomes negative near the wall for $\mathfrak{R} > 14$, indicating a backflow toward the vertex in this region. This is known as *flow separation*. The backflow is unstable, and a diverging laminar flow of the type assumed here cannot be maintained in practice; turbulence occurs when $\mathfrak{R}$ exceeds the value leading to separation.

10.3 ORDERS OF MAGNITUDE

The scaling of dependent and independent variables resulting in Eq. (10.12) was carried out with a particular end in mind. Both F and ϕ are quantities that are expected to be of the order of magnitude of unity. Furthermore, changes in F will be of order-of-magnitude unity, and they will occur over changes in ϕ of order unity; thus, derivatives of F are also expected to be of the order of magnitude of unity.* Each term in Eq. (10.12) is thus a quantity that is expected to be of order unity, multiplied by a parameter, α or $\mathfrak{R}$. The

---

*Note that this will not be true for large negative values of $\mathfrak{R}$, where changes in F all take place over a very small range of ϕ. This is an important point that will be taken up again in Sec. 10.6.

values of α and $\mathfrak{R}$ thus determine the relative importance of each of the terms in Eq. (10.12).

For example, consider the case in which $\mathfrak{R}$ is small. The term $\mathfrak{R} F \, dF/d\phi$ in Eq. (10.12) is then expected to be small relative to the other two terms. If we were to solve the equation by *neglecting* that term, we would expect the solution to be in error only by an amount of order $\mathfrak{R}$; for sufficiently small $\mathfrak{R}$, this error would be acceptable.

In the three sections that follow we will solve Eq. (10.12) for three limiting cases, $\mathfrak{R} \rightarrow 0$, $\alpha \rightarrow 0$, and $|\mathfrak{R}| \rightarrow \infty$. These three cases correspond to three common procedures for obtaining approximate solutions to the Navier–Stokes equations, known as the *creeping flow*, *lubrication*, and *boundary layer* approximations, respectively, which are discussed in detail in the next portion of the text. The approximations can be compared here to the exact solution in order to develop some feeling for the extent of applicability.

10.4 CREEPING FLOW, $\mathfrak{R} \rightarrow 0$

We first consider the case in which $\mathfrak{R}$ is small in magnitude. Since $\mathfrak{R}$ represents the ratio of inertial to viscous terms, this limit corresponds to a situation in which the inertial stresses (the $\rho v \cdot \nabla v$ terms) are small relative to the viscous stresses (the $\eta \nabla^2 v$ terms), as might occur in the flow of a highly viscous liquid such as a molten polymer. We assume that a solution with an error that is proportional to $\mathfrak{R}$ can be obtained by simply setting $\mathfrak{R}$ to zero in Eq. (10.12). We thus obtain the equation

$$\mathfrak{R} = 0: \quad \frac{d^3 F}{d\phi^3} + 4\alpha^2 \frac{dF}{d\phi} = 0 \tag{10.16}$$

Equation (10.16) is a linear equation with constant coefficients. The general solution is

$$F(\phi) = A + B \sin 2\alpha\phi + C \cos 2\alpha\phi \tag{10.17}$$

The conditions $F(+1) = F(-1) = 0$ give $B = 0$, $A = -C \cos 2\alpha$. Equation (10.17) is thus

$$\int_{-1}^{1} F(\phi) \, d\phi = 1 = \int_{-1}^{1} C(\cos 2\alpha\phi - \cos 2\alpha) \, d\phi = \frac{C(\sin 2\alpha - 2\alpha \cos 2\alpha)}{\alpha} \tag{10.18}$$

Equation (10.17) can thus be written

$$\mathfrak{R} = 0: \quad F(\phi) = \frac{\alpha(\cos 2\alpha\phi - \cos 2\alpha)}{\sin 2\alpha - 2\alpha \cos 2\alpha} \tag{10.19}$$

This equation is indeterminate for $\alpha = 0$, and L'Hôpital's rule must be applied to obtain the limiting value

$$\Re = 0: \quad \lim_{\alpha \to 0} F(\phi) = \tfrac{3}{4}(1 - \phi^2) \tag{10.20}$$

Equation (10.19) is plotted for several values of α in Fig. 10-4. The deviation from the parabola for $\alpha = 0$, Eq. (10.20), can be observed on the scale of the figure only for $\alpha > 0.35\ (20°)$.

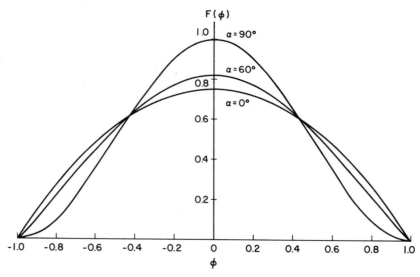

Figure 10-4. Dimensionless velocity function $F(\phi)$ for $\Re = 0$. The curves for $\alpha = 0$ and $\alpha = 20°$ differ by only a few percent, mostly near the wall ($\phi \longrightarrow \pm 1$), and cannot be distinguished on this scale.

We can obtain a rough estimate of the range of validity of the approximate solution by determining when the neglected term, $\Re F\, dF/d\phi$, will be comparable to the two terms that we have retained. The comparison can be made to either the term $d^3F/d\phi^3$ or the term $4\alpha^2\, dF/d\phi$, since they are equal in magnitude in the approximation $\Re = 0$. For $\alpha = \pi/4$ the maximum value of $F(\phi)$ in Eq. (10.19) occurs at the centerline and equals $\pi/4$, while the maximum value of $dF/d\phi$ occurs at the wall and equals $\pi^2/8$. Thus, within the accuracy of the approximation the maximum value of the product $F\, dF/d\phi$ is slightly less than unity, and the neglected term is indeed approximately equal to $\Re$. The maximum value of the term $4\alpha^2\, dF/d\phi$ is $\pi^4/32$, at $\phi = \pm 1$, which equals about 3. Thus, for $\alpha = \pi/4$, we must have $\Re \ll 3$ in order to use the approximate solution for $\Re = 0$ with a negligible error. As we saw in Sec. 10.2 in the discussion of Fig. 10-3, the solutions for $\Re = \pm 1$ are indistinguishable from the solution for $\Re = 0$ when $\alpha = \pi/4$.

Flows for which the inertial terms can be neglected are called *creeping flows*. The creeping flow approximation results in a linear set of differential equations.

10.5 LUBRICATION APPROXIMATION, $\alpha \to 0$

For small angles the terms involving α may be neglected, leading to a solution in which the error may be expected to be of order α. Since $\mathfrak{R}$ is proportional to α, setting α to zero in Eq. (10.12) leads to the approximate equation

$$\alpha = 0: \quad \frac{d^3 F}{d\phi^3} = 0 \tag{10.21}$$

This has a solution

$$F(\phi) = A + B\phi + C\phi^2 \tag{10.22}$$

From the no-slip boundary condition it follows that $A = -C$, $B = 0$, and it further follows from Eq. (10.14) that $C = -\frac{3}{4}$. Thus, the complete solution is

$$\alpha = 0: \quad F(\phi) = \tfrac{3}{4}(1 - \phi^2) \tag{10.23}$$

This is the result given in Eq. (10.20) for $\mathfrak{R} = 0$ in the limit $\alpha \to 0$, and we therefore see that the lubrication approximation is a special case of the creeping flow approximation.

The physical interpretation of the lubrication approximation* is obtained by using Eqs. (10.2) and (10.11) to write the velocity as

$$v_r = \frac{f(\theta)}{r} = \frac{3q}{4(r\alpha)}\left[1 - \left(\frac{r\theta}{r\alpha}\right)^2\right] \tag{10.24}$$

For a small angle the Cartesian coordinate distance y from the centerline is approximately equal to arc length,

$$y = r\theta \tag{10.25}$$

so that Eq. (10.24) can be written

$$v_r = \frac{3q}{4(H/2)}\left[1 - \frac{y^2}{(H/2)^2}\right] \tag{10.26}$$

Here, H is the width of the channel at any radial distance from the vertex and is a function of r:

$$H(r) = 2r \sin \alpha \approx 2r\alpha \tag{10.27}$$

Since q can be expressed in terms of the average velocity,

$$q = \langle v_r \rangle H \tag{10.28}$$

we can write Eq. (10.26) in the final form

$$v_r = \tfrac{3}{2}\langle v_r \rangle\left[1 - \left(\frac{2y}{H}\right)^2\right] \tag{10.29}$$

Equation (10.29) is identical to Eq. (8.15) for flow between parallel plates. The approximation $\alpha = 0$ is therefore equivalent to treating the flow *locally* (i.e., at each radial distance from the vertex) as though the flow were between

---

*The small-angle approximation is equivalent to an approximation commonly made in the analysis of lubrication problems, and hence the name.

parallel plates, but using the plate spacing H which is valid at that particular position.

The range of validity of the small-angle approximation cannot be estimated here, because when the solution is substituted into Eq. (10.12), the term $d^3F/d\phi^3$ vanishes and nothing is left for comparison with the terms $4\alpha^2\,dF/d\phi$ and $\mathfrak{R}F\,dF/d\phi$. We found in the preceding section, however, for the solution shown in Fig. 10-4 for $\mathfrak{R}=0$, that the parabola corresponding to $\alpha=0$ is a good approximation to the complete solution for $\mathfrak{R}=0$ for values of α up to about $20°$.

10.6 BOUNDARY LAYER APPROXIMATION, $|\mathfrak{R}| \rightarrow \infty$

If we divide Eq. (10.12) by $\mathfrak{R}$ and let $|\mathfrak{R}| \rightarrow \infty$, the equation simplifies to a single term:

$$|\mathfrak{R}| = \infty: \quad F\frac{dF}{d\phi} = 0 \tag{10.30}$$

This has a solution $F = $ constant, and it follows from the integral flow rate condition, Eq. (10.14), that the constant equals $\frac{1}{2}$:

$$|\mathfrak{R}| = \infty: \quad F(\phi) = \tfrac{1}{2} \tag{10.31}$$

This result does indeed correspond to the flow in Fig. 10.3 for large negative $\mathfrak{R}$ for the region away from the wall, but it cannot satisfy the no-slip condition $F(-1) = F(+1) = 0$ and it bears no resemblance at all to the flow for large positive $\mathfrak{R}$.

This situation illustrates a problem that can be encountered in the blind application of a limiting process. The mathematical limit $|\mathfrak{R}| \rightarrow \infty$ is *singular*, in that the highest derivative term is lost and the order of the differential equation drops from three to one. Thus, although there are three physical conditions to be imposed [Eqs. (10.13) and (10.14)], there is only one constant of integration.

The resolution lies in a physical understanding of the limiting process. Infinite $\mathfrak{R}$ corresponds to an inviscid fluid, since the viscous terms are taken as negligible relative to the inertial terms. Since an inviscid fluid does not transmit shear stresses, it need not satisfy the no-slip condition. We know from the problem studied in Chapter 9, however, that there is a region near the wall, penetrating into the fluid a distance proportional to $\sqrt{\eta/\rho}$, where the viscous terms *must* be considered. Thus, Eq. (10.31) may indeed describe the flow in the *inviscid core*, but the viscous terms cannot be neglected in the *boundary layer* near the wall.

The ordering arguments used to simplify the differential equation clearly break down in the boundary layer, since changes in F take place over an angle that is much smaller than α. Thus, the appropriate scaling variable for θ is not α, but the angle corresponding to the boundary layer thickness. This angle is estimated as follows:

From Eqs. (10.2) and (10.11), $v_r = f(\theta)/r = Fq/\alpha r$. For $F \approx \frac{1}{2}$, $v_r \approx q/2\alpha r$. Then, following a fluid particle,

$$v_r = \frac{dr}{dt} \approx \frac{q}{2\alpha r} \tag{10.32}$$

or

$$t \approx \frac{\alpha r^2}{q} + \text{constant} \tag{10.33}$$

If we consider outflow, we can take the constant to be zero. The boundary layer thickness was found in Chapter 9 to be proportional to the square root of the product of kinematic viscosity and time of flow past the wall; thus, using Eq. (10.33) as representative of the flow time,

$$\text{boundary layer thickness} = \delta \approx \sqrt{\frac{\eta t}{\rho}} = r\sqrt{\frac{\alpha \eta}{\rho |q|}} \tag{10.34}$$

The angle over which the boundary layer extends is therefore approximately δ/r, or

$$\text{boundary layer angle} = \frac{\delta}{r} = \sqrt{\frac{\alpha \eta}{\rho |q|}} \approx \alpha |\mathfrak{R}|^{-1/2} \tag{10.35}$$

It is therefore clear that the changes in F take place over an angle of order $\alpha |\mathfrak{R}|^{-1/2}$, and this is the angle that should be used to scale θ. Since we are interested in the region near the wall, it is convenient to measure the angle from the wall and to replace θ with* $\theta + \alpha$. We therefore define a scaled angle

$$\zeta = \frac{\theta + \alpha}{\alpha |\mathfrak{R}|^{-1/2}} = (\phi + 1)|\mathfrak{R}|^{1/2} \tag{10.36}$$

We can then replace ϕ in Eq. (10.12) with ζ and write

$$\mathfrak{R}|\mathfrak{R}|^{1/2} F \frac{dF}{d\zeta} + |\mathfrak{R}|^{3/2} \frac{d^3 F}{d\zeta^3} + 4\alpha^2 |\mathfrak{R}|^{1/2} \frac{dF}{d\zeta} = 0 \tag{10.37a}$$

or, dividing by $|\mathfrak{R}|^{1/2}$,

$$\frac{\mathfrak{R}}{|\mathfrak{R}|} F \frac{dF}{d\zeta} + \frac{d^3 F}{d\zeta^3} + \frac{4\alpha^2}{|\mathfrak{R}|} \frac{dF}{d\zeta} = 0 \tag{10.37b}$$

If we now let $|\mathfrak{R}| \longrightarrow \infty$, this simplifies to

$$\mathfrak{R} > 0: \quad F \frac{dF}{d\zeta} + \frac{d^3 F}{d\zeta^3} = 0 \tag{10.38a}$$

$$\mathfrak{R} < 0: \quad -F \frac{dF}{d\zeta} + \frac{d^3 F}{d\zeta^3} = 0 \tag{10.38b}$$

---

*This applies at the wall $\theta = -\alpha$; an identical procedure follows at the wall $\theta = +\alpha$ by replacing θ with $\theta - \alpha$.

Note that the problem is no longer singular, in that the third derivative is retained. The equations for converging and diverging flow differ in the sign of one term.

The condition $F = 0$ at $\phi = -\alpha$ is written in terms of ζ as

$$F(0) = 0 \tag{10.39}$$

As we move away from the wall to the edge of the thin boundary layer ϕ will not change much, but, because of the $|\Re|^{1/2}$ term, ζ gets very large. Thus, we can replace our other boundary conditions by the observation that as $\zeta \to \infty$, F approaches a constant value of approximately $\frac{1}{2}$:

$$\zeta \longrightarrow \infty: \quad F \longrightarrow \tfrac{1}{2} \quad \frac{dF}{d\zeta} \longrightarrow 0 \tag{10.40}$$

Equations (10.38) can be written

$$\Re > 0: \quad \frac{1}{2}\frac{dF^2}{d\zeta} + \frac{d^3F}{d\zeta^3} = 0 \tag{10.41a}$$

$$\Re < 0: \quad -\frac{1}{2}\frac{dF^2}{d\zeta} + \frac{d^3F}{d\zeta^3} = 0 \tag{10.41b}$$

Integrating once, and using Eq. (10.40),

$$\Re > 0: \quad \frac{1}{2}F^2 + \frac{d^2F}{d\zeta^2} = \text{constant} = \frac{1}{2}F^2(\infty) = \frac{1}{8} \tag{10.42a}$$

$$\Re < 0: \quad -\frac{1}{2}F^2 + \frac{d^2F}{d\zeta^2} = \text{constant} = -\frac{1}{2}F^2(\infty) = -\frac{1}{8} \tag{10.42b}$$

A second integration is carried out by multiplying by $dF/d\zeta$:

$$\Re > 0: \quad \frac{1}{2}F^2\frac{dF}{d\zeta} + \frac{dF}{d\zeta}\frac{d^2F}{d\zeta^2} - \frac{1}{8}\frac{dF}{d\zeta}$$
$$= \frac{1}{6}\frac{dF^3}{d\zeta} + \frac{1}{2}\frac{d}{d\zeta}\left(\frac{dF}{d\zeta}\right)^2 - \frac{1}{8}\frac{dF}{d\zeta} = 0 \tag{10.43a}$$

$$\Re < 0: \quad -\frac{1}{2}F^2\frac{dF}{d\zeta} + \frac{dF}{d\zeta}\frac{d^2F}{d\zeta^2} + \frac{1}{8}\frac{dF}{d\zeta}$$
$$= -\frac{1}{6}\frac{dF^3}{d\zeta} + \frac{1}{2}\frac{d}{d\zeta}\left(\frac{dF}{d\zeta}\right)^2 + \frac{1}{8}\frac{dF}{d\zeta} = 0 \tag{10.43b}$$

Integrating,

$$\Re > 0: \quad \frac{1}{6}F^3 + \frac{1}{2}\left(\frac{dF}{d\zeta}\right)^2 - \frac{1}{8}F$$
$$= \text{constant} = \frac{1}{6}F^3(\infty) - \frac{1}{8}F(\infty) = -\frac{1}{24} \tag{10.44a}$$

$$\Re < 0: \quad -\frac{1}{6}F^3 + \frac{1}{2}\left(\frac{dF}{d\zeta}\right)^2 + \frac{1}{8}F$$
$$= \text{constant} = -\frac{1}{6}F^3(\infty) + \frac{1}{8}F(\infty) = +\frac{1}{24} \tag{10.44b}$$

If Eqs. (10.44) are evaluated at $\zeta = 0$, with $F(0) = 0$, we have

$$\Re > 0: \quad \left(\frac{dF}{d\zeta}\right)^2 = -\tfrac{1}{12} \text{ at } \zeta = 0 \qquad (10.45a)$$

$$\Re < 0: \quad \left(\frac{dF}{d\zeta}\right)^2 = +\tfrac{1}{12} \text{ at } \zeta = 0 \qquad (10.45b)$$

Clearly, we cannot satisfy this equation with a real function F for $\Re > 0$. Thus, the hypothesis of the existence of a boundary layer has led to a contradiction, and *a solution of the type that we are seeking cannot exist for a diverging flow*. This observation is consistent with the numerical solution shown in Fig. 10-3. Henceforth, we will consider only converging flow, $\Re < 0$.

Equation (10.44b) can be rewritten

$$\pm\sqrt{3} \frac{dF}{\sqrt{F^3 - \tfrac{3}{4}F + \tfrac{1}{4}}} = \pm\sqrt{3} \frac{dF}{(\tfrac{1}{2} - F)\sqrt{F + 1}} = d\zeta \qquad (10.46)$$

or integrating,

$$\pm\sqrt{2} \ln\left\{\frac{\sqrt{1 + F} + \sqrt{\tfrac{3}{2}}}{\sqrt{1 + F} - \sqrt{\tfrac{3}{2}}} \cdot \frac{1 - \sqrt{\tfrac{3}{2}}}{1 + \sqrt{\tfrac{3}{2}}}\right\} = \zeta \qquad (10.47)$$

It readily follows that the positive sign is required to obtain the proper behavior as $\zeta \longrightarrow \infty$. Rearranging, we can solve for $F(\phi)$ as

$$F(\phi) = \frac{3}{2}\left[\frac{2.225e^{(\phi+1)\sqrt{|\Re|/2}} - 0.225}{2.225e^{(\phi+1)\sqrt{|\Re|/2}} + 0.225}\right]^2 - 1 \qquad (10.48)$$

This function is plotted in Fig. 10-5. Note that the wall boundary layer region

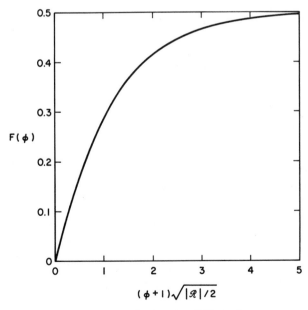

Figure 10-5. Dimensionless velocity function $F(\phi)$ for $\Re \longrightarrow -\infty$.

extends only an angular distance of order $7/\sqrt{|\Re|}$. Beyond this small distance $F(\phi)$ is approximately equal to $\frac{1}{2}$ and the fluid behaves as though it were inviscid.

The approach that we have taken here is a special case of a mathematical technique known as the theory of singular perturbations. As in the suddenly accelerated plate problem, we have computed behavior in a boundary layer. The type of approximation that we have used to obtain the solution here is extensively used in finding high Reynolds number boundary layer approximate solutions to the Navier–Stokes equations. We will return to such approximate solutions in Chapter 15.

10.7 POWER REQUIREMENT

We are frequently concerned with computing the power requirement for a given flow. We often speak loosely of computing "pressure drop," since for flow in a straight pipe the power requirement is simply the product of flow rate and pressure drop. In more complex flows, however, the pressure change by itself is not the quantity of significance. This can be nicely illustrated by considering Hamel flow.

We will compute the power requirement assuming a horizontal system, in which case the equivalent pressure $\mathcal{P}$ and pressure p are the same. If there is vertical motion over a large distance, the additional terms to overcome gravity must be added.

We will need the pressure p and the stress component τ_{rr} in order to compute the power. p is obtained from Eqs. (10.7) by integration. First, Eq. (10.7a) is rewritten

$$\frac{\partial p}{\partial r} = \frac{1}{r^3}\left(\rho f^2 + \eta \frac{d^2 f}{d\theta^2}\right) \tag{10.49}$$

This is integrated to obtain

$$p = p_\infty - \frac{1}{2r^2}\left(\rho f^2 + \eta \frac{d^2 f}{d\theta^2}\right) \tag{10.50}$$

The integration constant p_∞ is independent of r, but it could depend on θ. To determine this dependence we differentiate Eq. (10.50) with respect to θ to obtain

$$\frac{\partial p}{\partial \theta} = \frac{dp_\infty}{d\theta} - \frac{1}{2r^2}\left(2\rho f \frac{df}{d\theta} + \eta \frac{d^3 f}{d\theta^3}\right) = \frac{dp_\infty}{d\theta} + \frac{2\eta}{r^2}\frac{df}{d\theta} \tag{10.51}$$

The last substitution is carried out using Eq. (10.10). Comparison of Eq. (10.51) with Eq. (10.7b) demonstrates that $dp_\infty/d\theta$ is equal to zero and p_∞ is a true constant.

The stress component τ_{rr} is related to the velocity from Table 7-6 through the equation

$$\tau_{rr} = 2\eta \frac{\partial v_r}{\partial r} \tag{10.52}$$

Since $v_r = f(\theta)/r$, we then have

$$\tau_{rr} = -\frac{2\eta f(\theta)}{r^2} \tag{10.53}$$

The power requirement is equal to the rate at which work is done to force fluid through the system. At surface 1 in Fig. 10-6 the force per unit distance

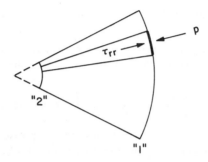

Figure 10-6. Schematic of stresses and surfaces used in power calculation.

into the paper on the arc sector $d\theta$ is $(p - \tau_{rr})r\,d\theta$. Thus, the rate of work per unit distance (henceforth, we will drop the term "per unit distance," but it will be understood) to move fluid across that sector of the surface is $v_r(p - \tau_{rr})r\,d\theta$. The total power requirement at surface 1 is therefore

$$P_1 = \int_{-\alpha}^{\alpha} v_r(p - \tau_{rr})r\,d\theta = \int_{-\alpha}^{\alpha} f(p - \tau_{rr})\,d\theta \tag{10.54}$$

Using Eqs. (10.50) and (10.53) we may write P_1 in terms of the function $f(\theta)$ as

$$P_1 = p_\infty q - \left(\frac{1}{2r_1^2}\right) \int_{-\alpha}^{\alpha} \left(\rho f^2 + \eta \frac{d^2 f}{d\theta^2} - 4\eta f\right) f\,d\theta \tag{10.55}$$

We obtain a similar result at surface 2, except that the sign is opposite, since here the control volume is pushing fluid into the surroundings. Thus,

$$P_2 = -p_\infty q + \left(\frac{1}{2r_2^2}\right) \int_{-\alpha}^{\alpha} \left(\rho f^2 + \eta \frac{d^2 f}{d\theta^2} - 4\eta f\right) f\,d\theta \tag{10.56}$$

The net power requirement is the algebraic sum of the two,

$$P = \left(\frac{1}{2r_2^2} - \frac{1}{2r_1^2}\right) \int_{-\alpha}^{\alpha} \left(\rho f^2 + \eta \frac{d^2 f}{d\theta^2} - 4\eta f\right) f\,d\theta \tag{10.57}$$

Equation (10.57) can be simplified a bit by integrating Eq. (10.10) once. Equation (10.10) can be written

$$\rho \frac{df^2}{d\theta} + \eta \frac{d^3f}{d\theta^3} + 4\eta \frac{df}{d\theta} = 0 \tag{10.58}$$

Integrating once, we obtain

$$\rho f^2 + \eta \frac{d^2f}{d\theta^2} + 4\eta f = \text{constant} = \eta f''(\alpha) \tag{10.59}$$

where $f''(\alpha)$ denotes $d^2f/d\theta^2$ evaluated at $\theta = \alpha$. When this is substituted into Eq. (10.57) and we use the fact that $\int_{-\alpha}^{\alpha} f(\theta)\, d\theta = q$, we obtain, finally,

$$P = \frac{\eta}{2}\left(\frac{1}{r_2^2} - \frac{1}{r_1^2}\right)\left[f''(\alpha)q - 8 \int_{-\alpha}^{\alpha} f^2(\theta)\, d\theta \right] \tag{10.60}$$

Our prior calculations have been in terms of the function $F = \alpha f/q$ and the independent variable $\phi = \theta/\alpha$. If we make these substitutions, then P can be written

$$P = \frac{\eta q^2}{2\alpha^3}\left(\frac{1}{r_2^2} - \frac{1}{r_1^2}\right)\left[8\alpha^2 \int_{-1}^{1} F^2(\phi)\, d\phi - F''(1) \right] \tag{10.61}$$

Finally, for purposes of presentation and comparison of results, it is useful to define a dimensionless power requirement P^* which is a function only of $\Re$ and α:

$$P = 4\eta q^2 \left[\frac{1}{(2r_2 \sin \alpha)^2} - \frac{1}{(2r_1 \sin \alpha)^2} \right] P^*(\Re, \alpha) \tag{10.62}$$

$$P^*(\Re, \alpha) \equiv \sin^2 \alpha \left[\frac{4}{\alpha} \int_{-1}^{1} F^2(\phi)\, d\phi - \frac{F''(1)}{2\alpha^3} \right] \tag{10.63}$$

The coefficient of P^* is written in terms of $r \sin \theta$ to facilitate thinking with regard to the configuration in Fig. 10-7, where $2r_1 \sin \theta$ and $2r_2 \sin \theta$ are the

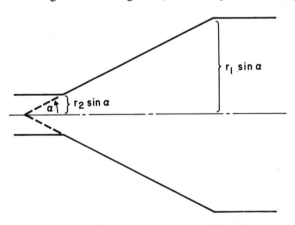

$r_1 \sin \alpha$

$r_2 \sin \alpha$

α

Figure 10-7. Schematic of a finite converging flow.

widths of the upstream and downstream channels (which, of course, are not considered to exist in this analysis.) Since the ratio of volumetric flow rate q to channel width is simply the mean velocity, P^* can be written alternatively as

$$P = 4\eta(V_2^2 - V_1^2)P^*(\mathfrak{R}, \alpha) \tag{10.64}$$

V_1 and V_2 are the mean velocities upstream and downstream of the converging section, respectively. We emphasize again that this result is only approximate, since we have not accounted for the sections where the profile changes from the plane Poiseuille flow up- and downstream of the converging section to the Hamel flow in the converging section. The closer the Hamel flow is to a parabola, the better we can expect the total power requirement to be given by Eq. (10.62) or (10.64). This relation is discussed further in Sec. 17.3.3.

The function P^* isolates the angle dependence of the converging section. We can obtain simple analytical expressions for the limiting cases $\mathfrak{R} \to 0$, $\alpha \to 0$, and $\mathfrak{R} \to -\infty$. For $\mathfrak{R} = 0$, the function $F(\theta)$ is given by Eq. (10.19). Substituting into Eq. (10.63) gives

$$\mathfrak{R} \longrightarrow 0: \quad P^* = \frac{[4\alpha(1 + \cos^2 2\alpha) - 4\sin 2\alpha \cos 2\alpha]\sin^2 \alpha}{(\sin 2\alpha - 2\alpha \cos 2\alpha)^2} \tag{10.65}$$

This function is plotted in Fig. 10-8.

For the lubrication approximation, $\alpha \to 0$, the function $F(\phi)$ is the parabola given by Eq. (10.20). When this is substituted into Eq. (10.63), we obtain the result

$$\alpha \longrightarrow 0: \quad P^* = \frac{3}{4\alpha} \tag{10.66}$$

Here, we have neglected a term proportional to α relative to the $\frac{1}{2}$ term, and we have used the small-angle approximation $\sin \alpha \approx \alpha$. Equation (10.66) is plotted in Fig. 10-8, and it is again evident that the lubrication approximation suffices up to an angle of about 20°.

For rapid converging flow, $\mathfrak{R} \to -\infty$, we can use the estimates from the boundary later analysis. We find there from Eq. (10.42b) that

$$\left.\frac{d^2F}{d\zeta^2}\right|_{\zeta=0} = -\frac{1}{8} \tag{10.67}$$

Since $d\zeta = |\mathfrak{R}|^{1/2}\, d\phi$ and $\phi = 1$ at $\zeta = 0$, this gives

$$F''(1) = -\frac{1}{8}|\mathfrak{R}| \tag{10.68}$$

In addition, outside the boundary layer $F \approx \frac{1}{2}$, so

$$\int_{-1}^{1} F^2(\phi)\, d\phi \approx \int_{-1}^{1} (\tfrac{1}{2})^2\, d\phi = \frac{1}{2} \tag{10.69}$$

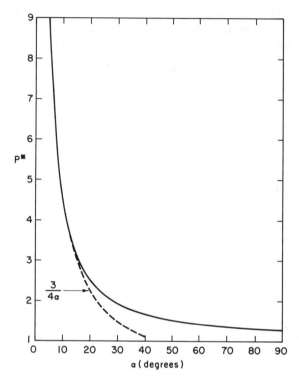

Figure 10-8. Dimensionless power requirement as a function of angle of convergence, $\Re = 0$. The dashed line is the result using the lubrication approximation for small α.

This term can be neglected relative to the term containing $|\Re|$, so we obtain

$$\Re \longrightarrow -\infty: \quad P^* = \frac{|\Re| \sin^2 \alpha}{16\alpha^3} \qquad (10.70)$$

Note that in this limit the power P is, in fact, independent of the viscosity, η.

For cases where the limits computed here are not applicable, the numerical solutions of Eq. (10.12) must be used.

10.8 CONCLUDING REMARKS

The important information to carry away from this chapter is the range of behavior to be expected of solutions to the Navier–Stokes equations when both viscous and inertial terms interact. It cannot be stated too often that exact solutions of the type exhibited in this chapter that include both inertial and viscous terms are rare, and students are not expected to be able to produce new nontrivial solutions routinely. Indeed, any new exact solution is a worthy subject of a research paper in a good technical journal. What can be expected of students, however, is a firm grasp of the way that inertial, viscous, and geometrical effects interact and influence flow behavior.

The three limiting cases applied to Hamel flow, the creeping flow, lubrication, and boundary layer approximations, are the subjects of subsequent chapters. These approximations represent simplifications of the mathematical structure at the expense of a severely limited range of applicability. It is only through such simplifications that most detailed flow problems of processing interest are solved. This is the time in the reading of the text to ensure that the relation between the approximate and exact solutions is clear; in subsequent applications we will not have an exact solution for comparison.

The final point made in the chapter, regarding the computation of power, is also quite important. The phrase "pressure drop" is firmly embedded in engineering culture, but the reading on a pressure gage may not have any significance in a given flow situation. The example considered here is illustrative of that important fact, since a knowledge of the pressure change between r_1 and r_2 cannot be translated into a calculation of the power requirement.

BIBLIOGRAPHICAL NOTES

The analytical solution for Hamel flow, and other exact solutions of the Navier–Stokes equations, can be found in

LANDAU, L. D., and LIFSHITZ, E. M., *Fluid Mechanics*, Addison-Wesley Publishing Company, Inc., Reading, Mass., 1959.

Other books with exact solutions are

BATCHELOR, G. K., *An Introduction to Fluid Mechanics*, Cambridge University Press, New York, 1967.

SCHLICHTING, H., *Boundary Layer Theory*, 6th ed., McGraw-Hill Book Company, New York, 1968.

All three of these books are well-written texts, and students of fluid mechanics owe it to themselves to become well acquainted with them. A recent exact solution that is not contained in standard textbooks is

ABBOTT, T. N. G., and WALTERS, K., *J. Fluid Mech.*, **40**, 205 (1970).

PROBLEMS

10.1. Show that a purely radial solution ($v_\theta = v_\phi = 0$) cannot exist for converging flow in a cone.

10.2. Plane stagnation flow is shown schematically in Fig. 14.2. The velocity field far from the stagnation point has the form $v_x = Ax$, $v_y = -Ay$, where x is distance from the stagnation point parallel to the wall and y is distance measured normal to the wall. Obtain an exact solution to the Navier–Stokes equations that is consistent with this asymptotic behavior; it will be sufficient to obtain an ordinary differential equation without solving the equation. (Hint: First show that $v_x = Ax$, $v_y = -Ay$ satisfies the Navier-Stokes equations but fails to satisfy a boundary condition. Then look for a generalization that retains this asymptotic form but satisfies all boundary conditions.)

Part V
Approximate Methods

Ordering and **11**
Approximation

11.1 INTRODUCTION

Because exact solutions of the Navier–Stokes equations are possible only in rare instances, it is necessary to establish systematic approaches for the development of approximate solutions. We saw in Chapter 10 that the differential equation describing Hamel flow could be simplified considerably under certain limiting conditions. In this chapter we will examine the logic that was used to obtain the simplifications, so that we may develop procedures that can be applied to the Navier–Stokes equations in other flow situations.

11.2 CHARACTERISTIC QUANTITIES

Approximation requires an estimate of the order of magnitude of each term in each equation, so that the important terms may be retained and the unimportant terms neglected. Such an estimate is obtained by making each term dimensionless. We define a dimensionless velocity vector $\tilde{\mathbf{v}}$ as $\mathbf{v}/V$, where V is a characteristic velocity of a moving surface, for example, or a mean velocity computed from the flow rate. It is presumed that the maximum velocity in the system is of the same order of magnitude as the characteristic velocity, so the maximum magnitude of the dimensionless velocity $\tilde{\mathbf{v}}$ will not be large compared to unity. We shall refer to terms whose magnitudes range between zero and a value that is not large relative to unity as being *of order unity*.

For simplicity we shall restrict ourselves to incompressible Newtonian fluids, and we shall not consider free surface flows for which the gravity term needs to be treated separately. We assume that there is a single characteristic velocity and a single characteristic distance. The characteristic distance is the length scale over which changes take place in the flow field. In pipe flow the characteristic distance is the diameter or the radius; in flow past a sphere it is the radius or diameter of the sphere. We thus define characteristic and dimensionless quantities as shown in Table 11-1. The assumption that the flow field is described by a single characteristic length is a restriction that we shall have to relax in later chapters.

TABLE 11-1

CHARACTERISTIC AND DIMENSIONLESS VARIABLES

Physical variable	Characteristic quantity	Dimensionless variable
Velocity, $\mathbf{v}$	V	$\tilde{\mathbf{v}} = \dfrac{\mathbf{v}}{V}$
Position, $\mathbf{x}$	L	$\tilde{\mathbf{x}} = \dfrac{\mathbf{x}}{L}$
Gradient operator, ∇	L^{-1}	$\tilde{\nabla} = \nabla L$
Time, t	T	$\tilde{t} = \dfrac{t}{T}$
Equivalent pressure, $\mathcal{P}$	Π	$\tilde{\mathcal{P}} = \dfrac{\mathcal{P}}{\Pi}$

The continuity equation for an incompressible fluid is

$$\nabla \cdot \mathbf{v} = 0 \tag{11.1}$$

Substituting the dimensionless variables defined in Table 11-1 then gives

$$\frac{V}{L}\tilde{\nabla} \cdot \tilde{\mathbf{v}} = 0 \tag{11.2a}$$

or

$$\tilde{\nabla} \cdot \tilde{\mathbf{v}} = 0 \tag{11.2b}$$

The continuity equation is unchanged in form by the nondimensionalization and does not contain any parameters.

The Navier–Stokes equations are

$$\rho \frac{\partial \mathbf{v}}{\partial t} + \rho \mathbf{v} \cdot \nabla \mathbf{v} = -\nabla \mathcal{P} + \eta \nabla^2 \mathbf{v} \tag{11.3}$$

Substitution of the dimensionless quantities then gives the form*

$$\left(\frac{\rho V}{T}\right)\frac{\partial \tilde{\mathbf{v}}}{\partial \tilde{t}} + \left(\frac{\rho V^2}{L}\right)\tilde{\mathbf{v}} \cdot \tilde{\nabla}\tilde{\mathbf{v}} = -\left(\frac{\Pi}{L}\right)\tilde{\nabla}\tilde{\mathcal{P}} + \left(\frac{\eta V}{L^2}\right)\tilde{\nabla}^2\tilde{\mathbf{v}} \tag{11.4a}$$

---

*A typical term in $\nabla^2 \mathbf{v}$ is $\partial^2 v_x / \partial y^2$. The substitution then follows as

$$\frac{\partial^2 v_x}{\partial y^2} = \frac{\partial^2 v_x}{\partial y \, \partial y} = \frac{\partial^2 (V\tilde{v}_x)}{\partial(L\tilde{y}) \, \partial(L\tilde{y})} = \frac{V}{L^2}\frac{\partial^2 \tilde{v}_x}{\partial \tilde{y} \, \partial \tilde{y}} = \frac{V}{L^2}\frac{\partial^2 \tilde{v}_x}{\partial \tilde{y}^2}$$

or, dividing by $\eta V/L^2$,

$$\left(\frac{\rho L^2}{\eta T}\right)\frac{\partial \tilde{\mathbf{v}}}{\partial \tilde{t}} + \left(\frac{\rho V L}{\eta}\right)\tilde{\mathbf{v}} \cdot \tilde{\boldsymbol{\nabla}}\tilde{\mathbf{v}} = -\left(\frac{\Pi L}{\eta V}\right)\tilde{\boldsymbol{\nabla}}\tilde{\mathcal{P}} + \tilde{\boldsymbol{\nabla}}^2\tilde{\mathbf{v}} \qquad (11.4b)$$

The coefficient of the $\tilde{\mathbf{v}} \cdot \tilde{\boldsymbol{\nabla}}\tilde{\mathbf{v}}$ term is readily seen to be the *Reynolds number*,

$$\text{Reynolds number:} \quad \text{Re} = \frac{\rho V L}{\eta} \qquad (11.5a)$$

so we may write Eq. (11.4) in the equivalent form

$$\left(\frac{\rho L^2}{\eta T}\right)\frac{\partial \tilde{\mathbf{v}}}{\partial \tilde{t}} + \text{Re}\,\tilde{\mathbf{v}} \cdot \tilde{\boldsymbol{\nabla}}\tilde{\mathbf{v}} = -\left(\frac{\Pi L}{\eta V}\right)\tilde{\boldsymbol{\nabla}}\tilde{\mathcal{P}} + \tilde{\boldsymbol{\nabla}}^2\tilde{\mathbf{v}} \qquad (11.4c)$$

It is sometimes convenient to define the *Strouhal number*,

$$\text{Strouhal number:} \quad \text{Sr} = \frac{VT}{L} \qquad (11.5b)$$

The Strouhal number is a dimensionless reciprocal time, or, better, a dimensionless frequency. The dimensionless Navier–Stokes equations can then be written in the form

$$\text{Re}\left(\text{Sr}^{-1}\frac{\partial \tilde{\mathbf{v}}}{\partial \tilde{t}} + \tilde{\mathbf{v}} \cdot \tilde{\boldsymbol{\nabla}}\tilde{\mathbf{v}}\right) = -\left(\frac{\Pi L}{\eta V}\right)\tilde{\boldsymbol{\nabla}}\tilde{\mathcal{P}} + \tilde{\boldsymbol{\nabla}}^2\tilde{\mathbf{v}} \qquad (11.4d)$$

11.3 CHARACTERISTIC PRESSURE

The characteristic pressure is not independent of the other characteristic quantities. We know from the engineering Bernoulli equation and the examples in Chapter 6 that for flows in which inertial effects dominate and losses are small, the pressure is of the order of ρV^2. Thus, we can write

$$\text{inertially dominated:} \quad \Pi = \rho V^2 \qquad (11.6)$$

Equation (11.4) can then be written

$$\text{inertially dominated:} \quad \text{Re}\left(\text{Sr}^{-1}\frac{\partial \tilde{\mathbf{v}}}{\partial \tilde{t}} + \tilde{\mathbf{v}} \cdot \tilde{\boldsymbol{\nabla}}\tilde{\mathbf{v}} + \tilde{\boldsymbol{\nabla}}\tilde{\mathcal{P}}\right) = \tilde{\boldsymbol{\nabla}}^2\tilde{\mathbf{v}} \qquad (11.7)$$

When the viscous terms are much more important than the inertial terms, we know that the pressure is of order ρl_V. Equivalently, we can say that the pressure is of the order of a characteristic viscous stress and write

$$\text{viscous dominated:} \quad \Pi = \frac{\eta V}{L} \qquad (11.8)$$

$$\text{viscous dominated:} \quad \text{Re}\left(\text{Sr}^{-1}\frac{\partial \tilde{\mathbf{v}}}{\partial \tilde{t}} + \tilde{\mathbf{v}} \cdot \tilde{\boldsymbol{\nabla}}\tilde{\mathbf{v}}\right) = -\tilde{\boldsymbol{\nabla}}\tilde{\mathcal{P}} + \tilde{\boldsymbol{\nabla}}^2\tilde{\mathbf{v}} \qquad (11.9)$$

When inertial and viscous terms are of comparable magnitude then either form may be used. Equation (11.8) is the more common estimate of order of magnitude in that case, so Eq. (11.9) is the dimensionless form of the Navier–Stokes equations most commonly encountered.

11.4 CHARACTERISTIC TIME

In some problems there is a characteristic time imposed on the system, such as the period of an oscillating surface. In others, however, the only characteristic time is the time required for a fluid element to move through a characteristic distance. In such a case the characteristic time is simply L/V, and we can write

$$\text{characteristic time} = \text{flow time:} \quad T = \frac{L}{V} \qquad (11.10)$$

Equation (11.9) is then

$$\text{characteristic time} = \text{flow time:} \quad \text{Re}\left(\frac{\partial \tilde{\mathbf{v}}}{\partial \tilde{t}} + \tilde{\mathbf{v}} \cdot \tilde{\nabla}\tilde{\mathbf{v}}\right) = -\tilde{\nabla}\tilde{\mathcal{P}} + \tilde{\nabla}^2\tilde{\mathbf{v}}$$

$$(11.11)$$

11.5 ORDERING ARGUMENTS

The dimensionless forms of the equations now allow us to make physically based simplifications. It is assumed that all dimensionless quantities are of order unity, and changes in dimensionless quantities take place over dimensionless length scales of order unity. Changes in quantities of order unity are also of order unity; thus, since derivatives are of the order of changes in dimensionless variables divided by the dimensionless characteristic distance, *all derivatives are of order unity*. Hence, if the proper characteristic quantities have been used for nondimensionalization, every term in the dimensionless equations is a term of order unity multiplied by a dimensionless group. The relative importance of each term is therefore determined by the magnitude of the dimensionless group.

As an illustrative example, consider the case of a long cylinder of radius R that is immersed in an infinite body of fluid. The cylinder oscillates about its axis with frequency ω:

$$v_\theta = V \sin \omega t \text{ at } r = R \qquad (11.12)$$

With reference to Eq. (11.9), we therefore have

$$\text{Re} = \frac{RV\rho}{\eta} \qquad \text{Sr} = \frac{V}{R\omega} \qquad (11.13)$$

Two limiting frequency ranges can then be considered:

$$\omega \ll \frac{V}{R}, \text{Sr} \gg 1: \quad \left|\text{Sr}^{-1}\frac{\partial \tilde{\mathbf{v}}}{\partial \tilde{t}}\right| \ll |\tilde{\mathbf{v}} \cdot \tilde{\nabla}\tilde{\mathbf{v}}| \qquad (11.14a)$$

$$\omega \gg \frac{V}{R}, \text{Sr} \ll 1: \quad \left|\text{Sr}^{-1}\frac{\partial \tilde{\mathbf{v}}}{\partial \tilde{t}}\right| \gg |\tilde{\mathbf{v}} \cdot \tilde{\nabla}\tilde{\mathbf{v}}| \qquad (11.14b)$$

In each case, one of the two inertial terms dominates the other, and we can write *approximations* to Eq. (11.9) as

$$\text{Sr} \gg 1: \quad \text{Re } \tilde{\mathbf{v}} \cdot \tilde{\nabla} \tilde{\mathbf{v}} = - \tilde{\nabla} \tilde{\mathscr{P}} + \tilde{\nabla}^2 \tilde{\mathbf{v}} \tag{11.15a}$$

$$\text{Sr} \ll 1: \quad \frac{\text{Re}}{\text{Sr}} \frac{\partial \tilde{\mathbf{v}}}{\partial \tilde{t}} = -\tilde{\nabla} \tilde{\mathscr{P}} + \tilde{\nabla}^2 \tilde{\mathbf{v}} \tag{11.15b}$$

Note that each of these approximate equations is a notable simplification. Equation (11.15a) is a pseudo-steady-state equation, because the time derivative term is no longer present. Equation (11.15b) is a linear equation, because the nonlinear portion of the inertial terms is negligible compared to the time rate-of-change term.

Equations (11.15) can both be further simplified in the case of small Reynolds number. If Re is small compared to unity, it follows from Eq. (11.15a) that the term on the left is small compared to the terms on the right; thus, we have

$$\text{Sr} \gg 1 \text{ and } \text{Re} \ll 1: \quad 0 = -\tilde{\nabla} \tilde{\mathscr{P}} + \tilde{\nabla}^2 \tilde{\mathbf{v}} \tag{11.16a}$$

If Re is small compared to Sr, the term on the left of Eq. (11.15b) is small compared to the terms on the right, and we have

$$\text{Sr} \ll 1 \text{ and } \text{Re} \ll \text{Sr}: \quad 0 = -\tilde{\nabla} \tilde{\mathscr{P}} + \tilde{\nabla}^2 \tilde{\mathbf{v}} \tag{11.16b}$$

Equations (11.16a) and (11.16b) are identical, but they apply in different ranges of the physical variables. The conditions for Eqs. (11.16a) and (11.16b) to hold are, respectively,

$$\text{Sr} \gg 1 \text{ and } \text{Re} \ll 1: \quad \omega \ll \frac{V}{R} \ll \frac{\eta}{\rho R^2} \tag{11.17a}$$

$$\text{Sr} \ll 1 \text{ and } \text{Re} \ll \text{Sr}: \quad \frac{V}{R} \ll \omega \ll \frac{\eta}{\rho R^2} \tag{11.17b}$$

It is essential in approximations like these that the limitations on applicability of the equations be defined carefully. When a solution to the approximate equations is obtained, it should be checked to establish that it does indeed satisfy the assumptions that were used in its derivation.

11.6 SOLUTION LOGIC

The logic for obtaining approximate solutions to the Navier–Stokes equations (as well as to other equations describing physical phenomena) is as follows:

1. Make the equations dimensionless so that all variables are of order unity.
2. Estimate the terms that can be neglected in the full (*exact*) equations because they are multiplied by small parameters. Delete these terms to obtain simplified (*approximate*) equations.

3. Obtain a solution to the approximate equations.
4. Estimate the magnitude of the neglected terms in the exact equations to ensure that they are truly negligible relative to the terms which were retained.

It is important to examine this logic. We are seeking an approximate solution to the exact equations. The actual solution process outlined above is shown in Fig. 11-1. Note that all operations are contained in the loop on the right, which does not include the sought-after solution. *Our approach is based on the presumption that an exact solution to the approximate equations is also an approximate solution to the exact equations.* This is certainly true in most cases, but it need not be so, and the approach is flawed logically.

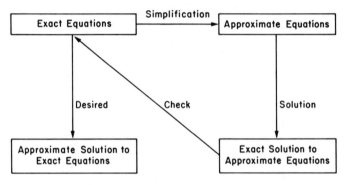

Figure 11-1. Logic of the process for obtaining approximate solutions.

The problem with the logic is nicely illustrated by the following example developed by Segal:

$$0.01x + y = 0.1 \qquad (11.18a)$$

$$x + 101y = 11 \qquad (11.18b)$$

If we assume that x and y are of comparable magnitude, we expect to be able to neglect the first term on the left of Eq. (11.18a) relative to the second. This leads to the approximate equations

$$y = 0.1 \qquad (11.19a)$$

$$x + 101y = 11 \qquad (11.19b)$$

This set of equations has a solution $y = 0.1$, $x = 11 - (101)(0.1) = 0.9$. The neglected term in Eq. (11.18a) is then $(0.01)(0.9) = 0.009$, which is negligible compared to the retained term, $y = 0.1$. Thus, the solution satisfies the consistency test outlined above. In fact, the exact solution to Eqs. (11.18a) and (11.18b) is $x = -90$, $y = 1$!

Counterexamples of this type are useful, in that they remind us to stay on our guard when seeking solutions to the mathematical equations describing physical phenomena.

11.7 INVISCID LIMIT

We have already seen that limiting behavior at high Reynolds numbers requires careful mathematical treatment. Equation (11.7) for inertially dominated flow can be rewritten

$$\text{Sr}^{-1}\frac{\partial \tilde{\mathbf{v}}}{\partial \tilde{t}} + \tilde{\mathbf{v}} \cdot \tilde{\nabla}\tilde{\mathbf{v}} + \tilde{\nabla}\tilde{\mathcal{P}} = \text{Re}^{-1}\tilde{\nabla}^2\tilde{\mathbf{v}} \qquad (11.20)$$

If we now set the term on the right to zero as an approximation for $\text{Re} \gg 1$, we obtain

$$\text{Re} \longrightarrow \infty: \quad \text{Sr}^{-1}\frac{\partial \tilde{\mathbf{v}}}{\partial \tilde{t}} + \tilde{\mathbf{v}} \cdot \tilde{\nabla}\tilde{\mathbf{v}} + \tilde{\nabla}\tilde{\mathcal{P}} = 0 \qquad (11.21)$$

This is the equation that would be obtained for an inviscid fluid, $\eta = 0$.

Equation (11.21) has been extensively studied, and its behavior in many geometries is discussed in texts on classical hydrodynamics. The limit in passing from Eq. (11.20) to (11.21) is *singular*, in that the highest derivative term has been lost from the equation. The mathematical consequence of this singularity is that one boundary condition cannot be satisfied. The unsatisfied boundary condition must be the no-slip condition, since an inviscid fluid cannot transmit a shear stress and can, therefore, slip past a wall without adhering. Classical hydrodynamics is therefore relevant only in cases where the flow is not affected in an important way by the presence of a no-slip surface. Inviscid flow is considered in Chapter 14.

The limit $\text{Re} \to \infty$ for cases in which the no-slip condition is important is discussed in Chapter 15 on boundary layer theory. That theory requires the solution to Eq. (11.21) as a part of the overall solution. The general approach is a part of a mathematical theory known as the theory of singular perturbations.

11.8 SINGULAR PERTURBATIONS*

The theory of singular perturbations and the way in which it can be applied to the Navier–Stokes equations can be illustrated by the following problem:

$$\epsilon\frac{d^2u}{dx^2} + u\frac{du}{dx} + f(x) = 0 \qquad (11.22)$$

$$u(0) = 0 \qquad u(1) = 1 \qquad (11.23)$$

This nonlinear equation is intended to serve as a prototype for a steady-state version of Eq. (11.20). ϵ is a small parameter, so $\epsilon \, d^2u/dx^2$ plays the role of the $\text{Re}^{-1}\tilde{\nabla}^2\tilde{\mathbf{v}}$ term; $u \, du/dx$ and f play the roles of the $\tilde{\mathbf{v}} \cdot \tilde{\nabla}\tilde{\mathbf{v}}$ and $\tilde{\nabla}\tilde{\mathcal{P}}$ terms,

*This section may be omitted on a first reading.

respectively. Equation (11.22) has an analytical solution only for the case in which $f(x)$ is zero.

An "inviscid" solution is obtained by taking the limit as $\epsilon \to 0$:

$$\epsilon = 0: \quad u\frac{du}{dx} + f(x) = \frac{1}{2}\frac{du^2}{dx} + f(x) = 0 \qquad (11.24)$$

Equation (11.24) is readily integrated once to give a solution

$$u(x) = \left[C_1 - 2\int_0^x f(\xi)\, d\xi \right]^{1/2} \qquad (11.25)$$

The single constant of integration C_1 cannot be chosen to satisfy both of the boundary conditions given by Eq. (11.23). We choose C_1 to satisfy the "outer" boundary condition, $u(1) = 1$. This leads to the *outer solution*,

$$u_{\text{outer}} = \left[1 + 2\int_x^1 f(\xi)\, d\xi \right]^{1/2} \qquad (11.26)$$

The outer solution does not satisfy the condition at $x = 0$ and cannot be applied near $x = 0$.

The region near $x = 0$ is accounted for by "stretching" the coordinates to account for changes in that region. We define the new "stretched" independent variable

$$y = \frac{x}{\epsilon} \qquad (11.27)$$

Equation (11.22) then becomes

$$\epsilon\frac{1}{\epsilon^2}\frac{d^2u}{dy^2} + \frac{1}{\epsilon}u\frac{du}{dy} + f(\epsilon y) = 0 \qquad (11.28a)$$

or, multiplying by ϵ,

$$\frac{d^2u}{dy^2} + u\frac{du}{dy} + \epsilon f(\epsilon y) = 0 \qquad (11.28b)$$

If we now take the limit $\epsilon \to 0$, the highest-order term is retained and we obtain an equation valid near $x = 0$:

$$\epsilon = 0: \quad \frac{d^2u}{dy^2} + u\frac{du}{dy} = \frac{d^2u}{dy^2} + \frac{1}{2}\frac{du^2}{dy} = 0 \qquad (11.29)$$

Equation (11.29) is integrated once to give

$$\frac{du}{dy} + \tfrac{1}{2}u^2 = C_2 \qquad (11.30)$$

This first-order equation can be put into the form

$$\int_0^u \frac{du}{C_2 - \tfrac{1}{2}u^2} = y \qquad (11.30)$$

The boundary condition $u = 0$ at $y = 0$ is included in the lower limit of integration. The integration can be carried out to give the *inner solution*,

$$u_{inner} = \sqrt{2C_2} \frac{\exp(2C_2 y) - 1}{\exp(2C_2 y) + 1} = \sqrt{2C_2} \frac{\exp(2C_2 x/\epsilon) - 1}{\exp(2C_2 x/\epsilon) + 1} \quad (11.32)$$

The constant of integration C_2 is evaluated by noting that the inner solution reaches its asymptotic value for large y quite rapidly, since the term x/ϵ becomes large for even small values of x. Thus, the inner and outer solutions must give the same value for the function $u(x)$ when y is large but x is small. The mathematical expression is then

$$\lim_{y \to \infty} u_{inner} = \lim_{x \to 0} u_{outer} \quad (11.33)$$

When Eqs. (11.26) and (11.32) are substituted into these limits, we obtain

$$\lim_{y \to \infty} u_{inner} = \sqrt{2C_2} = \lim_{y \to 0} u_{outer} = \left[1 + 2 \int_0^1 f(\xi) \, d\xi \right]^{1/2} \quad (11.34a)$$

or, solving for C_2,

$$C_2 = \tfrac{1}{2} + \int_0^1 f(\xi) \, d\xi \quad (11.34b)$$

The approximate solution to Eqs. (11.22) and (11.23) therefore has different forms that must be applied in different regions. In the *boundary layer* near $x = 0$, where $u(x)$ changes rapidly, the solution is given by Eq. (11.32), with C_2 given by Eq. (11.34b). Everywhere else the "inviscid" outer solution, Eq. (11.26), applies.

11.9 CONCLUDING REMARKS

The ordering analysis outlined here for simplification of the Navier–Stokes equations will be utilized in succeeding chapters for various limiting cases. The essence of an ordering approximation is the restriction of the physical range of applicability in order to simplify the equations and make solution possible. Systematic ordering of the relative magnitudes of terms is the single most important tool available to the analyst seeking solutions to mathematical representations of physical problems.

Selection of the correct characteristic quantities is essential for proper reduction of the equations. This point was illustrated in Section 10.6 on the boundary layer approximation for converging flow, where we found that the correct characteristic length was not the full channel opening but rather a boundary layer thickness proportional to $(\eta/\rho)^{1/2}$. In some of the chapters that follow we will find that specification of the characteristic quantities can be developed logically.

The ordering and dimensionless groups developed in this chapter presume the existence of a single characteristic length. Many flows of practical interest have *two* characteristic lengths, which introduces an additional dimensionless group (the ratio of lengths) and thus an additional ordering parameter. Lubrication and boundary layer flows are of this type. For such flows the procedures outlined in this chapter provide guidance, but the equations developed here cannot be applied directly.

BIBLIOGRAPHICAL NOTES

The counterexample in Section 11.6 is from a very instructive paper on scaling,

SEGAL, L. A., *SIAM Rev.*, **14**, 547 (1972).

The limiting cases Re $\longrightarrow$ 0 and Re $\longrightarrow \infty$ will be be considered in subsequent chapters. We will not consider cases where the Strouhal number is important; one such case of interest is the periodic formation of vortices in steady flow past submerged cylinders. There is a brief discussion of this flow, known as *von Kármán vortices*, in *Perry's Handbook;* see also

BERGER, E., and WILLE, R., *Ann. Rev. Fluid Mech.*, **4**, 313 (1972).

Creeping Flow 12

12.1 INTRODUCTION

Many flows of practical interest have a single characteristic length L and characteristic velocity V, and have no characteristic time other than $T = L/V$. In that case the dimensionless form of the Navier–Stokes equations for Newtonian fluids can be written in the form of Eq. (11.11):

$$\text{Re}\left(\frac{\partial \tilde{\mathbf{v}}}{\partial \tilde{t}} + \tilde{\mathbf{v}} \cdot \tilde{\boldsymbol{\nabla}}\tilde{\mathbf{v}}\right) = -\tilde{\boldsymbol{\nabla}}\tilde{\mathcal{P}} + \tilde{\nabla}^2\tilde{\mathbf{v}} \tag{12.1}$$

Here, Re is expressed in terms of the characteristic length and velocity,

$$\text{Re} = \frac{LV\rho}{\eta} \tag{12.2}$$

When Re is very small, the inertial terms on the left of Eq. (12.1) will be very small compared to the pressure and viscous stress terms on the right. For sufficiently small Re we may ignore the inertial terms and write the *creeping flow approximation* to the Navier–Stokes equations,

$$\text{Re} \ll 1: \quad 0 = -\tilde{\boldsymbol{\nabla}}\tilde{\mathcal{P}} + \tilde{\nabla}^2\tilde{\mathbf{v}} \tag{12.3a}$$

The dimensional form of this equation is

$$\text{Re} \ll 1: \quad 0 = -\boldsymbol{\nabla}\mathcal{P} + \eta\nabla^2\mathbf{v} \tag{12.3b}$$

The continuity equation is unchanged in this approximation.

The creeping flow approximation applies to flows that are "slow" in the sense of having a small characteristic velocity, V, but the condition Re $\ll 1$ can also be satisfied by a very small characteristic length, L, or a very large viscosity, η. The large viscosity of molten polymers means that Re will be small for these liquids even for flows with large characteristic velocities, and the creeping flow approximation is nearly always applicable for polymer melt flows.

The creeping flow equations, Eq. (12.3b), can be obtained directly from the Navier–Stokes equations in Tables 7-4, 7-7, and 7-10 by setting the density to zero. This illustrates the point that the approximation applies to flows in which the inertial (ρV^2) terms are negligible relative to the viscous ($\eta V/L$) terms. The mathematical consequence of the creeping flow approximation is the reduction of the nonlinear Navier–Stokes equations to a set of linear equations. The theory of linear partial differential equations is much better developed than the theory of nonlinear equations, so the possibility of solution is greatly enhanced. This broadening of the class of possible solutions is obtained, of course, by imposing the restrictive physical condition of a very small Reynolds number.

12.2 FLOW BETWEEN ROTATING DISKS

Flow between rotating disks is shown schematically in Fig. 12-1. A fluid is contained between two circular disks, one of which rotates relative to the other with angular velocity Ω. The spacing H between the disks is assumed to be very small relative to the disk outer radius, R. This configuration arises in many applications, including lubrication and polymer processing, and it is a useful design for viscosity measurement. We wish to determine the velocity field and the torque required to turn the moving disk. A closed-form analytical solution cannot be obtained for this problem for arbitrary Reynolds number, and we will obtain a solution that is valid only in the creeping flow limit.

The system is most conveniently described in cylindrical coordinates,

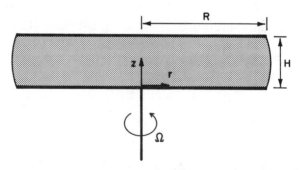

Figure 12-1. Schematic of flow between rotating disks.

with the origin at the center of the bottom plate. We will assume that neither the velocity nor the pressure is a function of angular position, so $\partial/\partial\theta = 0$. The continuity equation for an incompressible fluid and the creeping flow approximation to the Navier–Stokes equations are then obtained from Tables 7-1 and 7-7, respectively, as

$$\frac{1}{r}\frac{\partial}{\partial r}(rv_r) + \frac{\partial v_z}{\partial z} = 0 \tag{12.4}$$

$$r \text{ component:} \quad 0 = -\frac{\partial \mathcal{P}}{\partial r} + \eta\left[\frac{\partial}{\partial r}\left(\frac{1}{r}\frac{\partial}{\partial r}rv_r\right) + \frac{\partial^2 v_r}{\partial z^2}\right] \tag{12.5a}$$

$$\theta \text{ component:} \quad 0 = \eta\left[\frac{\partial}{\partial r}\left(\frac{1}{r}\frac{\partial}{\partial r}rv_\theta\right) + \frac{\partial^2 v_\theta}{\partial z^2}\right] \tag{12.5b}$$

$$z \text{ component:} \quad 0 = -\frac{\partial \mathcal{P}}{\partial z} + \eta\left[\frac{1}{r}\frac{\partial}{\partial r}\left(r\frac{\partial v_z}{\partial r}\right) + \frac{\partial^2 v_z}{\partial z^2}\right] \tag{12.5c}$$

The density was set equal to zero in Table 7-7 in order to obtain Eqs. (12.5) from the Navier–Stokes equations.

If we take the bottom plate as moving and the top plate as stationary, then the velocity at any radial position on the bottom plate is tangential, with a value $r\Omega$. Thus, the no-slip boundary condition requires that $\mathbf{v}$ vanish on the upper plate and that v_r and v_z vanish on the lower plate, with v_θ equal to $r\Omega$:

$$z = 0: \quad v_r = v_z = 0 \qquad v_\theta = r\Omega \tag{12.6a}$$

$$z = H: \quad v_r = v_z = v_\theta = 0 \tag{12.6b}$$

We have not written any boundary conditions in the r direction. We require finiteness at $r = 0$. The assumption $H/R \ll 1$ allows us to neglect the small region near the outer edge of the disks, and a boundary condition at $r = R$ is not required. This procedure is analogous to neglecting the small entry and exit regions for fully developed flow in a long pipe.

Equation (12.5b) for v_θ is uncoupled from the other equations and can be solved directly. It provides a nice demonstration of the way in which the boundary conditions can be used to deduce the form of the solution. Note that v_θ is proportional to r over the entire lower plate. The simplest way to ensure satisfaction of this boundary condition is to look for a solution that is *always* proportional to r:

$$v_\theta = rf(z) \tag{12.7}$$

$f(z)$ is an unknown function of z. It follows from Eqs. (12.6a) and (12.6b) that $f(z)$ must satisfy boundary conditions

$$f(0) = \Omega \qquad f(H) = 0 \tag{12.8}$$

Substitution of Eq. (12.7) into the θ component of the creeping flow equation, Eq. (12.5b), leads to an equation for $f(z)$,

$$\frac{d^2f}{dz^2} = 0 \qquad (12.9)$$

Thus, $f(z)$ must be a linear function of z. The boundary conditions (12.8) are satisfied by the linear function

$$f(z) = \Omega\left(1 - \frac{z}{H}\right) \qquad (12.10)$$

The velocity then follows from Eq. (12.7) as

$$v_\theta = r\Omega\left(1 - \frac{z}{H}\right) \qquad (12.11)$$

Equations (12.4), (12.5a), and (12.5c) for v_r, v_z, and $\mathcal{P}$ are clearly satisfied by

$$v_r = v_z = 0 \qquad \mathcal{P} = \text{constant} \qquad (12.12)$$

This means that all fluid particles move in circular paths in a plane. The constancy of the pressure in this rotating system may seem surprising at first, since the θ component of velocity increases with distance from the origin. There are no centrifugal terms to be balanced by a radial pressure gradient, however; the centrifugal terms are inertial (ρV^2) terms, and these are neglected in the creeping flow approximation.

To find the torque we must first evaluate the shear stress at the lower plate. From Table 7-6 and Eqs. (12.11) and (12.12), we have

$$\tau_{z\theta} = \eta\left(\frac{\partial v_\theta}{\partial z} + \frac{\partial v_z}{\partial \theta}\right) = -\frac{\eta r\Omega}{H} \qquad (12.13)$$

(Note that $\tau_{z\theta}$ is independent of z for this flow.) A differential segment of the lower disk is shown in Fig. 12-2. The differential force on the sector of area $r\,dr\,d\theta$ is

$$dF = \tau_{z\theta}r\,dr\,d\theta = -\frac{\eta\Omega}{H}r^2\,dr\,d\theta \qquad (12.14)$$

The force is directed tangentially. The differential torque is rdF:

$$dG = rdF = -\frac{\eta\Omega}{H}r^3\,dr\,d\theta \qquad (12.15)$$

The direction of the torque vector is normal to the plates and is the same everywhere. Thus, the differential torques can be summed to give the total torque:

$$G = \int dG = -\frac{\eta\Omega}{H}\int_0^R\int_0^{2\pi}r^3\,dr\,d\theta = -\frac{\pi\eta\Omega R^4}{2H} \qquad (12.16)$$

R is the outer radius of the disks. The torque required to turn the plate is $-G$.

Equation (12.16) is the basic equation for the parallel-disk viscometer. Torque is a straightforward quantity to measure. The viscosity of a Newtonian

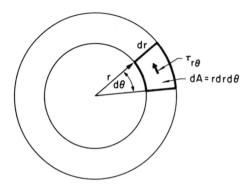

Figure 12-2. Schematic of a differential segment of the bottom disk for calculation of the torque.

fluid is obtained from the slope of a plot of torque versus Ω/H for various values of Ω and/or H.

It is still necessary to establish that the conditions for creeping flow have been satisfied. The characteristic length is clearly the plate spacing, H. The characteristic velocity is less obvious, but it cannot exceed the tip speed, $R\Omega$. Thus, the Reynolds number can be written

$$\text{Re} = \frac{LV\rho}{\eta} = \frac{HR\Omega\rho}{\eta} = \frac{H^2\Omega\rho}{\eta}\frac{R}{H} \qquad (12.17)$$

The condition $\text{Re} \ll 1$ could be a stringent one, because R/H will be quite large.

This estimate of the range of validity of the creeping flow solution is, in fact, too restrictive, and a better estimate can be obtained from direct examination of the Navier–Stokes equations. With the assumption that v_θ is the dominant velocity component, it then follows from inspection of the left hand side of the Navier–Stokes equations in Table 7-7 that the dominant term is $\rho v_\theta^2/r$. Thus, we can write

$$\text{inertial terms} \sim \frac{\rho v_\theta^2}{r} \sim \rho r\Omega^2 \qquad (12.18a)$$

The estimate of the viscous terms is not as straightforward. The dominant term is clearly $\eta\,\partial^2 v_\theta/\partial z^2$, but this term vanishes identically in the creeping flow approximation. From Eq. (12.13) it follows that $\eta\,\partial v_\theta/\partial z$ is of order $\eta r\Omega/H$. Even if the creeping flow approximation is slightly in error, we expect changes in $\eta\,\partial v_\theta/\partial z$ over the spacing to be bounded by an amount of the magnitude of the stress itself, so the derivative term will be bounded by a term of order $\eta r\Omega/H$ divided by the spacing, H. That is,

$$\text{viscous terms} \sim \eta\frac{\partial^2 v_\theta}{\partial z^2} \sim \frac{\eta r\Omega}{H^2} \qquad (12.18b)$$

We can therefore write

$$\frac{\text{inertial terms}}{\text{viscous terms}} \sim \frac{pr\Omega^2}{\eta r\Omega/H^2} = \frac{H^2\Omega\rho}{\eta} \ll 1 \qquad (12.19)$$

This condition is easily satisfied in practice for viscometers. We require separately that $R/H \gg 1$, for we have neglected any deviations from the circular flow pattern in the neighborhood of the liquid–air interface at the edge.

12.3 FLOW AROUND A SPHERE

12.3.1 Problem Formulation

We studied flow around a submerged sphere in Chapter 4, where we found that for Re < 1 the drag coefficient varies inversely with Reynolds number [Eq. (4.3)]. This was interpreted as a regime of negligible inertial forces. We will show here that this result, known as *Stokes' law*, can be derived from the creeping flow equations.

The flow is shown schematically in Fig. 12-3. A sphere of diameter D is submerged in an infinite body of fluid. The fluid moves past the sphere uniformly with a velocity U; that is, an observer situated at any point on the surface of the sphere will see a uniform flow with velocity U at a large distance from the sphere, although the velocity near the sphere will differ from the constant value. (We have changed nomenclature here from that used in Chapter 4. The subscript p to denote particle has not been retained, and the relative velocity between the sphere and the surrounding fluid has been changed from V_p to U. U is commonly used in the fluid mechanics literature to denote a velocity far from, and unaffected by, a solid surface.)

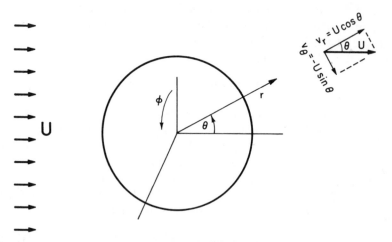

Figure 12-3. Schematic of uniform flow past a sphere.

The characteristic velocity for this flow is clearly the relative uniform velocity, U, and the characteristic length is the sphere diameter, D. These choices are unambiguous, because U and D are the only velocity and length specified in the problem. The Reynolds number is therefore as defined in Eq. (4.1a),

$$\text{Re} = \frac{DU\rho}{\eta} \qquad (12.20)$$

The condition for applicability of the creeping flow approximation is, of course, $\text{Re} \ll 1$.

The flow is best described in a spherical coordinate system, as shown in Fig. 12-3. We will assume ϕ-direction symmetry, $\partial/\partial\phi = 0$, and $v_\phi = 0$. Then the continuity equation is

$$\frac{1}{r^2}\frac{\partial}{\partial r}(r^2 v_r) + \frac{1}{r \sin\theta}\frac{\partial}{\partial\theta}(\sin\theta\, v_\theta) = 0 \qquad (12.21)$$

The r and θ components of the creeping flow approximation to the Navier–Stokes equations are obtained from Table 7-10 by setting $\rho = 0$, as follows:

$$0 = -\frac{\partial\mathcal{P}}{\partial r} + \eta\left[\frac{1}{r^2}\frac{\partial}{\partial r}\left(r^2\frac{\partial v_r}{\partial r}\right) + \frac{1}{r^2 \sin\theta}\frac{\partial}{\partial\theta}\left(\sin\theta\frac{\partial v_r}{\partial\theta}\right)\right.$$
$$\left. -\frac{2v_r}{r^2} - \frac{2}{r^2}\frac{\partial v_\theta}{\partial\theta} - \frac{2}{r^2}v_\theta \cot\theta\right] \qquad (12.22a)$$

$$0 = -\frac{1}{r}\frac{\partial\mathcal{P}}{\partial\theta} + \eta\left[\frac{1}{r^2}\frac{\partial}{\partial r}\left(r^2\frac{\partial v_\theta}{\partial r}\right) + \frac{1}{r^2 \sin\theta}\frac{\partial}{\partial\theta}\right.$$
$$\left.\cdot\left(\sin\theta\frac{\partial v_\theta}{\partial\theta}\right) + \frac{2}{r^2}\frac{\partial v_r}{\partial\theta} - \frac{v_\theta}{r^2 \sin^2\theta}\right] \qquad (12.22b)$$

The no-slip boundary condition at the sphere surface $r = R$ requires

$$r = R: \quad v_r = v_\theta = 0 \qquad (12.23a)$$

The boundary condition far from the sphere is obtained from the requirement that the flow approach the uniform velocity, U. When the velocity vector at large r is resolved into r and θ components, as shown in Fig. 12-3, we obtain the condition

$$r \longrightarrow \infty: \quad v_r = U \cos\theta \qquad v_\theta = -U \sin\theta \qquad (12.23b)$$

12.3.2 Solution of Equations

The solution of Eqs. (12.21), (12.22), and (12.23) is suggested by the form of the boundary condition at infinity. The boundary conditions usually provide insight into the structure of the solution. In this case we see that the θ dependence far from the sphere is given by a cosine and sine relation. It seems reasonable to look for a solution that retains this simple angle dependence over the entire field. Thus, we are motivated to seek a solution in the form

$$v_r = A(r) \cos \theta \qquad (12.24a)$$

$$v_\theta = B(r) \sin \theta \qquad (12.24b)$$

$A(r)$ and $B(r)$ are unknown functions of r, but comparison with Eqs. (12.23) shows that A and B must satisfy the following conditions at $r = R$ and $r \longrightarrow \infty$:

$$r = R: \quad A(R) = B(R) = 0 \qquad (12.25a)$$

$$r \longrightarrow \infty: \quad A(\infty) = U \qquad B(\infty) = -U \qquad (12.25b)$$

The remainder of this section is given over to the use of Eqs. (12.21) and (12.22) to find the functions $A(r)$ and $B(r)$. Readers who are not interested in the details may wish to skip to the solution, Eqs. (12.34) and (12.35).

Substitution of Eqs. (12.24) into the continuity equation, Eq. (12.21), yields

$$\frac{\cos \theta}{r^2} \left[2r A(r) + r^2 \frac{dA(r)}{dr} \right] + \frac{2 \cos \theta}{r} B(r) = 0 \qquad (12.26a)$$

Dividing by $2 \cos \theta / r$ eliminates the θ dependence and gives an equation involving only A and B,

$$A(r) + \frac{r}{2} \frac{dA}{dr} + B(r) = 0 \qquad (12.26b)$$

This equation shows that the assumed form of solution is consistent with continuity, and furthermore continuity reduces the number of unknown variables by showing how A and B must be related.

We now turn to the r component of the creeping flow equations, Eq. (12.22a). When the form of the solution given by Eqs. (12.24) is substituted into this component equation, we obtain an equation that can be written

$$\frac{\partial \mathcal{P}}{\partial r} = \eta \, (\text{function of } r) \cos \theta \qquad (12.27)$$

This relation requires that $\mathcal{P}$ be of the form

$$\mathcal{P} = \mathcal{P}_0 + \eta \Pi(r) \cos \theta \qquad (12.28)$$

$\Pi(r)$ is an unknown function of r. $\mathcal{P}_0$ must be independent of r, although it could depend on θ. Substitution of Eq. (12.28) into Eq. (12.22a) then gives a relation among Π, A, and B:

$$0 = -\frac{d\Pi(r)}{dr} + \frac{1}{r^2} \frac{d}{dr} r^2 \frac{dA(r)}{dr} - \frac{4A(r)}{r^2} - \frac{4B(r)}{r^2} \qquad (12.29)$$

We must now satisfy the θ component of the creeping flow equations, Eq. (12.22b). Note that $\mathcal{P}$, v_r, and v_θ are now expressed in terms of the functions $\Pi(r)$, $A(r)$, and $B(r)$, respectively, with specified θ dependence, so there are no remaining degrees of freedom. Thus, Eq. (12.22b) provides an important consistency check on the assumptions that we have made about the solution. When Eqs. (12.24) and (12.28) are substituted into Eq. (12.22b) we obtain, after some grouping of terms and multiplication by $r/\sin\theta$,

$$0 = \Pi(r) + \frac{1}{r}\frac{d}{dr}r^2\frac{dB(r)}{dr} - \frac{2B(r)}{r} - \frac{2A(r)}{r} \tag{12.30}$$

Thus, the assumed θ dependence for v_r and v_θ does reduce the partial differential equations to a set of linear *ordinary* differential equations for the three functions of r, $\Pi(r)$, $A(r)$, and $B(r)$. It should be noted that it is also established in the derivation of Eq. (12.30) that $\mathcal{P}_0$ must be a true constant and not a function of θ.

The three linear ordinary differential equations, Eqs. (12.24), (12.29), and (12.30), are most easily solved by eliminating Π and B to obtain a single equation for A. Equation (12.30) is differentiated with respect to r to obtain an equation for $d\Pi/dr$, which is then substituted into Eq. (12.29). Equation (12.26b) is then used to eliminate B. The resulting fourth-order equation for $A(r)$, after some simplification, is

$$0 = r\frac{dA}{dr} - r^2\frac{d^2A}{dr^2} - r^3\frac{d^3A}{dr^3} - \frac{r^4}{8}\frac{d^4A}{dr^4} \tag{12.31}$$

Two of the boundary conditions for this fourth-order equation are given directly in Eqs. (12.25), and the other two are obtained by evaluating Eq. (12.26b) at $r = R$ and $r \to \infty$ and using the conditions for B in Eqs. (12.25). The resulting four boundary conditions are thus

$$r = R: \quad A(R) = \frac{dA}{dr} = 0 \tag{12.32a}$$

$$r \longrightarrow \infty: \quad A(\infty) = U \quad \frac{dA}{dr} = 0 \tag{12.32b}$$

Equation (12.31) is a special form of linear homogeneous equation that is known as *Euler's equation.** This equation has solutions of the form r^n. Substituting $A = r^n$ into Eq. (12.31) gives

$$r^n(-n^4 - 2n^3 + 5n^2 + 6n) = 0$$

n must be a root of the polynomial in parentheses. The roots are $n = 0, -1, -3, +2$, so the general solution to Eq. (12.31) is

$$A(r) = C_1 + C_2 r^{-1} + C_3 r^{-3} + C_4 r^2$$

The conditions (12.32b) at $r \to \infty$ require that $C_4 = 0$, $C_1 = +U$, while the conditions (12.32a) at $r = R$ become

$$A(R) = U + \frac{C_2}{R} + \frac{C_3}{R^3} = 0$$

$$\frac{dA(R)}{dr} = -\frac{C_2}{R^2} - \frac{3C_3}{R^4} = 0$$

---

*There are many equations named after Euler. Equation (12.31) should not be confused with the flow equations for an inviscid fluid, which are also known as Euler's equation.

These latter two equations have a solution $C_2 = -3UR/2$, $C_3 = +UR^3/2$. The complete solution to Eqs. (12.31) and (12.32) is then

$$A(r) = U\left[1 - \frac{3}{2}\frac{R}{r} + \frac{1}{2}\left(\frac{R}{r}\right)^3\right] \tag{12.33a}$$

$B(r)$ and $\Pi(r)$ then follow from Eqs. (12.26b) and (12.30) as

$$B(r) = -U\left[1 - \frac{3}{4}\frac{R}{r} - \frac{1}{4}\left(\frac{R}{r}\right)^3\right] \tag{12.33b}$$

$$\Pi(r) = \frac{-3U}{2R}\left(\frac{R}{r}\right)^2 \tag{12.33c}$$

Finally, the r and θ components of the velocity and the equivalent pressure are obtained from Eqs. (12.24) and (12.28):

$$v_r = U\left[1 - \frac{3}{2}\frac{R}{r} + \frac{1}{2}\left(\frac{R}{r}\right)^3\right]\cos\theta \tag{12.34a}$$

$$v_\theta = -U\left[1 - \frac{3}{4}\frac{R}{r} - \frac{1}{4}\left(\frac{R}{r}\right)^3\right]\sin\theta \tag{12.34b}$$

$$\mathcal{P} = \mathcal{P}_0 - \frac{3\eta U}{2R}\left(\frac{R}{r}\right)^2\cos\theta \tag{12.35}$$

$\mathcal{P}_0$ is the uniform pressure in the fluid far from the sphere.

12.3.3 Form and Friction Drag

We are now able to calculate the force exerted by the moving fluid on the sphere. In doing this calculation it is helpful to make use of the relation $\mathcal{P} = p + \rho g h$, where p is the pressure and h is the height above an arbitrary datum. We need to work with the pressure and gravity terms separately in order to distinguish the buoyancy term. We take the flow direction to be at an angle α to the direction of gravity, as shown in Fig. 12-4, with $h = 0$ at the center of the sphere. Then $h = r\cos(\theta - \alpha) = r(\cos\alpha\cos\theta + \sin\alpha\sin\theta)$

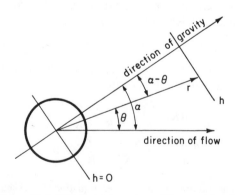

Figure 12-4. Relative orientation of flow and gravity.

and

$$p = \mathcal{P}_0 - \rho g r(\cos \alpha \cos \theta + \sin \alpha \sin \theta) - \frac{3\eta U}{2R}\left(\frac{R}{r}\right)^2 \cos \theta \quad (12.36)$$

We wish to calculate the net force on the sphere in the direction of flow. This is done by multiplying the stress at the surface by the differential surface area, taking the component in the direction of flow, and integrating over the entire surface of the sphere. When the velocity, Eqs. (12.34), is substituted into the stress equations in Table 7-9 we find that the only nonzero term is $\tau_{r\theta}$,

$$\tau_{r\theta} = \eta\left[r\frac{\partial}{\partial r}\left(\frac{v_\theta}{r}\right) + \frac{1}{r}\frac{\partial v_r}{\partial \theta}\right]$$

At $r = R$, $\tau_{r\theta}$ has the value

$$r = R: \quad \tau_{r\theta} = -\frac{3\eta U}{2R}\sin \theta \quad (12.37)$$

The resolution of the stresses is shown in Fig. 12-5. The component of $\tau_{r\theta}$ in the direction of flow is $-\tau_{r\theta}\sin \theta$, and the component of pressure in the flow direction is $-p\cos \theta$. The differential surface area in spherical coor-

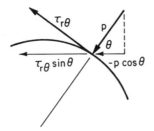

Figure 12-5. Schematic of the stresses acting on the surface of a sphere.

dinates is $R^2 \sin \theta\, d\theta\, d\phi$ for $0 \leq \theta < \pi$, and $-R^2 \sin \theta\, d\theta\, d\phi$ for $\pi \leq \theta < 2\pi$. Because of the symmetry about $\theta = 0$, we need only consider the half-plane $0 \leq \theta < \pi$, and we obtain the total force by multiplying the result for the half-plane by two. The net force in the flow direction is thus

$$F = 2\int_{\phi=0}^{\phi=2\pi}\int_{\theta=0}^{\theta=\pi}(-p\cos \theta - \tau_{r\theta}\sin \theta)R^2 \sin \theta\, d\theta\, d\phi$$

$$= \int_{\phi=0}^{\phi=2\pi}\int_{\theta=0}^{\theta=\pi}\left[-\mathcal{P}_0 + \rho g R(\cos \alpha \cos \theta + \sin \alpha \sin \theta) + \frac{3\eta U}{2R}\cos \theta\right]$$

$$\cdot R^2 \sin \theta \cos \theta\, d\theta\, d\phi + \int_{\phi=0}^{\phi=2\pi}\int_{\theta=0}^{\theta=\pi}\frac{3\eta U}{2R}\sin^2 \theta\, R^2 \sin \theta\, d\theta\, d\phi$$

$$= \frac{4\pi}{3}\rho g R^3 \cos \alpha + 2\pi\eta R U + 4\pi\eta R U \quad (12.38)$$

Note that the result is independent of the pressure at infinity, $\mathcal{P}_0$.

The first term on the right of Eq. (12.38) results from the gravitational force and will be recognized as the *buoyant force*; compare Eq. (4.28). This force will be exerted even in a stationary fluid. The second term $(2\pi\eta RU)$ arises from the pressure variation induced by the flow and is often called *form drag*. The final term $(4\pi\eta RU)$, which arises from the shear stress at the surface, is often called *friction drag* or *skin friction*. The total drag force induced by flow is

$$F_D = 6\pi\eta RU = 3\pi\eta DU \qquad (12.39)$$

This is *Stokes' law*, Eq. (4.4), which we have seen in Fig. 4-1 to be a good representation of experimental data for Re < 1.

12.3.4 Consistency of Solution

Finally, it is of some interest to estimate the extent of applicability of the solution in order to check for consistency of the approximation. The inertial terms are of order $\rho v_r \partial v_r / \partial r$. With Eq. (12.34a) we can write

$$\text{inertial:} \quad \rho v_r \frac{\partial v_r}{\partial r} = \rho \left\{ \left[1 - \frac{3}{2}\frac{R}{r} + \frac{1}{2}\left(\frac{R}{r}\right)^3 \right] \cos\theta \right\}$$
$$\cdot \left\{ U\left[\frac{3}{2}\frac{R}{r^2} - \frac{3}{2}\frac{R^3}{r^4} \right] \cos\theta \right\} \qquad (12.40)$$
$$\sim \frac{3\rho U^2 R}{2r^2}$$

Here we have retained only the dominant terms, noting that $R \leq r$ and $\cos\theta$ is of order unity. Similarly, the magnitude of the viscous terms is estimated from

$$\text{viscous:} \quad \eta\frac{1}{r^2}\frac{\partial}{\partial r}\left(r^2\frac{\partial v_r}{\partial r} \right) \sim \frac{3\eta UR^3}{r^5} \qquad (12.41)$$

Thus, the ratio of inertial to viscous terms is

$$\frac{\text{inertial}}{\text{viscous}} \sim \frac{3\rho U^2 R/2r^2}{3\eta UR^3/r^5} = \frac{1}{2}\frac{RU\rho}{\eta}\left(\frac{r}{R}\right)^3 = \frac{1}{4}\,\text{Re}\left(\frac{r}{R}\right)^3 \qquad (12.42)$$

The ratio of inertial to viscous stresses is small near the sphere as long as Re is small. Sufficiently far from the sphere, however, where $r \gg R$, the inertial terms become comparable to the viscous terms and the creeping flow assumption breaks down. The failure of the assumptions far from the sphere is not important physically, because for $r \gg R$ both the inertial and viscous terms are negligible and the velocity and pressure are close to the uniform values that they would have in the absence of the sphere. Thus, we only need a solution that is valid close to the sphere, and the creeping flow approximation provides this. The inconsistency of the creeping flow solution far from the sphere is of mathematical importance, however, for it restricts the procedures that can be used to construct "corrections" to the creeping flow solution for flows when Re is small but cannot be taken to be zero; specifically, the approach outlined in Sec. 17.2 cannot be used.

12.4 SQUEEZE FILM

The analysis of a squeeze film provides a rather nice example of solution of the creeping flow equations. The squeeze film process is shown schematically in Fig. 12-6. Fluid is contained between two disks of radius R. The upper disk moves toward the lower with a speed V. As the spacing H between the disks decreases, fluid is squeezed out at the edges. The problem is to compute the relationship between the imposed force, the speed and spacing of the plates, and time. Configurations of this type occur in lubrication and in the molding of objects from molten polymers, although the surfaces might be of a more complex shape in practical applications.

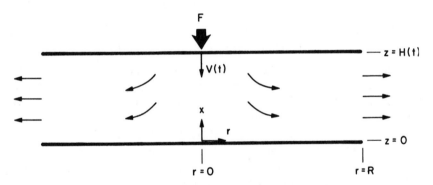

Figure 12-6. Schematic of a squeeze film.

The relevant length scale in the problem is a characteristic spacing between the plates; the initial spacing, H_0, is the maximum value and hence gives the most conservative estimate. The characteristic velocity is the maximum velocity of the upper plate, V_m. The Reynolds number is then

$$\mathrm{Re} = \frac{H_0 V_m \rho}{\eta} \tag{12.43}$$

The only characteristic time in the problem is the time required for the plates to come together, which is of order H_0/V_m. Thus, the characteristic time is constructed from the characteristic length and velocity, and the dimensionless Navier–Stokes equations are as given by Eq. (12.1). In the limit Re → 0 we therefore obtain Eq. (12.3b). It is important to note that the time derivative term is neglected, even though we are dealing with a transient flow. *The creeping flow limit for this time-varying flow introduces a pseudosteady-state assumption in a natural way,* and the problem can be analyzed as though the flow were steady at each instant in time.

For the problem under consideration here there is no rotation, so we assume that $v_\theta = 0$, $\partial/\partial\theta = 0$, and write the creeping flow equations in cylindrical coordinates as

$$0 = -\frac{\partial \mathcal{P}}{\partial r} + \eta\left[\frac{\partial}{\partial r}\frac{1}{r}\frac{\partial}{\partial r}(rv_r) + \frac{\partial^2 v_r}{\partial z^2}\right] \qquad (12.44a)$$

$$0 = -\frac{\partial \mathcal{P}}{\partial z} + \eta\left[\frac{1}{r}\frac{\partial}{\partial r}\left(r\frac{\partial v_z}{\partial r}\right) + \frac{\partial^2 v_z}{\partial z^2}\right] \qquad (12.44b)$$

The θ-component equation is simply $\partial\mathcal{P}/\partial\theta = 0$. The continuity equation is unchanged by the creeping flow approximation:

$$\frac{1}{r}\frac{\partial}{\partial r}(rv_r) + \frac{\partial v_z}{\partial z} = 0 \qquad (12.45)$$

No-slip boundary conditions at the bottom plate are

$$z = 0: \quad v_z = v_r = 0 \qquad (12.46a)$$

The fluid adheres to the top plate, so

$$z = H(t): \quad v_z = -V(t) \qquad v_r = 0 \qquad (12.46b)$$

Note that the location of the top plate may, in fact, be unknown. The negative sign is needed with $V(t)$ because the plate moves down in the figure, while the positive z direction is up. V may be constant or it may vary with time, depending on the problem formulation. We shall require a boundary condition in the r direction as well, but this is best discussed later.

Our usual practice in obtaining a solution to the flow problem is to examine the flow field and attempt to make a physical statement regarding the velocity. The boundary conditions are suggestive regarding v_z. At the upper plate the fluid moves downward uniformly at all radial positions, so v_z at the upper plate is independent of r. Similarly, at the lower plate v_z is also independent of r (it is, in fact, identically zero). It seems likely that the downward portion of the motion is always uniform, and we seek a solution in which v_z is independent of r:

$$v_z = \phi(z) \qquad (12.47)$$

(v_z must depend on t as well, but we can suppress this dependence for now since t does not enter the flow equations at all.) From the continuity equation, we then have

$$\frac{1}{r}\frac{\partial}{\partial r}(rv_r) = -\frac{d\phi(z)}{dz}$$

or, after integrating,

$$v_r = -\frac{r}{2}\frac{d\phi}{dz} + \frac{\text{constant}}{r} \qquad (12.48)$$

The constant of integration (which could be a function of z) must be zero, since v_r is surely finite at $r = 0$ for all z. Thus, it follows from the hypothesis Eq. (12.47) that

$$v_r = -\frac{r}{2}\frac{d\phi}{dz} \tag{12.49}$$

We now turn to the creeping flow equations, (12.44a) and (12.44b). Upon substitution of Eqs. (12.47) and (12.49) for the velocity components, these equations simplify to

$$0 = -\frac{\partial \mathcal{P}}{\partial r} - \frac{1}{2}\eta r \frac{d^3\phi}{dz^3} \tag{12.50a}$$

$$0 = -\frac{\partial \mathcal{P}}{\partial z} + \eta \frac{d^2\phi}{dz^2} \tag{12.50b}$$

These two equations are most easily solved for ϕ and $\mathcal{P}$ by first eliminating the equivalent pressure. We differentiate Eq. (12.50a) with respect to z and Eq. (12.50b) with respect to r to obtain

$$0 = -\frac{\partial^2 \mathcal{P}}{\partial r\,\partial z} - \frac{1}{2}\eta r \frac{d^4\phi}{dz^4} \tag{12.51a}$$

$$0 = -\frac{\partial^2 \mathcal{P}}{\partial r\,\partial z} + 0 \tag{12.51b}$$

The pressure term in Eq. (12.51a) is therefore zero, and it follows that $\phi(z)$ satisfies the equation

$$\frac{d^4\phi}{dz^4} = 0 \tag{12.52}$$

Following four integrations, we obtain

$$\phi(z) = a + bz + cz^2 + dz^3 \tag{12.53}$$

The boundary conditions are expressed in terms of $\phi(z)$ by substituting Eqs. (12.47) and (12.49) for the velocity components v_z and v_r into Eqs. (12.46):

$$\text{at } z = 0: \quad \phi = \frac{d\phi}{dz} = 0 \tag{12.54a}$$

$$\text{at } z = H(t): \quad \phi = -V(t) \qquad \frac{d\phi}{dz} = 0 \tag{12.54b}$$

Note that time, t, enters the solution through the time dependence of $H(t)$ and $V(t)$ in the boundary conditions. Evaluating a, b, c, and d then provides the details of the velocity field:

$$v_z = \phi(z) = -3V(t)\left[\frac{z}{H(t)}\right]^2\left[1 - \frac{2}{3}\frac{z}{H(t)}\right] \tag{12.55a}$$

$$v_r = -\frac{r}{2}\frac{d\phi}{dz} = \frac{3rzV(t)}{H^2(t)}\left[1 - \frac{z}{H(t)}\right] \tag{12.55b}$$

If either $H(t)$ or $V(t)$ is specified, this pair of equations completes the description; note that $V(t) = -dH/dt$.

The equivalent pressure is computed from Eqs. (12.50). From Eq. (12.50a), we have

$$\frac{\partial \mathcal{P}}{\partial r} = -\tfrac{1}{2}\eta r \frac{d^3\phi}{dz^3} = -\frac{6\eta Vr}{H^3} \tag{12.56a}$$

This is integrated once to give

$$\mathcal{P} = -\frac{3\eta Vr^2}{H^3} + \text{function of } z \tag{12.57a}$$

From Eq. (12.50b), we have

$$\frac{\partial \mathcal{P}}{\partial z} = \eta \frac{d^2\phi}{dz^2} \tag{12.56b}$$

which integrates once to

$$\mathcal{P} = \eta \frac{d\phi}{dz} + \text{function of } r = -\frac{6\eta Vz}{H^2}\left(1 - \frac{z}{H}\right) + \text{function of } r \tag{12.57b}$$

Comparison of Eqs. (12.57a) and (12.57b) establishes the functions of r and z to within a constant. The function of z in Eq. (12.57a) must equal the z-dependent term in Eq. (12.57b), except for an additive constant, and the function of r in Eq. (12.57b) must similarly equal the r-dependent term in Eq. (12.57a) to within an additive constant. We can therefore combine the two expressions for $\mathcal{P}$ and write

$$\mathcal{P} = \mathcal{P}_0 + \frac{3\eta V}{H}\left[2\frac{z}{H}\left(\frac{z}{H} - 1\right) - \frac{r^2}{H^2}\right] \tag{12.58}$$

The constant $\mathcal{P}_0$ must still be evaluated, since we will require $\mathcal{P}$ if we wish to compute the force. We have already used all the boundary conditions stated thus far, so we shall have to introduce an additional physical condition. This is done by considering the forces acting on the fluid as it exits from the space between the disks at $r = R$.

We take the datum for the gravitational term at the upper plate, $z = H$. At this elevation, then, $\mathcal{P} = p$. The fluid exiting at the edge of the disk is exposed to atmospheric pressure, so that a balance of normal stress at the surface requires that σ_{zz} equal atmospheric pressure; see the schematic in Fig. 12-7. There is no loss of generality in taking atmospheric pressure to be zero, so we may write

$$\text{at } z = H, r = R: \quad 0 = \sigma_{zz} = -p + 2\eta \frac{\partial v_z}{\partial z} \tag{12.59a}$$

From Eq. (12.55a) it follows that $\partial v_z/\partial z$ is zero at $z = H$. Substitution of Eq. (12.58) into Eq. (12.59a) then gives

$$\text{at } z = H, r = R: \quad 0 = \sigma_{zz} = -\mathcal{P}_0 + \frac{3\eta VR^2}{H^3} \tag{12.59b}$$

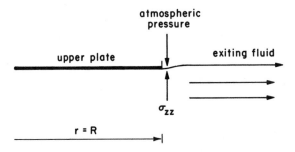

Figure 12-7. Schematic of flow at the outer edge of the upper plate of a squeeze film. It is assumed that the fluid exits parallel to the wall, without curvature, so that the z direction is normal to the free surface.

We can thus evaluate $\mathcal{P}_0$ and express the equivalent pressure as

$$\mathcal{P}(r, z) = \frac{3\eta V}{H}\left[2\frac{z}{H}\left(\frac{z}{H} - 1\right) + \frac{R^2 - r^2}{H^2}\right] \tag{12.60}$$

Note that this analysis assumes that there is no significant curvature of the fluid surface at the exit, so that surface tension effects may be neglected. It should also be noted that $\mathcal{P}_0$ can be evaluated by considering a normal stress balance at the free surface adjacent to the lower plate, including the gravitational term, and the identical result is obtained.

The parabolic distribution of pressure with radial position noted in Eq. (12.60) is quite important in practice. There is a considerable variation of stress over the surface of the plate, and the stress is most intense at the center. Thus, buckling might be a problem in the absence of adequate precautions.

The total force which must be imposed on the upper plate is the integral of the negative of σ_{zz}, evaluated at $z = H$; the negative is required because σ_{zz} is the stress exerted by the fluid in the positive z direction. At $z = H$ the term $2\eta\, \partial v_z/\partial z$ is zero, so $\sigma_{zz} = -\mathcal{P}$. Thus,

$$F = \int_0^R 2\pi r\mathcal{P}(r, H)\, dr = \int_0^R \frac{3\eta V}{H^3}(R^2 - r^2)2\pi r\, dr = \frac{3\pi\eta V R^4}{2H^3} \tag{12.61}$$

Let us suppose that the squeezing of the plates is carried out under constant force. We can then use Eq. (12.61) to find the variation of plate spacing with time. Noting that $V = -dH/dt$, we can rewrite Eq. (12.61) as

$$\frac{dH}{dt} = -\left(\frac{2F}{3\pi\eta R^4}\right)H^3 \tag{12.62}$$

This first-order equation has a solution

$$H(t) = \left(\frac{1}{H_0^2} + \frac{4Ft}{3\pi\eta R^4}\right)^{-1/2} \tag{12.63}$$

H_0 is the spacing at time zero. Equation (12.63) is known as the *Stefan equa-*

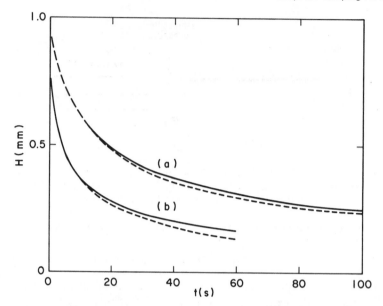

Figure 12-8. Squeeze film spacing as a function of time for two silicone fluids. The solid lines are the continuous experimental traces from a position transducer and the dashed lines are Eq. (12.63). (a) Dow Corning 200 fluid, $\eta = 13.9$ Pa · s, $F = 80$ N, $R = 63.5$ mm. (b) GE Viscasil, $\eta = 58$ Pa · s, $F = 77.8$ N, $R = 31.8$ mm. Data of S.J. Lee.

tion. Figure 12-8 shows a comparison between the Stefan equation and some data on the squeezing of two Newtonian fluids. It is evident that the equation agrees well with the data.

The range of applicability of the solution can be estimated from the Reynolds number. The characteristic spacing is the height, H, and the characteristic velocity is V. From Eq. (12.62), however, we can replace V with $2FH^3/3\pi\eta R^4$. Thus, we have

$$\mathrm{Re} = \frac{HV\rho}{\eta} = \frac{2\rho FH^4}{3\pi\eta^2 R^4} \ll 1 \qquad (12.64)$$

Alternatively, we can estimate the neglected terms in the full Navier–Stokes equations. In the z-component equation, for example, the retained terms are of order $\eta\partial^2 v_z/\partial z^2 \sim 6\eta V/H^2$. The term $\rho\partial v_z/\partial t$ is approximately $\rho \, dV/dt = \rho d^2H/dt^2$. Thus,

$$\frac{\rho\partial v_z/\partial t}{\eta\partial^2 v_z/\partial z^2} \approx \frac{\rho d^2H/dt^2}{6\eta V/H^2} = \frac{(6\rho FH^2 \, dH/dt)/3\pi\eta R^4}{(6\eta \, dH/dt)/H^2} = \frac{\rho FH^4}{3\pi\eta^2 R^4} \ll 1 \quad (12.65)$$

Here, we have used Eq. (12.62) to determine d^2H/dt^2. Similarly, the term $\rho v_z \, \partial v_z/\partial z$ is of order $6\rho V^2/H$. Hence,

$$\frac{\rho v_z \, \partial v_z/\partial z}{\eta \, \partial^2 v_z/\partial z^2} \approx \frac{6\rho(dH/dt)^2/H}{6\eta(dH/dt)/H^2} = \frac{2\rho FH^4}{3\pi\eta^2 R^4} \ll 1 \qquad (12.66)$$

There is consistency, then, between the various estimates of the range for which inertial terms can be neglected, and a conservative criterion is always

$$F \ll \frac{\eta^2}{\rho} \left(\frac{R}{H}\right)^4 \tag{12.67}$$

12.5 CONCLUDING REMARKS

The examples in this chapter have been selected to illustrate the scope of the creeping flow approximation and its application to both steady and time-dependent flows. The mathematical simplification, which leads to a set of linear equations, is obtained at the expense of restricting the scope of the problems that can be solved. This trade-off is frequently fruitful, for there are many low Reynolds number problems of physical interest, and the techniques for solving linear partial differential equations are well developed. It is always necessary in such a case to be sure to establish the consistency and range of applicability of the solution, however.

These three examples also illustrate the manner in which the form of the motion is deduced from the boundary conditions. Each problem involves velocities that are functions of two independent variables, but in each case the dependence on one independent variable could be established from the functional form at a boundary. This approach will not always provide the correct relation, but it is systematic and is often effective.

BIBLIOGRAPHICAL NOTES

Creeping flow is the subject of an exhaustive treatise,

HAPPEL, J., and BRENNER, H. B., *Low Reynolds Number Hydrodynamics*, Prentice-Hall, Inc., Englewood Cliffs, N.J., 1965.

See also sections in the texts on fluid mechanics cited at the end of Chapter 10.

The most complete treatments of the squeeze film, including data on non-Newtonian fluids and analyses for power-law and viscoelastic fluids, are in

BRINDLEY, G., DAVIES, J. M., and WALTERS, K., *J. Non-Newtonian Fluid Mech.*, **1**, 19 (1976).
LEIDER, P. J., and BIRD, R. B., *Ind. Eng. Chem. Fundamentals*, **13**, 336, 342 (1974).
TICHY, J. A., and WINER, W. O., *Trans. ASME.*, series F, **100**, 56 (1978).

PROBLEMS

12.1. Obtain the velocity profile and power requirement for creeping flow of an incompressible Newtonian liquid between converging infinite planes (Fig. 10.2). Define limits on the applicability of the solution.

12.2. An incompressible Newtonian liquid flows through a cone of half-angle α. Obtain the velocity distribution under creeping flow conditions. (Hint: Seek a radial solution, with $v_\theta = v_\phi = 0$. The problem reduces to solution of a linear ordinary differential equation.)

12.3. An incompressible Newtonian fluid flows radially outward between two disks (Fig. 12P3). Ignoring the flow rearrangement near the entrance, obtain the velocity profile and flow rate in terms of the pressure drop from R_1 to R_2. Ignore inertia, and define conditions under which the solution is valid.

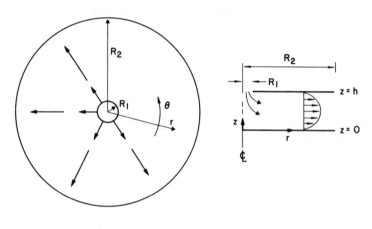

Top View   Side View

12.4. Use your solution to Problem 12.3 to estimate the time required to fill a disk mold, Fig. 12P4. The mold cavity is vented, so the pressure at the fluid-gas interface is always atmospheric. (Hint: Note that the creeping flow equations apply to quasi-steady state flows with negligible inertia. Thus, replace R_2 by R in your equation for the flow rate and obtain an independent equation for R from a macroscopic mass balance.)

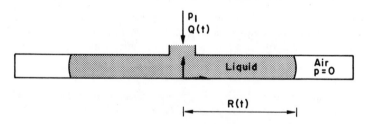

12.5. A simplified schematic of a compression molding process is shown in Fig. 12P5. A cylindrical charge of a very viscous incompressible Newtonian liquid is placed between two disks, and the disks are then brought together. Obtain an equation relating the spacing between the disks to the imposed force.

(Hint: This problem is nearly identical to the squeeze film, but fluid does not leave the edge of the disk. Thus, the boundary condition Eq. (12.59a) is not applicable, and a different boundary condition is required. You may assume that the radius of the fluid cylinder is always much larger than the disk spacing, in which case it can be shown that the curvature of the free surface can be neglected.)

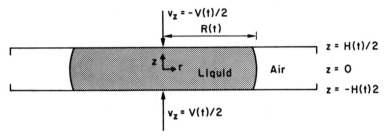

12.6. A cone-and-plate viscometer is shown schematically in Fig. 12P6. Assuming that inertia can be neglected, obtain the velocity distribution and torque. (Hint: Assume a purely tangential flow, with $v_r = v_\theta = 0$. Deduce the r-dependence of v_ϕ from the boundary conditions. The problem will reduce to a linear ordinary differential equation with variable coefficients.)

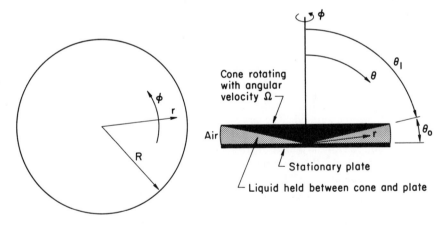

Top View   Side View

12.7. A column of liquid is drawn down in cross-section by imposition of a force, as shown in Fig. 12P7. The imposed force is sufficiently large that the gravitational body force may be neglected. The cross-sectional area of the cylinder of liquid is the same at all axial positions at any time. Inertia may be neglected.
a) Show that the assumption of "uniform drawdown" requires that v_z be

  independent of r and θ, and obtain an expression for v_r in terms of v_z.
  b) Show that p is independent of r and equals $-\eta \, dv_z/dz$. (Hint: Consider σ_{rr} at the outer radius of the liquid column.)
  c) Show that $v_z = \Gamma z$, where Γ may be a function of t.
  d) For a constant imposed force, show that $L/L_0 = (1 - Ft/3A_0\eta)^{-1}$, where L_0 and A_0 are the initial length and area of the liquid column.

(This analysis is a first step in describing the process of continuous filament drawing to form fibers.)

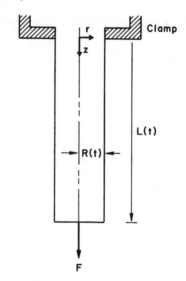

12.8. A small bubble is formed and grows in a large container of an isothermal Newtonian liquid. The pressure in the bubble (relative to the quiescent liquid far from the bubble) can be related thermodynamically to the superheat required for the bubble to grow. Obtain the relation between the bubble radius at any time, $R(t)$, and the bubble pressure. Inertia of the fluid being pushed away by the growing bubble may be neglected.

12.9. In commercial compression molding a cold fluid may be placed between hot plates, causing a viscosity gradient. As a first approximation to this problem, repeat the analysis for the squeeze film, Sec. 12.4, for the case in which the viscosity is a known function of position between the plates: $\eta = \eta(z)$. (v_r can be expressed explicitly in terms of an integral of a function of $\eta(z)$ for the case in which η is symmetric about the center plane. The problem is slightly easier if the origin is taken at the center plane, with each disk moving towards the center with velocity $V/2$.)

The Lubrication 13
Approximation

13.1 INTRODUCTION

The lubrication approximation is a simplification that applies to flow between "nearly parallel" surfaces. This approximation was first used by Reynolds in 1886 in a study of lubrication, hence the name. The common name of the procedure is unfortunate, however, for it implies an unduly restrictive range of applications; the lubrication approximation is fundamental to the study of polymer processing, where it forms the basis for the analysis of extrusion, coating, calendering, and molding operations. It is, therefore, one of the most important methods of approximate solution of the Navier–Stokes equations.

We shall first introduce the lubrication approximation in an intuitive manner, and this introduction may suffice for some readers. We shall then develop the formalism through a more careful ordering analysis and consider further applications.

13.2 INTUITIVE DEVELOPMENT

Consider the pressure-driven flow of an incompressible Newtonian fluid between converging planes with a very small half-angle α, shown in Fig. 13-1. If the walls were perfectly parallel ($\alpha = 0$) this would simply be the problem of plane Poiseuille flow studied in Sec. 8.2, with a parabolic velocity profile

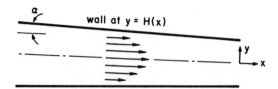

Figure 13-1. Schematic of pressure-driven flow in a plane channel with slowly converging walls.

given by Eq. (8.12):

$$v_x = \frac{H^2}{8\eta}\left(-\frac{d\mathcal{P}}{dx}\right)\left[1 - \left(\frac{2y}{H}\right)^2\right]$$ (13.1)

(We have retained $d\mathcal{P}/dx$ in place of $\Delta\mathcal{P}/L$ for reasons that will become obvious shortly.) Similarly, using Eq. (8.14) we may write the flow rate per unit width, q, as

$$q = \frac{H^3}{12\eta}\left(-\frac{d\mathcal{P}}{dx}\right)$$ (13.2)

If the angle α is small, we may reasonably expect the flow to be nearly the same as flow between parallel walls, so we expect Eqs. (13.1) and (13.2) to apply, *except that H is now a function of x.* The flow rate is a constant at all values of x, so it follows from Eq. (13.2) that $d\mathcal{P}/dx$ must vary as $H^{-3}(x)$ and cannot be a constant.

The power consumption depends on evaluation of σ_{xx}; compare Sec. 10.7. The viscous stress term $2\eta \, \partial v_x/\partial x$ is zero for flow between parallel walls, so the power consumption is simply equal to the product of flow rate and pressure drop. For flow between nearly parallel walls we may obtain $\partial v_x/\partial x$ by differentiation of Eq. (13.1); the result is proportional to dH/dx, which is equal to α for small α, and hence may be neglected. To within the small-angle approximation, therefore, the power consumption is again determined solely by the pressure drop. The pressure drop is obtained by rewriting Eq. (13.2) as an equation for $\mathcal{P}$,

$$\frac{d\mathcal{P}}{dx} = -\frac{12\eta q}{H^3(x)}$$ (13.3)

This is integrated to

$$\Delta\mathcal{P} = -12\eta q \int_{x_1}^{x_2} H^{-3}(x) \, dx$$ (13.4)

with a power consumption

$$P = q|\Delta\mathcal{P}| = 12\eta q^2 \int_{x_1}^{x_2} H^{-3}(x) \, dx$$ (13.5)

When the equation for the linear function $H(x)$ is substituted into Eq. (13.5), we simply obtain Eq. (10.66). This is plotted in Fig. 10-8, where we see that the assumption of nearly parallel flow is good up to a half-angle of about 15°.

It should be noted that nothing in the derivation of Eq. (13.5) in fact requires that $H(x)$ be linear. The conduit could have a nonlinear shape, as shown in Fig. 13-2, as long as the slope dH/dx is always small. In that case we can still compute $\Delta\mathcal{P}$ and P as long as the function $H(x)$ is known. We require separately, of course, that the length of the conduit be long compared to the average spacing between the walls, so that we may assume fully developed flow at all positions.

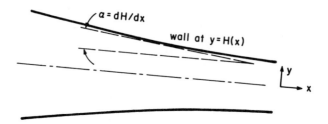

Figure 13-2. Schematic of a channel with slowly converging curved walls.

13.3 ORDERING ANALYSIS

We will develop the lubrication approximation for two-dimensional plane flows, but, with obvious changes appropriate to a different coordinate system, the results will clearly apply to axisymmetric three-dimensional flows as well. The geometry shown schematically in Fig. 13-3 is the basis for the

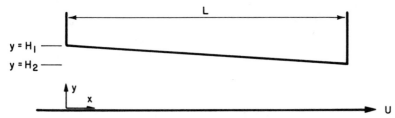

Figure 13-3. Schematic of a slider block. The bottom surface moves relative to the upper surface with velocity U. The spacing and change in spacing are both small relative to L.

ordering analysis. The lower surface moves at velocity U relative to the upper surface. The geometrical parameters satisfy the following inequalities:

$$\frac{H_1}{L} \ll 1 \tag{13.6a}$$

$$\frac{H_1 - H_2}{L} \ll 1 \tag{13.6b}$$

Equation (13.6a) is the assumption of a long flow channel, and Eq. (13.6b) is the "nearly parallel" assumption. We will be dealing at all times with flows for which the inertial terms are negligible,* as we shall show in Sec. 13.7. Our starting point is therefore the creeping flow equations for an incompressible Newtonian fluid, which for a two-dimensional planar flow have the form

$$\frac{\partial v_x}{\partial x} + \frac{\partial v_y}{\partial y} = 0 \tag{13.7}$$

$$0 = -\frac{\partial \mathcal{P}}{\partial x} + \eta\left(\frac{\partial^2 v_x}{\partial x^2} + \frac{\partial^2 v_x}{\partial y^2}\right) \tag{13.8a}$$

$$0 = -\frac{\partial \mathcal{P}}{\partial y} + \eta\left(\frac{\partial^2 v_y}{\partial x^2} + \frac{\partial^2 v_y}{\partial y^2}\right) \tag{13.8b}$$

To carry out an ordering analysis we will need to express the flow equations in dimensionless form. The significant feature of this problem is that it contains *two* characteristic lengths. The characteristic length in the y direction is clearly a spacing between the surfaces; to be specific we may take H_1, since H_1 and H_2 do not differ significantly. There are also changes in the x direction, however, and these take place over a distance of order L; thus, L is the characteristic length in the x direction. The dimensionless coordinates are therefore

$$\tilde{x} = \frac{x}{L} \qquad \tilde{y} = \frac{y}{H_1} \tag{13.9}$$

The characteristic velocity in the flow direction is clearly the linear velocity of the lower surface, U. There will also be some flow in the y direction, however, since the walls are not parallel. This y-direction flow will be characterized by a velocity that is different from U, and we shall denote it as V. V is an unknown at this stage of the development and must still be determined. Recognizing this fact, we write the dimensionless velocity components as

$$\tilde{v}_x = \frac{v_x}{U} \qquad \tilde{v}_y = \frac{v_y}{V} \tag{13.10}$$

Finally, the dimensionless equivalent pressure is written

$$\tilde{\mathcal{P}} = \frac{\mathcal{P}}{\Pi} \tag{13.11}$$

The characteristic pressure, Π, must also be determined.

It is convenient to consider the dimensionless equations one at a time. The dimensionless continuity equation, Eq. (13.7), is

$$\frac{U}{L}\frac{\partial \tilde{v}_x}{\partial \tilde{x}} + \frac{V}{H_1}\frac{\partial \tilde{v}_y}{\partial \tilde{y}} = 0 \tag{13.12a}$$

---

*Recall that the inertial terms vanish identically in perfectly parallel flows of the type studied in Chapter 8.

or

$$\left(\frac{UH_1}{VL}\right)\frac{\partial \tilde{v}_x}{\partial \tilde{x}} + \frac{\partial \tilde{v}_y}{\partial \tilde{y}} = 0 \qquad (13.12\text{b})$$

The dimensionless group UH_1/VL must be of order unity. This follows by considering the consequences of any other choice. If UH_1/VL is large compared to unity, then the $\partial \tilde{v}_x/\partial \tilde{x}$ term in the dimensionless equation (13.12) will dominate and $\partial \tilde{v}_y/\partial \tilde{y}$ can be neglected; in that case, the equation simplifies to $\partial \tilde{v}_x/\partial \tilde{x} = 0$, which contradicts the necessity of allowing v_x to vary with x as the spacing changes. Similarly, if UH_1/VL is very small compared to unity, we may neglect the $\partial \tilde{v}_x/\partial \tilde{x}$ term relative to $\partial \tilde{v}_y/\partial \tilde{y}$; in that case, the equation simplifies to $\partial \tilde{v}_y/\partial \tilde{y} = 0$, and the boundary condition requiring $\tilde{v}_y$ to vanish at the wall then requires that $\tilde{v}_y$ be zero everywhere, which is a contradiction in a changing cross section. The continuity equation therefore *defines* V by the requirement that UH_1/VL be of order unity:

$$V = \frac{UH_1}{L} \qquad (13.13)$$

Note that, consistent with one's intuition, $V \ll U$ for this nearly one-dimensional flow.

We now turn to Eq. (13.8a), the x component of the momentum equation. In dimensionless form this is written

$$0 = -\frac{\Pi}{L}\frac{\partial \tilde{\mathscr{P}}}{\partial \tilde{x}} + \frac{\eta U}{H_1^2}\left[\left(\frac{H_1}{L}\right)^2 \frac{\partial^2 \tilde{v}_x}{\partial \tilde{x}^2} + \frac{\partial^2 \tilde{v}_x}{\partial \tilde{y}^2}\right] \qquad (13.14)$$

One simplification is immediately obvious. Since $H_1/L \ll 1$, the x-derivative term in the brackets may be neglected relative to the y-derivative term. This is consistent with our intuitive understanding that rates of change in the y direction are much larger than rates of change in the x (flow) direction. We may thus rewrite Eq. (13.14) as

$$\left(\frac{\Pi H_1^2}{\eta UL}\right)\frac{\partial \tilde{\mathscr{P}}}{\partial \tilde{x}} = \frac{\partial^2 \tilde{v}_x}{\partial \tilde{y}^2} \qquad (13.15)$$

The two terms in Eq. (13.15) must be of comparable magnitude, since neither term can dominate without introducing a contradiction; indeed, for parallel walls Eq. (13.15) is simply a statement of the balance between the pressure drop and shear stress terms. Thus, the dimensionless group $\Pi H_1^2/\eta UL$ must be of order unity, and we obtain an expression for the characteristic pressure:

$$\Pi = \frac{\eta UL}{H_1^2} \qquad (13.16)$$

Finally, we consider the y component of the momentum equation, Eq. (13.8b). In dimensionless form, using Eq. (13.16) for Π, this is written

$$0 = -\frac{\eta UL}{H_1^3}\frac{\partial \tilde{\mathscr{P}}}{\partial \tilde{y}} + \frac{\eta U}{LH_1}\left[\left(\frac{H_1}{L}\right)^2 \frac{\partial^2 \tilde{v}_y}{\partial \tilde{x}^2} + \frac{\partial^2 \tilde{v}_y}{\partial \tilde{y}^2}\right] \qquad (13.17)$$

As before, we may neglect the x-derivative term in the brackets relative to the y-derivative term. Thus, we can write Eq. (13.17) after some simplification as

$$\frac{\partial \tilde{\mathcal{P}}}{\partial \tilde{y}} = \left(\frac{H_1}{L}\right)^2 \frac{\partial^2 \tilde{v}_y}{\partial \tilde{y}^2} \simeq 0 \qquad (13.18)$$

All the characteristic quantities have been defined, so there are no more degrees of freedom. We must therefore conclude from Eq. (13.18) that, to within the approximation that $H_1/L \ll 1$, $\partial \tilde{\mathcal{P}}/\partial \tilde{y}$ is negligible and $\tilde{\mathcal{P}}$ is a function only of $\tilde{x}$. This is a primary result—that we may neglect variations in the pressure over the width of the channel.

We can summarize the ordering analysis by rewriting Eqs. (13.15) and (13.18) in dimensional form:

$$\mathcal{P} = \mathcal{P}(x) \qquad (13.19)$$

$$\frac{d\mathcal{P}}{dx} = \eta \frac{\partial^2 v_x}{\partial y^2} \qquad (13.20)$$

These are the equations that describe flow between parallel walls, except that $d\mathcal{P}/dx$ need not be a constant and v_x may depend on x as well as on y.

13.4 LUBRICATION EQUATIONS

Equations (13.19) and (13.20) are the basic equations for the lubrication approximation. Because of their very simple structure they can be solved directly and expressed in alternative, more useful forms. Because $d\mathcal{P}/dx$ is independent of y, Eq. (13.20) can be integrated twice to give

$$v_x = \frac{1}{2\eta} \frac{d\mathcal{P}}{dx} y^2 + C_1 y + C_2 \qquad (13.21)$$

The "constants" of integration C_1 and C_2 are independent of y, but they will depend on x. If we take the origin of the y coordinate at the moving surface, we have boundary conditions

$$\text{at } y = 0: \quad v_x = U \qquad (13.22a)$$

$$\text{at } y = H(x): \quad v_x = 0 \qquad (13.22b)$$

C_1 and C_2 can then be evaluated to give

$$v_x = U\left[1 - \frac{y}{H(x)}\right] - \frac{1}{2\eta} \frac{d\mathcal{P}}{dx} y H(x)\left[1 - \frac{y}{H(x)}\right] \qquad (13.23)$$

Equation (13.23) is simply the equation describing the velocity distribution between two flat plates with an imposed pressure gradient. In this case, however, *we do not know the pressure gradient*. This is a typical situation in applications of the lubrication approximation. We may not know the flow rate between the plates either, but we do know that it must be the same at all values of x. Defining q as the flow rate per unit width, we have

$$q = \int_0^{H(x)} v_x \, dy = \text{constant} \tag{13.24}$$

and, carrying out the integration of Eq. (13.23),

$$q = \frac{UH(x)}{2} - \frac{H^3(x)}{12\eta} \frac{d\mathcal{P}}{dx} \tag{13.25}$$

Equation (13.25) is sometimes taken as the starting point for the lubrication approximation. It can be looked upon as an equation for $d\mathcal{P}/dx$ and rearranged to

$$\frac{d\mathcal{P}}{dx} = 12\eta \left[\frac{U}{2H^2(x)} - \frac{q}{H^3(x)} \right] \tag{13.26}$$

or, integrating once,

$$\mathcal{P}(x) = \mathcal{P}_0 + 6\eta U \int_0^x \frac{dx}{H^2(x)} - 12\eta q \int_0^x \frac{dx}{H^3(x)} \tag{13.27}$$

$\mathcal{P}_0$ is a constant of integration that represents the pressure at $x = 0$. Finally, a useful expression relating the flow rate, q; the relative velocity, U; and the overall pressure change, $\mathcal{P}_0 - \mathcal{P}(L)$; can be obtained by setting $x = L$ in Eq. (13.27) and rearranging:

$$q = \frac{\mathcal{P}_0 - \mathcal{P}(L)}{12\eta \int_0^L H^{-3}(x) \, dx} + \frac{U}{2} \frac{\int_0^L H^{-2}(x) \, dx}{\int_0^L H^{-3}(x) \, dx} \tag{13.28}$$

For the case in which $U = 0$, this is simply Eq. (13.4). Note that although Fig. 13-3 is drawn with two plane surfaces, $H(x)$ can in fact be any function of x as long as dH/dx is small compared to unity.

13.5 COATING

We analyzed the problem of wire coating in Sec. 8.5. The treatment there was somewhat oversimplified in that we took the die to be of uniform cross section and assumed that the reservoir pressure was atmospheric. The real situation is more likely to be like that shown in Fig. 13-4, with a possible pressure drop between the reservoir and the die exit. We will analyze this flow here for the coating of a sheet rather than a wire, in keeping with the two-dimensional equations developed in this chapter. The case of a wire die is identical, except that the equations for axisymmetric flow are used. The sheet case applies to the wire as well when the maximum spacing between the wire and the die wall is small compared to the radius of the wire.

The coating thickness H_t is related to the flow rate by the equation

$$q = UH_t \tag{13.29}$$

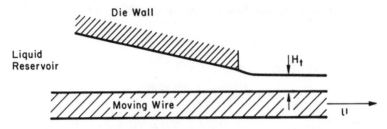

Figure 13-4. Schematic of a wire coating die.

We can therefore write Eq. (13.28) as

$$H_t = \frac{\mathscr{P}_0 - \mathscr{P}(L)}{12\eta U \int_0^L H^{-3}(x)\,dx} + \frac{1}{2}\frac{\int_0^L H^{-2}(x)\,dx}{\int_0^L H^{-3}(x)\,dx} \qquad (13.30)$$

If there is no net pressure drop through the die $[\mathscr{P}_0 = \mathscr{P}(L)]$, the coating thickness depends only on die geometry. In general, however, a pressure drop across the die will be employed, and for a given sheet or wire speed the pressure drop determines the thickness. The pressure drop required for a given coating thickness is obtained by rewriting Eq. (13.30) in the form

$$\mathscr{P}_0 - \mathscr{P}(L) = 12\eta U \int_0^L \frac{1}{H^2(x)}\left[\frac{H_t}{H(x)} - \frac{1}{2}\right]dx \qquad (13.31)$$

It is evident from Eq. (13.31) that the coating thickness can never be less than one-half the exit spacing of a converging die, or else a negative pressure drop would be required.

It is of interest to examine the velocity profile in the die. If we replace q in Eq. (13.26) with UH_t and substitute for $d\mathscr{P}/dx$ in Eq. (13.23), we can express the velocity at any position as

$$v_x = U\left(1 - \frac{y}{H}\right)\left[1 - 3\left(1 - \frac{2H_t}{H}\right)\frac{y}{H}\right] \qquad (13.32)$$

It is understood in Eq. (13.32) that H is a function of x. The term in brackets will be negative over a portion of the cross section whenever $H > 3H_t$, indicating a negative velocity and a region of backflow near the wall, as shown schematically in Fig. 13-5. The region of zero net flow (flow forward exactly compensated by reverse flow) occurs in the region $H_0 \le y \le H$, where H_0 is defined by the equation

$$0 = \int_{H_0}^H v_x\,dy \qquad (13.33)$$

When Eq. (13.32) is substituted into Eq. (13.33) and the integration carried out, H_0 is found to satisfy the equation

$$H_0(x) = \frac{H(x)H_t}{H(x) - 2H_t} \qquad (13.34)$$

Equation (13.34) is plotted in Fig. 13-5 for a plane wall. Fluid in the region $y \leq H_0(x)$ is swept out and forms the coating, while fluid in the region $H_0(x) \leq y \leq H(x)$ simply recirculates. The recirculation will result in a long residence time in the die for a portion of the fluid and could lead to degradation of the coating material in some cases. Recirculation can be avoided by ensuring that the spacing between the moving surface and the die wall never exceeds $3H_t$.

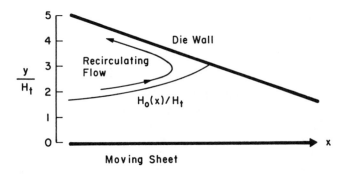

Figure 13-5. Recirculation in a coating die with a plane wall.

13.6 SLIDER BLOCK

The application of the lubrication approximation to a classical problem in lubrication is illustrated by reference to Fig. 13-3. We suppose that the stationary block is completely immersed in the fluid. The pressure in the fluid outside the space between the moving sheet and the block, but near $x = 0$ and $x = L$, is then simply hydrostatic. Thus, $\mathcal{P}(L) = \mathcal{P}_0$, and we may take $\mathcal{P}_0 = 0$ without any loss of generality. It then follows from substitution of Eq. (13.28) into (13.27) that

$$\mathcal{P}(x) = 6\eta U \int_0^x \frac{H(x) - \mathcal{H}}{H^3(x)} \, dx \tag{13.35}$$

where

$$\mathcal{H} = \frac{\int_0^L H^{-2}(x) \, dx}{\int_0^L H^{-3}(x) \, dx} \tag{13.36}$$

For the special case of a plane surface (a *wedge*),

$$\text{wedge:} \quad \mathcal{P}(x) = \frac{6\eta U}{H_1^2 - H_2^2} \frac{[H_1 - H(x)][H(x) - H_2]}{H^2(x)} \tag{13.37}$$

The upward stress σ_{yy} consists of a pressure term and a term $2\eta \, \partial v_y / \partial y$. From the continuity equation, $\partial v_y / \partial y$ is equal to $-\partial v_x / \partial x$, and the latter is proportional to dH/dx and can thus be neglected. The force per unit width acting upward on the block is therefore

$$F_N = \int_0^L \mathcal{P}(x)\, dx \tag{13.38}$$

For a wedge this force is

$$\text{wedge:} \quad F_N = \frac{6\eta U L^2}{H_1^2 - H_2^2}\left[\ln\left(\frac{H_1}{H_2}\right) - \frac{2(H_1 - H_2)}{H_1 + H_2}\right] \tag{13.39}$$

The normal force is nonzero only for $H_1 \neq H_2$. The force F_N takes on a maximum for $H_1 \simeq 2.2 H_2$; in that case

$$\text{wedge:} \quad F_{N,\text{max}} \simeq 0.16 \frac{\eta U L^2}{H_2^2} \tag{13.40}$$

Clearly, a very large normal force is exerted by the thin film of liquid for small H_2. This is the basis of effective lubrication.

The force required to pull the plate past the block is obtained by integrating the shear stress at the plate:

$$F_s = -\int_0^L \eta \frac{\partial v_x}{\partial y}\bigg|_{y=0} dx \tag{13.41}$$

(The negative sign is required because $\eta \partial v_x/\partial y$ is the stress exerted by the fluid on the plate, while we require the equal and opposite value.) For the special case of a wedge the integration gives

$$\text{wedge:} \quad F_s = \frac{\eta U L}{H_1 - H_2}\left[4\ln\frac{H_1}{H_2} - \frac{6(H_1 - H_2)}{H_1 + H_2}\right] \tag{13.42}$$

At the ratio $H_1/H_2 = 2.2$, corresponding to the maximum upward force, we have

$$\text{wedge:} \quad F_{s,\text{max}} \simeq 0.75 \frac{\eta U L}{H^2} \tag{13.43}$$

We can then compute the *coefficient of friction*, the ratio of imposed shear force to obtained normal force, as

$$\text{coefficient of friction} = \frac{F_{s,\text{max}}}{F_{N,\text{max}}} \simeq 5\frac{H_2}{L} \tag{13.44}$$

This can be made a very small value for a sufficiently thin liquid film.

13.7 NEGLECT OF INERTIA

Our starting point in the derivation of the lubrication equations was taken to be the creeping flow equations. It is helpful to examine the inertial terms to ensure that they are indeed negligible.

The x component of the steady-state Navier–Stokes equations is

$$\rho\left(v_x\frac{\partial v_x}{\partial x} + v_y\frac{\partial v_x}{\partial y}\right) = -\frac{\partial \mathcal{P}}{\partial x} + \eta\left(\frac{\partial^2 v_x}{\partial x^2} + \frac{\partial^2 v_x}{\partial y^2}\right) \qquad (13.45)$$

In dimensionless form, using V and Π as defined by Eqs. (13.13) and (13.16), respectively, we obtain

$$\frac{\rho U^2}{L}\left(\tilde{v}_x\frac{\partial \tilde{v}_x}{\partial \tilde{x}} + \tilde{v}_y\frac{\partial \tilde{v}_x}{\partial \tilde{y}}\right) = \frac{\eta UL}{H^2}\left[-\frac{\partial \tilde{\mathcal{P}}}{\partial \tilde{x}} + \left(\frac{H}{L}\right)^2\frac{\partial^2 \tilde{v}_x}{\partial \tilde{x}^2} + \frac{\partial^2 \tilde{v}_x}{\partial \tilde{y}^2}\right] \qquad (13.46)$$

Thus, the inertial terms will be negligible if

$$\frac{\rho U^2}{L} \ll \frac{\eta UL}{H^2} \qquad (13.47a)$$

or, equivalently,

$$\left(\frac{\rho UH}{\eta}\right)\frac{H}{L} = \mathrm{Re}\,\frac{H}{L} \ll 1 \qquad (13.47b)$$

This includes a much wider range of flow conditions than the stronger requirement for general creeping flow, $\mathrm{Re} \ll 1$.

13.8 CONCLUDING REMARKS

The lubrication approximation illustrates the way in which order-of-magnitude estimates can be used to simplify flow problems when the flow is "almost" one-dimensional, providing two characteristic length scales and hence two characteristic velocities. The same type of ordering will be used again in Chapter 15 on boundary layer flows. The use of the lubrication approximation has led to considerable insight into both lubrication flows and polymer processing operations. Successful application to polymer processing has been achieved because processing applications such as extrusion, molding, coating, and calendering involve the flow of a very viscous liquid between surfaces that are in close proximity and in relative motion with respect to one another.

BIBLIOGRAPHICAL NOTES

For applications in polymer processing, see

McKelvey, J. M., *Polymer Processing*, John Wiley & Sons, Inc., New York, 1962.

Middleman, S., *Fundamentals of Polymer Processing*, McGraw-Hill Book Company, New York, 1977.

Pearson, J. R. A., *Mechanical Principles of Polymer Melt Processing*, Pergamon Press, Oxford, 1965.

PROBLEMS

13.1. Estimate the pressure drop in the slowly-varying cylindrical contraction shown in Fig. 13P1.

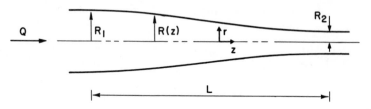

13.2. Repeat Problems 12.3 and 12.4 for a case in which the spacing H between the disks is a slowly varying function of radius, $H(r)$.

13.3. An end-fed sheeting die is shown schematically in Fig. 13P3. Fluid flows axially because of an axial pressure gradient. There is also a side flow through the die because of the pressure difference between the tube and the die exit. Estimate the flow distribution along the length of the die. You may assume that the axial tube flow is locally fully-developed Poiseuille flow, and that the die flow rate at each position is fully-developed plane Poiseuille flow. (A nearly identical analysis applies to slow leakage through a porous-walled tube, as in flow in the kidney.)

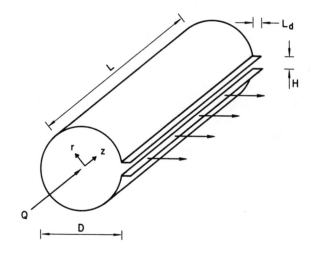

Stream Function, Vorticity, and Potential Flow

14

14.1 INTRODUCTION

Three auxiliary functions, the *stream function*, the *vorticity*, and the *potential function*, are often used to represent and interpret solutions of fluid mechanics problems. The stream function is applicable to certain incompressible flows. Vorticity is a measure of local rotation and is broadly used. The potential function is important in the inviscid limit. We shall briefly introduce these concepts in this chapter.

14.2 STREAM FUNCTION

Consider a two-dimensional plane flow of an incompressible fluid with $v_z = 0$. The continuity equation is

$$\frac{\partial v_x}{\partial x} + \frac{\partial v_y}{\partial y} = 0 \tag{14.1}$$

The *stream function* $\psi(x, y)$ is defined as the function such that

$$v_x = -\frac{\partial \psi}{\partial y} \qquad v_y = +\frac{\partial \psi}{\partial x} \tag{14.2}$$

Clearly, the continuity equation is automatically satisfied when Eqs. (14.2) are substituted into Eq. (14.1). The equivalent relations in cylindrical coordinates, with $v_z = 0$, are

$$v_r = -\frac{1}{r}\frac{\partial \psi}{\partial \theta} \qquad v_\theta = +\frac{\partial \psi}{\partial r} \tag{14.3}$$

The Navier–Stokes equations can be rewritten in terms of ψ.

The significance of the stream function is that lines of constant value of ψ in a steady flow are *streamlines* (compare Sec. 5.6). Streamlines are often more easily visualized than are velocity components. For the converging flow described in Sec. 10.1, for example, we might anticipate that the stream-lines will be straight lines converging on the vertex. In that case we expect ψ to be a function only of angle θ, and thus we are led immediately through Eq. (14.3) to Eq. (10.2b), $v_r = f(\theta)/r$, where $f(\theta) = -\partial\psi/\partial\theta$.

The difference in value of the stream function between two streamlines indicates the portion of the flow contained between those streamlines:

$$\int_{x_1}^{x_2} v_x \, dy = -\int_{x_1}^{x_2} \frac{\partial \psi}{\partial y} \, dy = \psi(x_1, y) - \psi(x_2, y) \tag{14.4}$$

14.3 VORTICITY

Vorticity is the name given to the local rate of rotation in a fluid, and the symbol ω is often used. Consider the element of fluid in Fig. 14-1 that is rotating with angular velocity ω, and hence with a linear velocity $r\omega$. Then

$$v_x = -r\omega \sin \theta = -r\omega \frac{y}{r} = -y\omega \tag{14.5a}$$

$$v_y = +r\omega \cos \theta = +r\omega \frac{x}{r} = +x\omega \tag{14.5b}$$

It therefore follows that

$$\omega = \frac{1}{2}\left(\frac{\partial v_y}{\partial x} - \frac{\partial v_x}{\partial y}\right) \tag{14.6}$$

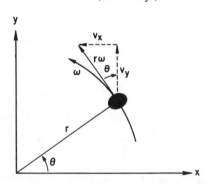

Figure 14-1. Schematic of a fluid element rotating with angular velocity ω.

Equation (14.6) serves to define the vorticity for a two-dimensional flow. The generalization requires that we consider the *components* of the vorticity, ω_{xy}, ω_{xz}, and so on. Equation (14.6) then defines ω_{xy}; note that $\omega_{xy} = -\omega_{yx}$. The other components are defined analogously. It should be noted that some authors define ω_{yx} by Eq. (14.6).

An *irrotational flow* is one for which the vorticity is zero.

14.4 TWO-DIMENSIONAL POTENTIAL FLOW

Two-dimensional irrotational flows, or *potential flows*, have special properties that have led to their being widely studied. If we substitute the stream function, Eq. (14.2), into the definition of vorticity, Eq. (14.6), we obtain

$$\omega = \frac{1}{2}\left(\frac{\partial^2 \psi}{\partial x^2} + \frac{\partial^2 \psi}{\partial y^2}\right) \tag{14.7}$$

For an irrotational flow, $\omega = 0$, and we obtain

$$\omega = 0: \quad \frac{\partial^2 \psi}{\partial x^2} + \frac{\partial^2 \psi}{\partial y^2} = 0 \tag{14.8}$$

Equation (14.8) is known as *Laplace's equation*, and a well-developed mathematical theory of the properties of its solutions exists.

Another important feature follows by noting that a potential flow must simultaneoulsy satisfy the condition of zero vorticity and the equation of continuity:

$$\omega = 0: \quad \frac{\partial v_x}{\partial y} = \frac{\partial v_y}{\partial x} \tag{14.9}$$

$$\text{continuity:} \quad \frac{\partial v_x}{\partial x} = -\frac{\partial v_y}{\partial y} \tag{14.10}$$

It then follows by differentiation of Eq. (14.9) with respect to y and Eq. (14.10) with respect to x that

$$\frac{\partial^2 v_x}{\partial x^2} + \frac{\partial^2 v_x}{\partial y^2} = 0 \tag{14.11}$$

This sum will be recognized as the viscous stress terms in the x component of the Navier–Stokes equations. The corresponding terms in the y-component equation similarly vanish. We are thus led to the conclusion that *a potential flow solution to the inviscid fluid equations* ($\eta = 0$) *is also a solution to the Navier–Stokes equations.* An inviscid fluid need not satisfy the no-slip boundary condition, so such potential flow solutions may not have any physical significance.

Finally, we can show that, for a potential flow, Bernoulli's equation is equivalent to the Navier–Stokes equations. The Bernoulli equation for a flow without losses was derived in Chapter 5 and can be written in the form

$$\mathcal{P} + \tfrac{1}{2}\rho(v_x^2 + v_y^2) = \text{constant} \tag{14.12}$$

As derived in Chapter 5, the constant in Eq. (14.12) might vary from one streamline to the next, since the derivation was along a single streamline. We assume here that the constant is independent of position and is the same on every streamline.

If we differentiate Eq. (14.12) with respect to x, we obtain

$$\frac{\partial \mathcal{P}}{\partial x} + \rho v_x \frac{\partial v_x}{\partial x} + \rho v_y \frac{\partial v_y}{\partial x} = 0 \tag{14.13}$$

We can replace the term $\partial v_y / \partial x$ for a potential flow by use of Eq. (14.9) and rewrite Eq. (14.13) as

$$\rho \left(v_x \frac{\partial v_x}{\partial x} + v_y \frac{\partial v_x}{\partial y} \right) = -\frac{\partial \mathcal{P}}{\partial x} \tag{14.14}$$

Equation (14.13) is the x component of the Navier–Stokes equation in the inviscid limit $\eta = 0$. Similarly, the y component is obtained by differentiation of Eq. (14.12) with respect to y.

We can thus summarize this section by noting that a potentially useful tool has been developed. Potential flows may be found through solutions of Laplace's equation, which is one of the most thoroughly studied equations in mathematics. Such flows are solutions not only to the inviscid flow equations, but to the full Navier–Stokes equations. Potential flows cannot be expected to satisfy no-slip boundary conditions, however, so they will not be useful solutions to the Navier–Stokes equations near solid surfaces.

14.5 POTENTIAL FUNCTION

Many phenomena in physics representing fluxes of quantities can be described as gradients of scalars; in such a case the scalar is called a *potential*. Heat flux, for example, is proportional to the gradient of the temperature. Two-dimensional irrotational flows can be represented as the gradient of a potential, hence the name potential flow. The potential function is usually represented by the symbol Φ, so we write

$$\mathbf{v} = -\nabla \Phi \tag{14.15a}$$

or, equivalently,

$$v_x = -\frac{\partial \Phi}{\partial x} \qquad v_y = -\frac{\partial \Phi}{\partial y} \tag{14.15b}$$

In polar coordinates Eq. (14.15a) is expressed as

$$v_r = -\frac{\partial \Phi}{\partial r} \qquad v_\theta = -\frac{1}{r}\frac{\partial \Phi}{\partial \theta} \tag{14.15c}$$

Substitution of Eq. (14.15) into the condition $\omega = 0$, Eq. (14.9), establishes that the flow is indeed an irrotational flow. Substitution into the continuity equation leads to an equation for the potential function,

$$\frac{\partial^2 \Phi}{\partial x^2} + \frac{\partial^2 \Phi}{\partial y^2} = 0 \tag{14.16}$$

Thus, we again obtain Laplace's equation. It can be shown that lines of constant value of Φ are everywhere orthogonal to lines of constant value of ψ.

The theory of functions of a complex variable is a tool that is often used to study Laplace's equation, and complex variable theory is thus commonly used for the solution of potential flow fields. Readers who have studied complex variables will recognize in particular the applicability of the procedure of *conformal mapping*. Applications of complex variable theory are beyond the scope of this introductory text, however.

14.6 STAGNATION FLOW

As an example of the use of potential flow theory, consider the following quadratic function:

$$\Phi = -\tfrac{1}{2}A(x^2 - y^2) \tag{14.17}$$

where A is a constant, and the factor of $\tfrac{1}{2}$ is introduced for convenience. The velocity components are then found from Eq. (14.15b),

$$v_x = -\frac{\partial \Phi}{\partial x} = Ax \tag{14.18a}$$

$$v_y = -\frac{\partial \Phi}{\partial y} = -Ay \tag{14.18b}$$

The stream function is found from Eq. (14.2), as follows:

$$\psi = -\int v_x \, dy = -Axy + \text{function of } x \tag{14.19a}$$

$$\psi = +\int v_y \, dx = -Axy + \text{function of } y \tag{14.19b}$$

Evidently, the function of x and function of y must simply be a constant, and the constant can be set to zero, since only differences in stream function values have any significance. Thus, we have

$$\psi = -Axy \tag{14.20}$$

The streamlines, which are hyperbolas, are plotted in Fig. 14-2. The flow is that approaching a plane wall at $y = 0$. The fluid does not penetrate the wall ($v_y = 0$ at $y = 0$), but the fluid does slide along the wall because of the nonzero v_x. Thus, the velocity field described by Eqs. (14.18) cannot be followed by a real fluid right at the surface. As discussed in Sec. 10.6, how-

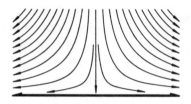

Figure 14-2. Streamlines for plane stagnation flow of an ideal fluid.

ever, and in more detail in Chapter 15, outside a boundary layer adjacent to the solid surface the fluid can be expected to follow the potential flow solution. Equations (14.18) are frequently used in mass and heat transfer studies for flow normal to a solid body.

The pressure distribution is obtained from the Bernoulli equation, (14.12), together with Eqs. (14.18), as

$$\mathcal{P} = \mathcal{P}_0 - \tfrac{1}{2}\rho A^2(x^2 + y^2) \tag{14.21}$$

The lines of constant pressure are concentric circles, with the maximum at the *stagnation point*, or point of zero velocity, $x = y = 0$. Because of the existence of a stagnation point, this type of flow field is often referred to as *plane stagnation flow*.

It should be noted that a stagnation flow can exist, if at all, only in a small neighborhood of the stagnation point. As the distance gets larger, the velocity grows and the pressure decreases without bound, which is impossible. There have been some experimental constructions of stagnation flows by shooting submerged jets of liquid against plane surfaces. The flow near the forward stagnation point of a cylinder is also approximated by Eqs. (14.18), as discussed in the next section.

14.7 FLOW PAST A CYLINDER

An example that we shall find useful in Chapter 15 is the potential given by

$$\Phi = U_\infty\!\left(r + \frac{R^2}{r}\right)\cos\theta \tag{14.22}$$

The corresponding velocity and stream function are, respectively,

$$v_r = -U_\infty\!\left(1 - \frac{R^2}{r^2}\right)\cos\theta \tag{14.23a}$$

$$v_\theta = +U_\infty\!\left(1 + \frac{R^2}{r^2}\right)\sin\theta \tag{14.23b}$$

$$\psi = U_\infty\!\left(r - \frac{R^2}{r}\right)\sin\theta \tag{14.24}$$

The streamlines are shown in Fig. 14-3. The flow corresponds to uniform motion past a cylinder of radius R, with the flow far from the cylinder at velocity U_∞. The flow has fore- and aft symmetry, and there is both a forward and a rear stagnation point where the velocity is zero. We emphasize again that v_θ is nonzero on the cylinder surface, so the no-slip boundary condition is not satisfied.

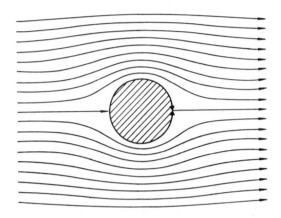

Figure 14-3. Streamlines for uniform flow of an ideal fluid past a cylinder.

The pressure is obtained from the Bernoulli equation as

$$\mathcal{P} = \mathcal{P}_0 - \frac{1}{2}\rho U_\infty^2 \left[\left(1 - \frac{R^2}{r^2}\right)^2 + 4\left(\frac{R}{r}\right)^2 \sin^2\theta \right] \qquad (14.25)$$

Note that the difference between the stagnation pressure and the pressure in the uniform flow far from the surface ($r \to \infty$) is simply $\rho U_\infty^2/2$. We have seen this result in Sec. 6.6 in the analysis of the pitot tube.

The pressure on the surface, $r = R$, is of particular interest:

$$r = R: \quad \mathcal{P} = \mathcal{P}_\infty - 2\rho U_\infty^2 \sin^2\theta \qquad (14.26)$$

The pressure is predicted to decrease from the stagnation point to $\theta = 90°$, and then to increase again. Experimental measurements of the pressure on the surface have been made, and Eq. (14.26) is followed approximately at very high Reynolds numbers up to about $\theta = 120°$; the fore and aft symmetry is not retained beyond this point experimentally, however, for reasons that are best discussed in the next chapter in the context of boundary layer theory.

It is useful to note that if we embed an xy coordinate system with its origin at the forward stagnation point, then the stream function, Eq. (14.24), can be expanded about $x = y = 0$ as

$$\psi = -\left(\frac{2U_\infty}{R}\right)xy + \text{order of } (x^3, x^2y, xy^2, y^3) \qquad (14.27)$$

Thus, the plane stagnation flow in the preceding section is an approximation to the flow in the neighborhood of the stagnation point of the cylinder, with $A = 2U_\infty/R$.

Finally, the fore and aft symmetry of the pressure means that the net pressure force exerted by the liquid on the cylinder is zero. We referred to such a drag term in Sec. 12.3.3 as *form drag*. (There is a nonzero form drag in a real fluid because of the experimental absence of symmetry.) The viscous stress term is identically zero in potential flow, so the theory cannot predict any friction drag. Thus, potential flow theory predicts that there is no drag at all on a submerged cylinder in a uniform flow or, in fact, on any body with fore and aft symmetry. This absence of form drag in potential flow theory is known as *d'Alembert's paradox*.

14.8 CONCLUDING REMARKS

The concept of the stream function is a useful one in formulating and representing solutions to flow problems. Potential flow theory has been introduced here primarily because of the role that it plays in the boundary layer approximation for high Reynolds number flows, to be discussed in Chapter 15. The theory has been quite useful in its own right as well, however; stagnation flow is often used in heat and mass transfer calculations, for example. The flow of a liquid around gas bubbles, where the no-slip boundary condition is not appropriate at the interface, is well described by potential flow, and the theory has been successfully applied to the motion of dilute solids in fluidized bed reactors.

BIBLIOGRAPHICAL NOTES

The standard references on inviscid flow, including potential flow theory, are

LAMB, H., *Hydrodynamics*, 6th ed., Dover Publications, New York, 1945.

MILNE-THOMPSON, L. M., *Theoretical Hydrodynamics*, 3rd ed., Macmillan Publishing Co., Inc., New York, 1955.

There is an excellent treatment in the translation of a classic,

PRANDTL, L., and TIETJENS, O. G., *Fundamentals of Hydro- and Aeromechanics*, Dover Publications, New York, 1957.

The Boundary Layer Approximation $\textbf{15}$

15.1 INTRODUCTION

Flows in which inertial terms are important are the most difficult to treat analytically. The limiting case Re $\longrightarrow \infty$ corresponds to an inviscid fluid. Inviscid fluid flow describes the large Reynolds number motion of real fluids away from solid surfaces, but motion near surfaces cannot be described in this limit because an inviscid fluid cannot satisty the no-slip boundary condition.

The boundary layer approximation was developed by Prandtl in 1904 to deal with flows in the neighborhood of a solid surface, where the inviscid fluid equations fail. It is assumed that the flow away from the surface is closely approximated by a potential flow, for which the Bernoulli equation applies (Chapter 14), and that the transition from the no-slip surface to the inviscid flow region takes place over a very thin fluid layer; we have seen from the exact solutions of the Navier–Stokes equations in Chapters 9 and 10 that this layer is proportional to $\sqrt{\eta/\rho}$. The resulting equations are much simpler than the full Navier–Stokes equations and can be solved for flows near surfaces of many different shapes. The boundary layer approximation forms the basis for most studies of heat and mass transfer near solid surfaces. It is assumed at all times that the potential flow solution for an inviscid fluid about an object of the same shape is available. Potential flow is discussed briefly in Chapter 14, where solutions for flow near a plane stagnation point and about a circular cylinder are given. Potential flow solutions are usually obtained using the theory of functions of a complex variable.

The boundary layer approximation is similar in concept to the lubrication approximation, in that there are two length scales that need to be taken into account, and the ordering analysis parallels the one in Section 13.3.

15.2 ORDERING ANALYSIS

The flow is shown schematically in Fig. 15-1. We consider only two-dimensional flows, with no variation in the z direction. The distance $\delta(x)$ over which the flow adjusts from the no-slip condition at the solid surface to the potential flow in the "free stream" is assumed to be very small compared to any linear

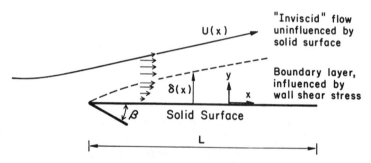

Figure 15-1. Schematic of boundary layer flow.

dimension in the system. We shall have an opportunity to verify this assumption later in the chapter. The starting point is the steady two-dimensional continuity and Navier–Stokes equations for an incompressible Newtonian fluid:

$$\frac{\partial v_x}{\partial x} + \frac{\partial v_y}{\partial y} = 0 \tag{15.1}$$

$$\rho\left(v_x \frac{\partial v_x}{\partial x} + v_y \frac{\partial v_x}{\partial y}\right) = -\frac{\partial \mathcal{P}}{\partial x} + \eta\left(\frac{\partial^2 v_x}{\partial x^2} + \frac{\partial^2 v_x}{\partial y^2}\right) \tag{15.2a}$$

$$\rho\left(v_x \frac{\partial v_y}{\partial x} + v_y \frac{\partial v_y}{\partial y}\right) = -\frac{\partial \mathcal{P}}{\partial y} + \eta\left(\frac{\partial^2 v_y}{\partial x^2} + \frac{\partial^2 v_y}{\partial y^2}\right) \tag{15.2b}$$

These equations must be made dimensionless for ordering purposes. The characteristic length in the x (flow) direction is L, which is a linear dimension of the solid body and reflects the distance over which changes occur in the direction of mean flow. Changes in the y direction occur over a distance δ, which characterizes the *boundary layer*, or transition region, thickness. Thus, the dimensionless spatial variables are

$$\tilde{x} = \frac{x}{L} \qquad \tilde{y} = \frac{y}{\delta} \qquad (15.3)$$

Recall that $\delta/L \ll 1$.

The characteristic velocity in the flow direction is U, the velocity parallel to the surface in the inviscid region. The characteristic velocity in the y direction is clearly much smaller than U, but its magnitude is not well defined and needs to be found as a part of the derivation; we shall denote this characteristic velocity as V. The dimensionless velocity variables are then

$$\tilde{v}_x = \frac{v_x}{U} \qquad \tilde{v}_y = \frac{v_y}{V} \qquad (15.4)$$

The characteristic pressure is denoted Π, and the dimensionless equivalent pressure is written

$$\tilde{\mathcal{P}} = \frac{\mathcal{P}}{\Pi} \qquad (15.5)$$

It can be anticipated that Π will be of the order of the pressure in the inviscid flow, $\frac{1}{2}\rho U^2$, but we shall let this result come naturally from the derivation.

It should be noted that U and δ refer to quantities that may vary with x position, as shown in Fig. 15-1. For purposes of the ordering analysis we shall assume that we are using the maximum values of these quantities over the length L, and we shall treat them as constants.

As in Section 13.3, it is convenient to deal with the dimensionless equations one at a time. The dimensionless continuity equation, Eq, (15.1), is

$$\frac{U}{L}\frac{\partial \tilde{v}_x}{\partial \tilde{x}} + \frac{V}{\delta}\frac{\partial \tilde{v}_y}{\partial \tilde{y}} = 0 \qquad (15.6a)$$

or

$$\left(\frac{U\delta}{VL}\right)\frac{\partial \tilde{v}_x}{\partial \tilde{x}} + \frac{\partial \tilde{v}_y}{\partial \tilde{y}} = 0 \qquad (15.6b)$$

Neither term in the continuity equation can dominate the other or a contradiction will result. If the group $U\delta/VL$ is very large compared to unity, the term $\partial \tilde{v}_y/\partial \tilde{y}$ can be neglected; in that case we have $\partial \tilde{v}_x/\partial \tilde{x} = 0$ and we conclude that $\tilde{v}_x$ is independent of $\tilde{x}$. Yet we know from the exact solutions in Chapters 9 and 10 that the flow in the x direction depends on the time that a fluid particle has spent in motion past the wall or, equivalently in this case, on the value of the x coordinate. Similarly, if the group $U\delta/VL$ is small compared to unity, we may neglect the term on the left; in that case we conclude that $\tilde{v}_y$ is independent of $\tilde{y}$. Since $\tilde{v}_y$ must equal zero at the surface, we would then be forced to conclude that $\tilde{v}_y$ is identically zero, and we have no a priori basis for such a conclusion. Thus, we conclude that $U\delta/VL$ must be of order unity to ensure that both terms in the continuity equation are of comparable magnitude, and we thus define V:

$$V = \frac{U\delta}{L} \tag{15.7}$$

Since $\delta/L \ll 1$, it follows, as expected, that $V \ll U$.

We next turn to the x component of the Navier–Stokes equations, Eq. (15.2a). Together with Eq. (15.7), we can write this in dimensionless form as

$$\frac{\rho U^2}{L}\left(\tilde{v}_x \frac{\partial \tilde{v}_x}{\partial \tilde{x}} + \tilde{v}_y \frac{\partial \tilde{v}_x}{\partial \tilde{y}}\right) = -\frac{\Pi}{L}\frac{\partial \tilde{\mathscr{P}}}{\partial \tilde{x}} + \frac{\eta U}{\delta^2}\left[\left(\frac{\delta}{L}\right)^2 \frac{\partial^2 \tilde{v}_x}{\partial \tilde{x}^2} + \frac{\partial^2 \tilde{v}_x}{\partial \tilde{y}^2}\right] \tag{15.8a}$$

or

$$\left(\frac{\rho U \delta^2}{\eta L}\right)\left(\tilde{v}_x \frac{\partial \tilde{v}_x}{\partial \tilde{x}} + \tilde{v}_y \frac{\partial \tilde{v}_x}{\partial \tilde{y}}\right) = -\left(\frac{\Pi \delta^2}{\eta U L}\right)\frac{\partial \tilde{\mathscr{P}}}{\partial \tilde{x}} + \left[\left(\frac{\delta}{L}\right)^2 \frac{\partial^2 \tilde{v}_x}{\partial \tilde{x}^2} + \frac{\partial^2 \tilde{v}_x}{\partial \tilde{y}^2}\right] \tag{15.8b}$$

The dimensionless group $\rho U \delta^2/\eta L$ must be of order unity. If the group is large compared to unity, the shear stress terms are negligible compared to the inertial terms and we are led to the inviscid fluid, which is not a consistent approximation near the surface. If the group is small compared to unity, the inertial terms are negligible compared to the shear stress terms; this is the creeping flow approximation and contradicts the assumption that inertia is important. Thus, the order of magnitude of δ is established:

$$\delta = \left(\frac{\eta L}{\rho U}\right)^{1/2} = L \, \mathrm{Re}_L^{-1/2} \tag{15.9}$$

(The symbol Re_x is often used in boundary layer flow to denote a Reynolds number based on distance in the flow direction:

$$\mathrm{Re}_x = \frac{xU\rho}{\eta} \tag{15.10}$$

Hence, $\mathrm{Re}_L = LU\rho/\eta$.) Equation (15.9) is consistent with the results obtained from exact solutions in Chapters 9 and 10, where it was found that the influence of the wall extended into the fluid a distance proportional to the square root of the product of η/ρ and the time of flow past the surface; L/U is the characteristic time for a particle to flow a distance L in the inviscid free stream.

The characteristic pressure is chosen to make the dimensionless group $\Pi\delta^2/\eta UL$ of order unity, to ensure that the pressure gradient term is of the same order as the inertial and viscous stress terms:

$$\Pi = \frac{\eta U L}{\delta^2} = \rho U^2 \tag{15.11}$$

The term $\eta UL/\delta^2$ is the same as Eq. (13.16), and reflects the fact that the pressure force in a nearly parallel flow ($\Pi\delta$) must be of the order of the viscous resistance ($\eta UL/\delta$). The equality to ρU^2 follows from Eq. (15.9), and is a consequence of the fact that the pressure and inertial terms balance one another in the inviscid region, where the Bernoulli equation applies.

Our examination of the x-component equation has served primarily to define the characteristic quantities. There is one important simplification that is possible as well. The dimensionless derivatives $\partial^2 \tilde{v}_x/\partial \tilde{x}^2$ and $\partial^2 \tilde{v}_x/\partial \tilde{y}^2$ are both of order unity, since the characteristic quantities have been chosen to make all variables and changes of order unity. Thus, the term $(\delta/L)^2 \, \partial^2 \tilde{v}_x/\partial \tilde{x}^2$ is negligible compared to $\partial^2 \tilde{v}_x/\partial \tilde{y}^2$ and may be neglected.

Finally, we turn to the y component of the Navier–Stokes equations, Eq. (15.2b). This equation may be written in dimensionless form, with Eqs. (15.7), (15.9), and (15.11) as

$$\frac{\rho U^2 \delta}{L^2}\left(\tilde{v}_x \frac{\partial \tilde{v}_y}{\partial \tilde{x}} + \tilde{v}_y \frac{\partial \tilde{v}_y}{\partial \tilde{y}}\right) = -\frac{\rho U^2}{\delta}\frac{\partial \tilde{\mathcal{P}}}{\partial \tilde{y}} + \frac{\rho U^2 \delta}{L^2}\left[\left(\frac{\delta}{L}\right)^2 \frac{\partial^2 \tilde{v}_y}{\partial \tilde{x}^2} + \frac{\partial^2 \tilde{v}_y}{\partial \tilde{y}^2}\right] \quad (15.12a)$$

or, rearranging,

$$\frac{\partial \tilde{\mathcal{P}}}{\partial \tilde{y}} = \left(\frac{\delta}{L}\right)^2 \left[\left(\frac{\delta}{L}\right)^2 \frac{\partial^2 \tilde{v}_y}{\partial \tilde{x}^2} + \frac{\partial^2 \tilde{v}_y}{\partial \tilde{y}^2} - \tilde{v}_x \frac{\partial \tilde{v}_y}{\partial \tilde{x}} - \tilde{v}_y \frac{\partial \tilde{v}_y}{\partial \tilde{y}}\right] \approx 0 \quad (15.12b)$$

The term in brackets is of order unity, and it is multiplied by $(\delta/L)^2$. Thus, we are led to the conclusion that $\partial \tilde{\mathcal{P}}/\partial \tilde{y}$ is of order $(\delta/L)^2$ and is hence negligible compared to unity. *We may therefore neglect any changes in $\mathcal{P}$ in the y direction.* This is one of the primary simplifications of the boundary layer approximation.

We may summarize the boundary layer approximation by rewriting the equations that we have derived in dimensional form:

$$\mathcal{P} = \mathcal{P}(x) \quad (15.13)$$

$$\rho\left(v_x \frac{\partial v_x}{\partial x} + v_y \frac{\partial v_x}{\partial y}\right) = -\frac{d\mathcal{P}}{dx} + \eta \frac{\partial^2 v_x}{\partial y^2} \quad (15.14)$$

[Recall that the term $\partial^2 v_x/\partial x^2$ is of order $(\delta/L)^2$ and is neglected relative to $\partial^2 v_x/\partial y^2$.] Since $\mathcal{P}$ is independent of y, the equivalent pressure in the boundary layer is the same as the pressure in the inviscid free stream at the same value of x, and is therefore presumed known either from the solution to the potential flow or from experiment.*

15.3 MATCHING THE FREE STREAM

The flow in the free stream, outside the boundary layer, satisfies Bernoulli's equation:

$$\mathcal{P} + \tfrac{1}{2}\rho v^2 = \text{constant} \quad (15.15)$$

---

*This result is directly applicable to the analysis of the pitot tube in Sec. 6.6. Refer to Fig. 6-6. It was assumed that the pressure acting on the side holes in the outer tube is the same as the pressure p_3 in the free stream. We now know that a boundary layer will form along the side wall of the tube, but the assumption regarding the pressure is still valid.

The boundary layer is a very thin region physically, even though significant changes are taking place; for a body with a linear dimension $L = 10^{-2}$ m, and with a fluid with the properties of water moving past at a speed of 10 m/s, for example, then according to Eq. (15.9) the boundary layer thickness is of order 10^{-4} m. Thus, the flow in the region where the boundary layer blends into the potential flow will be quite close to the potential flow that would exist at the solid surface.

This is an important point. The potential flow has no y component at the surface, but it does have an x component because of failure to satisfy the no-slip condition. Thus, the velocity in Bernoulli's equation, Eq. (15.15), is simply the x component, U, and we can write

$$\mathcal{P} + \tfrac{1}{2}\rho U(x)^2 = \text{constant} \qquad (15.16)$$

U is a function only of x because it is the solution to the potential flow problem at $y = 0$.

Equation (15.16) can be differentiated with respect to x to give

$$\frac{d\mathcal{P}}{dx} = -\rho U \frac{dU}{dx} \qquad (15.17)$$

Equation (15.14) can then be written in an alternative form that is frequently employed,

$$\rho\left(v_x \frac{\partial v_x}{\partial x} + v_y \frac{\partial v_x}{\partial y} - U \frac{dU}{dx}\right) = \eta \frac{\partial^2 v_x}{\partial y^2} \qquad (15.18)$$

$U(x)$ is always assumed to be known. For a plane stagnation flow, for example, $U(x)$ is given by Eq. (14.18a).

Equation (15.18) is second order with respect to y and therefore requires two boundary conditions. The no-slip condition, $v_x = 0$ at $y = 0$, is clearly one condition. The other condition is that v_x must approach the potential flow velocity, $U(x)$, when y is large compared to the boundary layer thickness, δ. This latter condition can be expressed in a form that is useful analytically in a manner similar to that used in Sec. 10.6. As shown by Eq. (15.9), the boundary layer thickness is of order $\eta^{1/2}$. To make the x and y coordinates comparable, therefore, the y-coordinate distance must be "stretched" by a factor $\mathrm{Re}_L^{1/2}$, and we can think in terms of the variable $y\,\mathrm{Re}_L^{1/2}$. We then have the result that v_x must approach $U(x)$ for $y\,\mathrm{Re}_L^{1/2}$ large compared to L. But even though y is still quite small at the edge of the boundary layer, we are dealing with the asymptotic case of very large Reynolds number, so we may replace the condition $y\,\mathrm{Re}_L^{1/2} \gg L$ by the condition $y\,\mathrm{Re}_L^{1/2} \to \infty$. We thus obtain the boundary condition*

$$v_x \longrightarrow U(x) \text{ as } y\,\mathrm{Re}_L^{1/2} \longrightarrow \infty \qquad (15.19)$$

*Equation (15.19) is often simply written

$$v_x \longrightarrow U(x) \quad \text{as} \quad y \longrightarrow \infty$$

The end result of any calculation will be the same, since Re_L is always a finite constant, but some of the physical meaning of the condition is lost.

The fact that the boundary condition is moved "off to infinity" is of major significance in enabling solutions to the boundary layer equations to be found.

The continuity equation contains a term $\partial v_y/\partial y$, so one y boundary condition is needed for v_y, and this is the no-slip condition at $y = 0$. The remaining boundary conditions normal to the surface are, therefore,

$$v_x = v_y = 0 \text{ at } y = 0 \tag{15.20}$$

There is a term $\partial v_x/\partial x$ in Eq. (15.18) and the continuity equation, so a boundary condition in the x direction is also needed. In the case of an object with a *sharp leading edge*, like that shown in Fig. 15-1, the potential flow can be assumed to exist undisturbed right up until the leading edge, so we have

$$\text{sharp leading edge:} \quad v_x = U(0) \text{ at } x = 0 \tag{15.21}$$

15.4 FLOW OVER A FLAT PLATE

The most elementary application of the boundary layer approximation is to the flow over a flat plate at zero angle of incidence, as shown in Fig. 15-2.

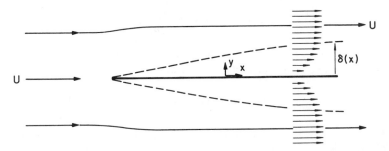

Figure 15-2. Schematic of flow over a flat plate at zero angle of incidence.

If the fluid were inviscid, it would simply slip past the plate with no change in velocity, so in the potential flow outside the boundary layer the velocity U is a constant, independent of x, and there is no pressure gradient. In that case Eq. (15.18) simplifies even further to

$$\rho\left(v_x \frac{\partial v_x}{\partial x} + v_y \frac{\partial v_x}{\partial y}\right) = \eta \frac{\partial^2 v_x}{\partial y^2} \tag{15.22}$$

The structure of Eq. (15.22) is quite similar to that of Eq. (9.2a) for flow past a suddenly accelerated flat plate, and the physical analogy seems to be a realistic one from the point of view of a fluid particle. For the flat plate accelerated from rest, we found that the absence of any true characteristic length for an infinite plate dictated a solution in terms of the single combination of variables $y(\rho/\eta t)^{1/2}$. Here, the "time" that a typical fluid element has been exposed to the presence of the plate is of order x/U, so we might expect a solution in terms of the single variable $y(\rho U/\eta x)^{1/2}$. A dimensional-analysis argument identical to the one used in Sec. 9.4 does indeed lead to such an independent variable.

There is an alternative, less formal approach that leads to the same result. The search for the simplest possible solution to Eq. (15.22) suggests consideration of the possibility that solutions are *similar*; that is, as in flow in a nearly parallel channel (Chapter 13), we look for a solution in which the dimensionless velocity depends only on dimensionless distance from the plate:

$$\frac{v_x}{U} = \text{function of } \frac{y}{\delta(x)} \tag{15.23}$$

We use $\delta(x)$, the boundary layer thickness at distance x from the leading edge, to make y dimensionless, because this is the distance over which the flow is changing. We know from Eq. (15.9) that the boundary layer thickness is of order $(\eta L/\rho U)^{1/2}$. On a "semi-infinite" flat plate the only distance scale of significance is the distance from the origin to a particular point; thus, it is reasonable to expect $\delta(x)$ to be of order $(\eta x/\rho U)^{1/2}$. In that case it follows from Eq. (15.23) that we should seek solutions of the form

$$\frac{v_x}{U} = \text{function of } y\left(\frac{\rho U}{\eta x}\right)^{1/2} \tag{15.24}$$

which is the same result as that obtained from the dimensional analysis argument. We define the variable ζ as the argument in Eq. (15.24):

$$\zeta = y\left(\frac{\rho U}{\eta x}\right)^{1/2} \tag{15.25}$$

It will be necessary to perform one integration of the function in Eq. (15.24) in order to find v_y. For this reason, and the resulting convenience, it is customary to express the ratio v_x/U as the *derivative* of a function $f(\zeta)$:

$$v_x = Uf'(\zeta) \tag{15.26}$$

Here, f' denotes $df/d\zeta$; the prime notation is used in all writing on boundary layer theory, and we shall use it for consistency. The following manipulations, analogous to Eqs. (9.11), are then useful:

$$\frac{\partial f}{\partial y} = \frac{df}{d\zeta}\frac{\partial \zeta}{\partial y} = \left(\frac{\rho U}{\eta x}\right)^{1/2} f' \tag{15.27a}$$

$$\frac{\partial^2 f}{\partial y^2} = \frac{\rho U}{\eta x} f'' \tag{15.27b}$$

$$\frac{\partial}{\partial x} f' = f'' \frac{\partial \zeta}{\partial x} = -\frac{1}{2} x^{-3/2} y \left(\frac{\rho U}{\eta}\right)^{1/2} f'' \tag{15.27c}$$

We can now obtain v_y from the continuity equation, as follows:

$$\frac{\partial v_y}{\partial y} = -\frac{\partial v_x}{\partial x} = -U \frac{\partial}{\partial x} f'(\zeta) = \frac{1}{2} \left(\frac{\eta U}{\rho x}\right)^{1/2} y \frac{\partial^2 f}{\partial y^2} \qquad (15.28)$$

$$v_y = \int_0^y \frac{\partial v_y}{\partial y} \, dy = \frac{1}{2} \left(\frac{\eta U}{\rho x}\right)^{1/2} \int_0^y y \frac{\partial^2 f}{\partial y^2} \, dy = \frac{1}{2} \left(\frac{\eta U}{\rho x}\right)^{1/2} \left(y \frac{\partial f}{\partial y} - f\right)$$

$$= \frac{1}{2} \left(\frac{\eta U}{\rho x}\right)^{1/2} [\zeta f'(\zeta) - f(\zeta)] \qquad (15.29)$$

We have taken $f(0) = 0$ in the integration by parts in Eq. (15.29); it is readily established that any arbitrary value of $f(0)$ can be used without loss of generality.

Substitution of Eqs. (15.26), (15.27), and (15.29) into the boundary layer equation, Eq, (15.22), leads to a nonlinear ordinary differential equation:

$$ff'' + 2f''' = 0 \qquad (15.30)$$

Boundary conditions follow from Eqs. (15.25), (15.26), and (15.29). The solid surface, $y = 0$, corresponds to $\zeta = 0$. The no-slip boundary conditions, Eqs. (15.20), then become

$$f(0) = f'(0) = 0 \qquad (15.31a)$$

The definition of ζ is such that $y \, \mathrm{Re}^{1/2} \rightarrow \infty$ corresponds to $\zeta \rightarrow \infty$; thus, the condition that v_x approach $U(x)$ outside the boundary layer becomes

$$f'(\infty) = 1 \qquad (15.31b)$$

We now have three boundary conditions for the third-order equation, but we have not yet used Eq. (15.21) at $x = 0$. Note, however, that $x = 0$ also corresponds to $\zeta = \infty$, and in that case Eq. (15.21) *also* reduces to Eq. (15.31b). It is this collapsing of the four boundary conditions to three that enables us to reduce the problem to solution of an ordinary differential equation. Partial differential equations can often be reduced to ordinary differential equations by variable changes, but it is only in exceptional cases that the boundary conditions transform as well!

Equation (15.30), with boundary conditions (15.31), was first solved by Blasius in 1908 by a tedious series method. The equation is easily solved numerically on modern computers by assuming a value of $f''(0)$, integrating Eq. (15.30) to large values of ζ, and adjusting the value of $f''(0)$ until the condition $f'(\infty) = 1$ is satisfied; the correct value of $f''(0)$ is 0.332. The solution is shown in Fig. 15-3, together with some experimental data obtained by Nikuradse. Clearly, the theory and experiment are in good agreement. Note that the approach to the condition $f'(\infty) = 1$ is asymptotic. If we define the boundary layer thickness as that distance for which $v_x = 0.99U$,

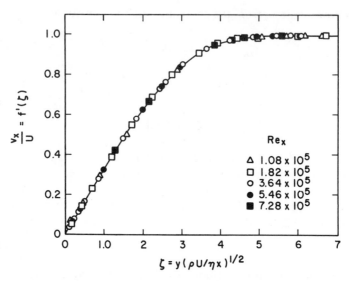

Figure 15-3. Numerical solution to Eqs. (15.30) and (15.31), with experimental data of Nikuradse.

then y equals δ at a value of ζ approximately equal to 5; setting $y = \delta$ and $\zeta = 5$ in Eq. (15.25) then gives

$$\delta(x) = 5\left(\frac{\eta x}{\rho U}\right)^{1/2} = 5x\, \mathrm{Re}_x^{-1/2} \qquad (15.32)$$

This result is consistent with the ordering analysis in Eq. (15.9), where the numerical factor of 5 could not be obtained.

The shear stress at the wall of the plate can be calculated by the use of Eq. (15.27b):

$$\tau_w = \eta \frac{\partial v_x}{\partial y}\bigg|_{y=0} = \eta U \left(\frac{\rho U}{\eta x}\right)^{1/2} f''(0) = 0.332 \eta U \left(\frac{\rho U}{\eta x}\right)^{1/2} \qquad (15.33)$$

τ_w is about half again as large as the crudest estimate, $\eta U/\delta(x)$. Equation (15.33) is in good agreement with experimental measurements of frictional drag.

The solution obtained here is only found experimentally over the first portion of the plate. There is a transition to turbulence when Re_x exceeds a value in the range 2×10^5 to 6×10^5. Turbulent boundary layers do not grow according to the $x^{1/2}$ relation in Eq. (15.32). It is interesting to note that the transition to turbulence occurs at a Reynolds number computed on the basis of boundary layer thickness, $\rho U \delta(x)/\eta$, of about 3000, which is the same order of magnitude as the transition to turbulence in parallel flow in a closed conduit.

15.5 FLOW PAST A WEDGE

The potential flow past a wedge of angle β (Fig. 15-1) is given by

$$U(x) = Ax^m \qquad (15.34a)$$

$$m = \frac{\beta}{2\pi - \beta} \qquad (15.34b)$$

The flat plate is the limiting case $\beta = 0$, while the plane stagnation flow in Sec. 14.6 corresponds to $\beta = \pi$, $m = 1$. Flow past a wedge of any angle has the interesting property that the time for a fluid particle in the potential flow to move from the leading edge to a downstream position x is proportional to $x/U(x)$:

$$t = \int_0^x \frac{dx}{U} = \int_0^x \frac{dx}{Ax^m} = \frac{1}{1-m}\frac{x}{Ax^m} = \frac{1}{1-m}\frac{x}{U(x)} \qquad (15.35)$$

Thus, the "accelerated flat plate" logic used in the preceding section, with t replaced by x/U, might be expected to work for all wedges. In that case we would again define ζ by Eq. (15.25), with the understanding that $U = Ax^m$, and seek a solution in the form

$$v_x = U(x)f'(\zeta) \qquad (15.36)$$

This transformation does work, and an ordinary differential equation is obtained:

$$f''' + \frac{m+1}{2}ff'' + m(1 - f'^2) = 0 \qquad (15.37)$$

Boundary conditions are given by Eqs. (15.31). The case $m = 0$ is Eq. (15.30). Equation (15.37) has been solved numerically for a range of values of m.

The stagnation flow, $m = 1$, has one particularly interesting feature. x/U is a constant, so ζ is independent of x and the boundary layer thickness is the same everywhere. If we use the estimate $A = 2U_\infty/R$ [Eq. (14.27)], where R is the radius of a cylinder, then the boundary layer thickness everywhere in the neighborhood of the forward stagnation point is of order $\delta/R = \mathrm{Re}^{-1/2}$, where Re is based on cylinder diameter.

15.6 INTEGRAL MOMENTUM APPROXIMATION

The integral momentum, or von Kármán–Polhausen, approximation was described in Sec. 9.5. This approximation is often used to obtain solutions to the boundary layer equations, and we shall demonstrate it by application to the problem of flow over a flat plate at zero angle of incidence. (Some common terminology is of interest here. The preceding two sections have dealt with *exact solutions*, meaning, of course, exact solutions to a set of approximate equations; compare Sec. 11.6. Solutions such as the one to be described here are called *approximate solutions*. Exact solutions are philosophically preferable, because the approximation in deriving the starting equations is usually better understood than the approximations involved in solution. As a practical matter, however, approximate solutions are usually required.)

There are several approaches to the integral momentum method, as outlined in Secs. 9.5 and 7.6, and all lead to equivalent results. We shall use the method of direct integration of the differential equation. Our starting point is Eq. (15.22), which we integrate over y from $y = 0$ to $y = \infty$:

$$\rho \int_0^\infty \left(v_x \frac{\partial v_x}{\partial x} + v_y \frac{\partial v_x}{\partial y} \right) dy = \eta \int_0^\infty \frac{\partial^2 v_x}{\partial y^2} \, dy = -\eta \left. \frac{\partial v_x}{\partial y} \right|_{y=0} = -\tau_w \quad (15.38)$$

τ_w is the shear stress at the wall. We have used the fact that $\partial v_x/\partial y$ is zero for $y \rightarrow \infty$, since v_x approaches the constant velocity U in the potential flow outside the boundary layer.

The integral of $v_y \, \partial v_x/\partial y$ can be expressed entirely in terms of v_x and U through an integration by parts:

$$\int_0^\infty v_y \frac{\partial v_x}{\partial y} \, dy = \left. v_y v_x \right|_0^\infty - \int_0^\infty v_x \frac{\partial v_y}{\partial y} \, dy = \left. U v_y \right|_{y=\infty} + \int_0^\infty v_x \frac{\partial v_x}{\partial x} \, dy \quad (15.39)$$

Here, we have made use of the no-slip condition at $y = 0$, the condition $v_x = U$ as $y \rightarrow \infty$, and the continuity equation. Also, by integration of the continuity equation,

$$\left. v_y \right|_{y=\infty} = -\int_0^\infty \frac{\partial v_x}{\partial x} \, dy \quad (15.40)$$

Substitution of Eqs. (15.39) and (15.40) with Eq. (15.38) thus leads to an alternative form,

$$\tau_w = -\rho \int_0^\infty \left(2v_x \frac{\partial v_x}{\partial x} - U \frac{\partial v_x}{\partial x} \right) dy \quad (15.41)$$

Since $2v_x \, \partial v_x/\partial x = \partial(v_x)^2/\partial x$, and U is independent of x, Eq. (15.41) can also be written as

$$\tau_w = -\rho \int_0^\infty \left(\frac{\partial v_x^2}{\partial x} - \frac{\partial U v_x}{\partial x} \right) dy \quad (15.42)$$

Finally, interchanging the order of integration with respect to y and differentiation with respect to x, we obtain the equivalent of Eq. (9.20):

$$\tau_w = \rho \frac{d}{dx} \int_0^\infty v_x(U - v_x) \, dy \quad (15.43)$$

We now make use of our physical understanding of boundary layer flows. Outside the boundary layer v_x is nearly equal to U, although we know that the approach is asymptotic. For purposes of approximation *we assume that v_x is indentically equal to U for $y \geq \delta$*. In that case the integrand in Eq. (15.43) is zero for $y \geq \delta$, and we can replace the upper limit of ∞ with δ:

$$\tau_w = \rho \frac{d}{dx} \int_0^\delta v_x (U - v_x) \, dy \tag{15.44}$$

We further assume that the dimensionless velocity, v_x/U, is a function only of dimensionless distance from the wall, y/δ:

$$v_x = U\phi \left(\frac{y}{\delta}\right) \tag{15.45}$$

We know that this assumption is valid for flow past a plate, and in fact for all wedge flows, since we have seen that the exact solution is a similarity solution. Equation (15.45) is always used in this approximation, however, even in cases where a similarity solution cannot exist.

The no-slip condition and the requirement that $v_x = U$ at $y = \delta$ impose two conditions on the function ϕ:

$$\phi(0) = 0 \qquad \phi(1) = 1 \tag{15.46}$$

The wall shear stress is expressed in terms of ϕ by

$$\tau_w = \eta \frac{\partial v_x}{\partial y}\bigg|_{y=0} = \frac{\eta U}{\delta} \phi'(0) \tag{15.47}$$

The prime denotes differentiation with respect to the argument, y/δ. Equation (15.44) can then be written in terms of the function ϕ as

$$\frac{\eta U}{\delta} \phi'(0) = \rho U^2 \frac{d}{dx} \int_0^\delta \phi\left(\frac{y}{\delta}\right)\left[1 - \phi\left(\frac{y}{\delta}\right)\right] dy \tag{15.48}$$

Finally, it is convenient to define a new variable, $\xi = y/\delta$, and rewrite Eq. (15.48) as

$$\frac{\eta U}{\delta} \phi'(0) = \rho U^2 \frac{d}{dx} \delta \int_0^1 \phi(\xi)[1 - \phi(\xi)] \, d\xi \tag{15.49}$$

Note that the term $\int_0^1 \phi(\xi)[1 - \phi(\xi)] \, d\xi$ is a constant, independent of x, whose value is fixed when the function ϕ is specified.

Equation (15.49) is a differential equation for the boundary layer thickness, $\delta(x)$. The equation can be rewritten

$$2\delta \frac{d\delta}{dx} = \frac{d\delta^2}{dx} = \frac{2\eta}{\rho U} \frac{\phi'(0)}{\int_0^1 \phi(\xi)[1 - \phi(\xi)] \, d\xi} \tag{15.50}$$

With the assumption that the boundary layer does not begin to form until the leading edge of the plate, and hence $\delta(0) = 0$, Eq.(15.50) can be integrated at once to obtain

$$\delta(x) = \left\{ \frac{2\phi'(0)}{\int_0^1 \phi(\xi)[1 - \phi(\xi)]\, d\xi} \right\}^{1/2} \left(\frac{\eta x}{\rho U}\right)^{1/2} \tag{15.51}$$

Similarly, the wall shear stress is

$$\tau_w = \eta \frac{\partial v_x}{\partial y}\Big|_{y=0} = \left\{\frac{1}{2}\phi'(0) \int_0^1 \phi(\xi)[1 - \phi(\xi)]\, d\xi \right\}^{1/2} \eta U \left(\frac{\rho U}{\eta x}\right)^{1/2} \tag{15.52}$$

Except for the numerical coefficients, which depend on the still-unspecified function $\phi(\xi)$, Eqs. (15.51) and (15.52) are identical in form to the boundary layer thickness and wall shear stress obtained from the exact solution, Eqs. (15.32) and (15.33).

The function $\phi(\xi)$ must be completely specified in order to evaluate the coefficients in Eqs. (15.51) and (15.52) and to obtain, if desired, the complete velocity distribution from Eq. (15.45). If we should choose to approximate the velocity by a cubic, which has four coefficients, then the two conditions given by Eq. (15.46) are not sufficient to completely specify ϕ, and two more are required. A third is obtained by noting that since $v_x = v_y = 0$ at $y = 0$, the boundary layer equation, Eq. (15.22), requires that $\partial^2 v_x/\partial y^2 = 0$, or

$$\phi''(0) = 0 \tag{15.53}$$

A fourth condition is continuity of the shear stress at $y = \delta$, which requires that $\partial v_x/\partial y = 0$ at $y = \delta$, or

$$\phi'(1) = 0 \tag{15.54}$$

The numerical coefficients are relatively insensitive to the function used. If we write the boundary layer thickness and wall shear stress in the forms

$$\delta(x) = a\left(\frac{\eta x}{\rho U}\right)^{1/2} \tag{15.55a}$$

$$\tau_w = b\eta U \left(\frac{\rho U}{\eta x}\right)^{1/2} \tag{15.55b}$$

then the values of a and b given in Table 15-1 for various functions $\phi(\xi)$ do

TABLE 15-1

COEFFICIENTS OF $\delta(x)$ AND τ_w FOR VARIOUS
APPROXIMATING FUNCTIONS FOR THE INTEGRAL
MOMENTUM SOLUTION OF FLOW OVER A FLAT PLATE

$\phi(\xi)$	a	b
Exact solution	5 (asymptotic)	0.332
ξ	3.5	0.289
$\frac{3}{2}\xi - \frac{1}{2}\xi^3$	4.6	0.323
$\sin(\pi\xi/2)$	4.8	0.327

not differ significantly from the exact solution. Note that the approximating function $\phi(\xi)$ need not be a polynomial.

15.7 ENTRY LENGTH FOR LAMINAR FLOW

One of the simple results that can be obtained from boundary layer theory is a rough order-of-magnitude estimate of the entry length in laminar flow. The system is shown schematically in Fig. 15-4. Fluid enters the conduit with a

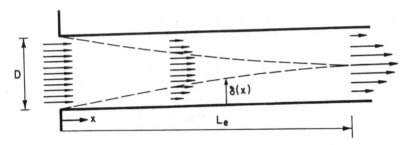

Figure 15-4. Schematic of flow development in the entry region of a conduit.

flat velocity profile; the profile evolves into the parabolic profile of Poiseuille flow and is essentially fully developed a distance L_e downstream of the entrance. The developing profile can be visualized as the growth of a boundary layer along the wall, with the centerline velocity playing the role of the free stream velocity, U. The flat plate theory should be very roughly applicable here, since the centerline velocity does not change much in the developing region; at the entry the centerline velocity equals the mean velocity, while in the fully developed region it equals twice the mean velocity for flow in a tube [Eq. (8.39)] and $\frac{3}{2}$ the mean velocity for plane Poiseuille flow [Eq. (8.15)].

According to flat plate boundary layer theory, the boundary layer thickness is approximately

$$\delta(x) = 5\left(\frac{\eta x}{\rho U}\right)^{1/2} \tag{15.32}$$

We can estimate L_e as the distance from the leading edge at which the boundary layers growing from opposite sides meet at the centerline. Thus, $x = L_e$ for $\delta = D/2$, or

$$\frac{D}{2} \approx 5\left(\frac{\eta L_e}{\rho U}\right)^{1/2} \tag{15.56}$$

This can be rearranged to

$$\frac{L_e}{D} \approx 0.01 \, \text{Re} \tag{15.57}$$

where Re is based on the mean velocity and the diameter. Given the rather crude estimates used here, this is good agreement with the experimental entry length of approximately 0.055Re, Eq. (3.14a). Order-of-magnitude estimates of this type, using theories which apply to idealizations of the true situation, are common and important engineering tools. More detailed solutions, which attempt to account for the changing centerline velocity, give coefficients of Re in the range 0.028 to 0.068.

15.8 FLOW PAST A CYLINDER AND BOUNDARY LAYER SEPARATION

The derivation of the boundary layer equations was carried out in Cartesian coordinates for flow over a plane surface. Because we are dealing with a very thin region, however, the equations also apply to flow over a curved surface, as long as the curvature does not change suddenly. For example, we can apply the boundary layer equations to flow past a circular cylinder, as shown in Fig. 15-5. Here, the x coordinate is the arc length along the surface,

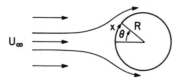

Figure 15-5. Schematic of boundary layer flow past a cylinder.

measured from the forward stagnation point, and the y coordinate is distance normal to the surface. The relation between x and the angle θ is

$$x = R\theta \qquad (15.58)$$

The potential flow past a cylinder is given by Eqs. (14.23). The component v_θ, evaluated at the surface $r = R$, corresponds to $U(x)$ in the boundary layer approximation. Using Eq. (15.58) for θ, we therefore have

$$U(x) = 2U_\infty \sin\left(\frac{x}{R}\right) \qquad (15.59)$$

The boundary layer equations can be solved analytically for this flow, using a very tedious series method for an exact solution or using the integral momentum method, but we shall not present the details here. The exact solution is *not* a similarity solution; this is not surprising, because the time that a fluid particle in the free stream is influenced by the solid surface is not proportional to $x/U(x)$. The velocity profile is shown for various angles in Fig. 15-6, and the wall shear stress in Fig. 15-7.

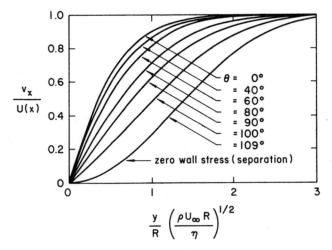

Figure 15-6. Velocity profiles as a function of angle for boundary layer flow past a cylinder.

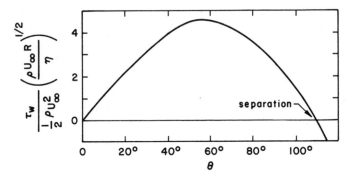

Figure 15-7. Wall shear stress as a function of angle for boundary layer flow past a cylinder.

This flow illustrates an important feature of flows past objects in which there is an *adverse pressure gradient* (i.e., the pressure increases with distance). According to the pressure distribution given by Eq. (14.26), such an adverse gradient occurs here for angles greater than 90° in flow past the cylinder. The increasing pressure causes a deceleration of fluid in the free stream, since the Bernoulli equation requires a trade-off between pressure and kinetic energy terms. The fluid in the boundary layer has less inertia than the fluid in the free stream, and thus the deceleration in the boundary layer is greater. Under some conditions, as shown for the cylinder in Figs. 15-6 and 15-7 for $\theta = 109°$, the deceleration in the boundary layer is sufficient to cause a zero velocity gradient and zero shear stress at the surface. Beyond this point there will be a backflow at the surface, the transition region will move from the surface out into the free stream, and the assumption of a thin boundary layer will break down. This phenomenon of *boundary layer separation* leads to the formation of a wake behind the body, where the pressure and velocity distribution is quite different from that given by the potential flow solution.

Boundary layer separation causes a significant increase in the drag on a body because of the additional energy dissipation in the wake. Aerodynamic "streamlining" of airplanes and automobiles is carried out to delay or prevent the onset of separation of the boundary layer. The very dramatic drop in the drag coefficient for cylinders and spheres at Reynolds numbers above 2×10^5, shown in Fig. 4-8, is a result of a transition to turbulence in the boundary layer and a resulting shift in the point of separation toward the rear of the body. The delay in separation more than offsets the increased wall shear stress in turbulent flow.

15.9 CONCLUDING REMARKS

The boundary layer approximation represents one of the major methods for obtaining solutions to the Navier–Stokes equations, and the results are particularly valuable in studies of mass and heat transfer near solid surfaces. The estimate of the order of magnitude of the boundary layer thickness, Eq. (15.9), is often useful in making preliminary estimates and should be committed to memory.

Exact solutions of the boundary layer equations, both for the two-dimensional flows studied here and for three-dimensional flows, have been obtained for a large number of shapes, usually through series expansions. Many more cases have been studied using the integral momentum method, and results are also available for boundary layer flows of non-Newtonian fluids.

We have not dealt here with boundary layer separation in any detail, but it is an important concept, for naive application of boundary layer results for the calculation of mass and heat transfer coefficients can lead to large errors in cases where separation is important.

BIBLIOGRAPHICAL NOTES

The basic text is

SCHLICHTING, H., *Boundary Layer Theory*, 6th ed., McGraw-Hill Book Company, New York, 1968.

See also the chapter by Howarth,

HOWARTH, L., "Laminar Boundary Layers," *Encyclopedia of Physics* (*Handbuch der Physik*), Vol. 8, No. 1, Springer-Verlag, Berlin, 1958.

There are some excellent photographs of flow visualization experiments on boundary layer flows and separation in

PRANDTL, O., and TIETJENS, O. G., *Applied Hydro- and Aeromechanics*, Dover Publications, New York, 1957.

and in chapters by Abernathy and Hazen in

Illustrated Experiments in Fluid Mechanics, MIT Press, Cambridge, Mass., 1972.

The latter is based on a fine series of films produced by the National Committee for Fluid Mechanics Films.

The boundary layer approximation can be approached from the point of view of the mathematical theory of singular perturbations, introduced in Sec. 11.8. See Sec. 4.2 of

COLE, J. D., *Perturbation Methods in Applied Mathematics*, Blaisdell/Ginn, Waltham, Mass., 1968.

For the extension to non-Newtonian fluids, see

ACHARYA, A., MASHELKAR, R. A., and ULBRECHT, J., *Rheol. Acta*, **15**, 454 (1976).
ACRIVOS, A., SHAH, M. J., and PETERSON, E. E., *Chem. Eng. Sci.*, **20**, 101 (1965).
DENN, M. M., *Chem. Eng. Sci.*, **22**, 395 (1967).
SCHOWALTER, W. R., *AIChE J.*, **6**, 27 (1960).

There are interesting historical discussions of boundary layer theory and reminiscences of Prandtl in

FLÜGGE-LOTZ, I., and FLÜGGE, W., *Ann. Rev. Fluid Mech.*, **5**, 1 (1973).
TANI, I., *Ann. Rev. Fluid Mech.*, **9**, 87 (1977).

PROBLEMS

15.1. Equations (15.29) and (15.40) indicate that there is a finite value of v_y outside the boundary layer over a flat plate, although such a component would not exist in an inviscid fluid.

  a) Using Eq. (15.29), show that this result is consistent with the continuity equation.

  b) Explain the finite v_y *qualitatively* by means of a macroscopic mass balance over the control volume shown in Fig. 15P1.

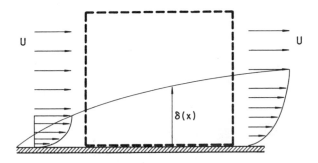

15.2. Compute the maximum thickness of a laminar boundary layer over a flat plate, and the distance from the leading edge at which it occurs for (a) air, (b) water, and (c) glycerine flowing at a velocity of 3 m/s at atmospheric pressure and 20°C.

15.3. Comment on the "drag crisis" in Fig. 4.1 at Re $\sim 2 \times 10^5$ for flow past a sphere in terms of the criterion for the onset of turbulence in the boundary layer.

15.4. Apply the integral momentum method to obtain the boundary layer thickness over a wedge of arbitrary angle (Eqs. (15.34)). Show that the boundary layer thickness is constant for stagnation flow, $m = 1$.

15.5. Derive the boundary layer equations for a power law fluid (Eq. 8.86) and show that a similar solution can be obtained for flow over a flat plate at zero angle of incidence. (A similar solution can in fact be obtained for flow over a wedge of arbitrary angle.)

Part VI
Advanced Topics

Turbulence **16**

16.1 INTRODUCTION

The exact and approximate solutions to the Navier–Stokes equations that we have obtained thus far have all been for laminar flow conditions. Indeed, the procedures that we have developed are *necessarily* restricted to laminar flow, inasmuch as the methods require an understanding of the general structure of the flow field which is to be calculated.

Turbulent flows are chaotic and require special techniques for analysis. Such flows are characterized by rapid, apparently random fluctuations in the flow variables. The response of a velocity probe at a point in a turbulent flow field is shown in Fig. 16-1. The heavy line denotes the mean value about which the actual velocity fluctuates. The fluctuations are rapid, and a probe with sufficiently slow response will be unable to follow the fluctuations and will record only the average. When we measure a macroscopic property, such as the friction factor in Fig. 3-1, we are measuring a mean property.

We shall restrict ourselves in this chapter to incompressible Newtonian fluids, although it should be evident that many parts of the preliminary development are expressed in terms of the stresses and apply to any incompressible fluid. We are therefore seeking an approximate solution to the Navier–Stokes equations under turbulent conditions, where we must take explicit account of the random fluctuations in velocity and pressure. The most fruitful approach is to recognize that in many applications we are interested only in average quantities, and thus we will obtain relations defin-

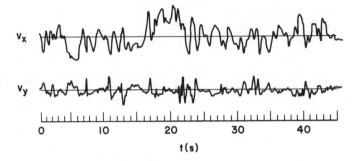

Figure 16-1. Velocity at a point as a function of time in turbulent flow in an oil channel, Re = 7000. x is the direction of mean flow and y is the direction normal to the wall. Reproduced from Wallace, Eckelmann, and Brodkey, *J. Fluid Mech.*, **54**, 39 (1972), copyright by Cambridge University Press, by permission.

ing the *average* behavior over a time scale that is long compared to the time characteristic of fluctuations about the mean. We shall therefore derive the time average of the Navier–Stokes and continuity equations in the next section. This averaging is straightforward, but it does require some care because of the nonlinear nature of the equations. The remainder of the chapter will then be given over to obtaining approximate solutions to these time-averaged equations.

16.2 TIME AVERAGING

16.2.1 Basic Relations

The average $\bar{\xi}$ of any quantity ξ, for which we have N measurements, $\xi_1, \xi_2, \ldots, \xi_N$, is defined as

$$\bar{\xi} = \frac{1}{N} \sum_{i=1}^{N} \xi_i \tag{16.1}$$

If ξ varies continuously in time, and we can make continuous measurements $\xi(\tau)$ over a time interval $t - T \leq \tau \leq t + T$, then the analogous definition of the average is

$$\bar{\xi}(t) = \frac{1}{2T} \int_{t-T}^{t+T} \xi(\tau)\, d\tau \tag{16.2}$$

Note that $\bar{\xi}$ might be a function of time, t. Equation (16.2) defines a *running average*.

A *stationary* process is one in which the average, $\bar{\xi}$, does not change over time as long as T is large compared to the time scale of fluctuations. In a steady turbulent flow, such as flow in a pipe for $Re > 2100$, the average quantities, such as flow rate and pressure drop, do not change in time, so we are interested in stationary processes. *We shall restrict ourselves to stationary processes, or processes which are steady in the sense that averages over time do not themselves change with time.* This will include most pipe and boundary layer flows of processing importance, but may of course exclude some cases of interest. For stationary process we can simply treat the time scale of measurement as unbounded, and define the average by

$$\bar{\xi} = \lim_{T \to \infty} \frac{1}{2T} \int_{t-T}^{t+T} \xi(\tau) \, d\tau \tag{16.3}$$

We now turn explicitly to flow variables. Consider the velocity component v_x, which is a function of position and time, $v_x(x, y, z, t)$. The time-averaged velocity component, $\overline{v_x}$, is a function only of position in a steady flow:

$$\overline{v_x}(x, y, z) = \lim_{T \to \infty} \frac{1}{2T} \int_{t-T}^{t+T} v_x(x, y, z, \tau) \, d\tau \tag{16.4}$$

The mean value at a point is a useful frame of reference, and we generally consider fluctuations about the mean. Fluctuations about the mean are conventionally written with a prime,*

$$v'_x(x, y, z, t) = v_x(x, y, z, t) - \overline{v_x}(x, y, z) \tag{16.5}$$

Note that the time average of a fluctuation must be zero. The integral of a sum is the sum of the integrals, so we may write the time average of Eq. (16.5) as

$$\overline{v'_x} = \overline{v_x} - (\overline{v_x}) = \overline{v_x} - \overline{v_x} = 0 \tag{16.6}$$

We have made use here of the fact that the time average is a constant in time, and the time average of a constant is the same constant; hence, the time average of any time average is simply the time average itself $(\overline{\overline{v_x}} = \overline{v_x})$.

We will have occasion to consider terms of the form $\overline{v_y} \, \partial v'_x / \partial y$. The time average of such a term is zero, as the following shows:

$$\overline{\overline{v_y} \frac{\partial v'_x}{\partial y}} = \lim_{T \to \infty} \frac{1}{2T} \int_{t-T}^{t+T} \overline{v_y} \frac{\partial v'_x}{\partial y} \, d\tau = \overline{v_y} \lim_{T \to \infty} \frac{1}{2T} \int_{t-T}^{t+T} \frac{\partial v'_x}{\partial y} \, d\tau$$

$$= \overline{v_y} \frac{\partial}{\partial y} \lim_{T \to \infty} \frac{1}{2T} \int_{t-T}^{t+T} v'_x \, d\tau = \overline{v_y} \frac{\partial \overline{v'_x}}{\partial y} = 0 \tag{16.7}$$

We can take $\overline{v_y}$ out of the integration because it is a constant, independent of

---

*This nomenclature is unfortunate, because we have previously used the prime to denote a derivative, but it is standard in the turbulence literature and is therefore used here.

time, and we can take $\partial/\partial y$ outside because we can interchange the order of spatial differentiation and time integration.* The derivative of $\overline{v'_x}$ is zero because $\overline{v'_x}$ itself is identically zero. This manipulation is easily generalized to show that *the time average of any term linear in a fluctuating quantity is zero.*

16.2.2 Time-Averaged Continuity Equation

The continuity equation for an incompressible fluid is

$$\frac{\partial v_x}{\partial x} + \frac{\partial v_y}{\partial y} + \frac{\partial v_z}{\partial z} = 0 \tag{16.8}$$

Introducing the time-averaged and fluctuating velocity components, $v_x = \overline{v_x} + v'_x$, $v_y = \overline{v_y} + v'_y$, $v_z = \overline{v_z} + v'_z$, we have

$$\frac{\partial \overline{v_x}}{\partial y} + \frac{\partial v'_x}{\partial y} + \frac{\partial \overline{v_y}}{\partial y} + \frac{\partial v'_y}{\partial y} + \frac{\partial \overline{v_z}}{\partial z} + \frac{\partial v'_z}{\partial z} = 0 \tag{16.9}$$

If the sum of terms on the left is always zero, then its time average must also always be zero, so we can write

$$\lim_{T \to \infty} \frac{1}{2T} \int_{t-T}^{t+T} \left(\frac{\partial \overline{v_x}}{\partial x} + \frac{\partial v'_x}{\partial x} + \frac{\partial \overline{v_y}}{\partial y} + \frac{\partial v'_y}{\partial y} + \frac{\partial \overline{v_z}}{\partial z} + \frac{\partial v'_z}{\partial z} \right) d\tau = 0 \tag{16.10}$$

or, carrying out the integration (averaging) term by term,

$$\frac{\partial \overline{v_x}}{\partial x} + \frac{\partial \overline{v'_x}}{\partial x} + \frac{\partial \overline{v_y}}{\partial y} + \frac{\partial \overline{v'_y}}{\partial y} + \frac{\partial \overline{v_z}}{\partial z} + \frac{\partial \overline{v'_z}}{\partial z} = 0 \tag{16.11}$$

But the time averages of the fluctuating quantities are zero, so Eq. (16.11) simplifies to

$$\frac{\partial \overline{v_x}}{\partial x} + \frac{\partial \overline{v_y}}{\partial y} + \frac{\partial \overline{v_z}}{\partial z} = 0 \tag{16.12}$$

By subtracting Eq. (16.12) from Eq. (16.9) we also obtain a useful relation between the fluctuating velocity components:

$$\frac{\partial v'_x}{\partial x} + \frac{\partial v'_y}{\partial y} + \frac{\partial v'_z}{\partial z} = 0 \tag{16.13}$$

16.2.3 Time-Averaged Momentum Equation

The momentum equation is most conveniently expressed in terms of the stresses. From Table 7-2 we obtain the x component in terms of the equivalent pressure as

$$\rho \left(\frac{\partial v_x}{\partial t} + v_x \frac{\partial v_x}{\partial x} + v_y \frac{\partial v_x}{\partial y} + v_z \frac{\partial v_x}{\partial z} \right) = -\frac{\partial \mathcal{P}}{\partial x} + \frac{\partial \tau_{xx}}{\partial x} + \frac{\partial \tau_{yx}}{\partial y} + \frac{\partial \tau_{zx}}{\partial z} \tag{16.14}$$

*There are some technical considerations in such an interchange that are dealt with in courses in calculus.

When this equation is time-averaged, we have

$$\rho\left(\overline{\frac{\partial v_x}{\partial t}} + \overline{v_x\frac{\partial v_x}{\partial x}} + \overline{v_y\frac{\partial v_x}{\partial y}} + \overline{v_z\frac{\partial v_x}{\partial z}}\right) = -\overline{\frac{\partial \mathcal{P}}{\partial x}} + \overline{\frac{\partial \tau_{xx}}{\partial x}} + \overline{\frac{\partial \tau_{yx}}{\partial y}} + \overline{\frac{\partial \tau_{zx}}{\partial z}} \quad (16.15)$$

The first term on the left, $\overline{\partial v_x/\partial t}$, must be zero, since in a randomly fluctuating system there is an equal likelihood of positive and negative rates of change and the two will cancel each other over a sufficiently long time. The averages of the products need some care; note that *the average of a product is not the product of the averages*. For a Newtonian fluid the stress averages are simply expressed in terms of the derivatives of the average velocities; for example,

$$\overline{\tau_{yx}} = \eta\left(\frac{\partial \overline{v_y}}{\partial x} + \frac{\partial \overline{v_x}}{\partial y}\right) \quad (16.16)$$

The last relation follows because of the linearity of the constitutive equation.

We now consider the product terms in Eq. (16.15) by examining a typical term. Using the definitions of the mean and fluctuating quantities, we can write

$$\overline{v_y\frac{\partial v_x}{\partial y}} = \overline{(\overline{v_y} + v_y')\frac{\partial}{\partial y}(\overline{v_x} + v_x')} = \overline{\overline{v_y}\frac{\partial \overline{v_x}}{\partial y}} + \overline{\overline{v_y}\frac{\partial v_x'}{\partial y}} + \overline{v_y'\frac{\partial \overline{v_x}}{\partial y}} + \overline{v_y'\frac{\partial v_x'}{\partial y}}$$

$$(16.17)$$

The first term in the sum is the average of a product of averages; the two terms in the product are independent of time, so the averaging does not change anything. Thus,

$$\overline{\overline{v_y}\frac{\partial \overline{v_x}}{\partial y}} = \overline{v_y}\frac{\partial \overline{v_x}}{\partial y} \quad (16.18)$$

The second term in the sum has already been shown to be zero in Eq. (16.7). The third term is also the time average of a quantity that is linear in a fluctuating component and is therefore also zero by an identical proof. The last term in the sum must be presumed to be nonzero in general, since, even though $\partial v_x'/\partial y$ and v_y' have average values which are zero, there is no reason to presume that their product has a zero average. A simple counterexample is given by the case in which v_y' equals $\partial v_x'/\partial y$; the product is then always positive and cannot have an average value which is zero.

The other nonlinear terms in Eq. (16.15) are treated in the same way, and the time-averaged x component of the momentum equation can now be written:

$$\rho\left(\overline{v_x}\frac{\partial \overline{v_x}}{\partial x} + \overline{v_y}\frac{\partial \overline{v_x}}{\partial y} + \overline{v_z}\frac{\partial \overline{v_x}}{\partial z}\right) = -\frac{\partial \overline{\mathcal{P}}}{\partial x} + \frac{\partial \overline{\tau_{xx}}}{\partial x} + \frac{\partial \overline{\tau_{yx}}}{\partial y} + \frac{\partial \overline{\tau_{zx}}}{\partial z}$$

$$- \rho\left(\overline{v_x'\frac{\partial v_x'}{\partial x}} + \overline{v_y'\frac{\partial v_x'}{\partial y}} + \overline{v_z'\frac{\partial v_x'}{\partial z}}\right) \quad (16.19)$$

The last group of terms can be rearranged by using Eq. (16.13) and the following identities:

$$v'_x \frac{\partial v'_x}{\partial x} = \frac{1}{2} \frac{\partial v'_x v'_x}{\partial x} \tag{16.20a}$$

$$v'_y \frac{\partial v'_x}{\partial y} = \frac{\partial v'_y v'_x}{\partial y} - v'_x \frac{\partial v'_y}{\partial y} \tag{16.20b}$$

$$v'_z \frac{\partial v'_x}{\partial z} = \frac{\partial v'_z v'_x}{\partial z} - v'_x \frac{\partial v'_z}{\partial z} \tag{16.20c}$$

Equation (16.19) then takes the form

$$\rho\left(\overline{v}_x \frac{\partial \overline{v}_x}{\partial x} + \overline{v}_y \frac{\partial \overline{v}_x}{\partial y} + \overline{v}_z \frac{\partial \overline{v}_x}{\partial z}\right) = -\frac{\partial \overline{\mathcal{P}}}{\partial x} + \frac{\partial}{\partial x}(\overline{\tau}_{xx} - \rho\overline{v'_x v'_x})$$

$$+ \frac{\partial}{\partial y}(\overline{\tau}_{yx} - \rho\overline{v'_y v'_x}) + \frac{\partial}{\partial z}(\overline{\tau}_{zx} - \rho\overline{v'_z v'_x}) \tag{16.21a}$$

In a similar way, the y and z components of the momentum equation become

$$\rho\left(\overline{v}_x \frac{\partial \overline{v}_y}{\partial x} + \overline{v}_y \frac{\partial \overline{v}_y}{\partial y} + \overline{v}_z \frac{\partial \overline{v}_y}{\partial z}\right) = -\frac{\partial \overline{\mathcal{P}}}{\partial y} + \frac{\partial}{\partial x}(\overline{\tau}_{xy} - \rho\overline{v'_x v'_y})$$

$$+ \frac{\partial}{\partial y}(\overline{\tau}_{yy} - \rho\overline{v'_y v'_y}) + \frac{\partial}{\partial z}(\overline{\tau}_{zy} - \rho\overline{v'_z v'_y}) \tag{16.21b}$$

$$\rho\left(\overline{v}_x \frac{\partial \overline{v}_z}{\partial x} + \overline{v}_y \frac{\partial \overline{v}_z}{\partial y} + \overline{v}_z \frac{\partial \overline{v}_z}{\partial z}\right) = -\frac{\partial \overline{\mathcal{P}}}{\partial z} + \frac{\partial}{\partial x}(\overline{\tau}_{xz} - \rho\overline{v'_x v'_z})$$

$$+ \frac{\partial}{\partial y}(\overline{\tau}_{yz} - \rho\overline{v'_y v'_z}) + \frac{\partial}{\partial z}(\overline{\tau}_{zz} - \rho\overline{v'_z v'_z}) \tag{16.21c}$$

16.3 REYNOLDS STRESSES

16.3.1 Definition

Equations (16.21) have a structure like the Navier–Stokes equations, except that each of the time-averaged viscous stress terms has a corresponding inertial term associated with the turbulent fluctuations (e.g., τ_{xy} in the Navier–Stokes equations is replaced by $\overline{\tau}_{xy} - \rho\,\overline{v'_x v'_y}$). These fluctuating terms play a role analogous to stresses, and they are sometimes referred to as *turbulent stresses*, or *Reynolds stresses*. The analogy between the time-averaged products of fluctuating velocities and the viscous stresses is reinforced by defining a "turbulent stress" as

$$\begin{aligned}
&\tau_{xx}^{(t)} = -\rho\overline{v'_x v'_x} && \tau_{xy}^{(t)} = \tau_{yx}^{(t)} = -\rho\overline{v'_x v'_y} && \tau_{xz}^{(t)} = \tau_{zx}^{(t)} = -\rho\overline{v'_x v'_z} \\
&\tau_{yy}^{(t)} = -\rho\overline{v'_y v'_y} && \tau_{yz}^{(t)} = \tau_{zy}^{(t)} = -\rho\overline{v'_y v'_z} && \tau_{zz}^{(t)} = -\rho\overline{v'_z v'_z}
\end{aligned} \tag{16.22}$$

The time-averaged momentum equations can then be written

$$\rho\left(\overline{v_x}\frac{\partial \overline{v_x}}{\partial x} + \overline{v_y}\frac{\partial \overline{v_x}}{\partial y} + \overline{v_z}\frac{\partial \overline{v_x}}{\partial z}\right) = -\frac{\partial \overline{\mathcal{P}}}{\partial x} + \frac{\partial}{\partial x}(\overline{\tau_{xx}} + \tau_{xx}^{(t)})$$

$$+ \frac{\partial}{\partial y}(\overline{\tau_{yx}} + \tau_{yx}^{(t)}) + \frac{\partial}{\partial z}(\overline{\tau_{zx}} + \tau_{zx}^{(t)}) \tag{16.23a}$$

$$\rho\left(\overline{v_x}\frac{\partial \overline{v_y}}{\partial x} + \overline{v_y}\frac{\partial \overline{v_y}}{\partial y} + \overline{v_z}\frac{\partial \overline{v_y}}{\partial z}\right) = -\frac{\partial \overline{\mathcal{P}}}{\partial y} + \frac{\partial}{\partial x}(\overline{\tau_{xy}} + \tau_{xy}^{(t)})$$

$$+ \frac{\partial}{\partial y}(\overline{\tau_{yy}} + \tau_{yy}^{(t)}) + \frac{\partial}{\partial z}(\overline{\tau_{zy}} + \tau_{zy}^{(t)}) \tag{16.23b}$$

$$\rho\left(\overline{v_x}\frac{\partial \overline{v_z}}{\partial x} + \overline{v_y}\frac{\partial \overline{v_z}}{\partial y} + \overline{v_z}\frac{\partial \overline{v_z}}{\partial z}\right) = -\frac{\partial \overline{\mathcal{P}}}{\partial z} + \frac{\partial}{\partial x}(\overline{\tau_{xz}} + \tau_{xz}^{(t)})$$

$$+ \frac{\partial}{\partial y}(\overline{\tau_{yz}} + \tau_{yz}^{(t)}) + \frac{\partial}{\partial z}(\overline{\tau_{zz}} + \tau_{zz}^{(t)}) \tag{16.23c}$$

Equations (16.12) and (16.23) are not sufficient to provide a solution to any turbulent flow problem. The mean stresses $\overline{\tau_{ij}}$ are easily related to the mean velocity through the appropriate constitutive equation, as in Eq. (16.16) for a Newtonian fluid. We need to be able to say something about the Reynolds stresses $\tau_{ij}^{(t)}$, however, if we are to obtain a complete set of equations. Determining the behavior of quantities such as $\overline{v_x' v_y'}$ is a central problem of modern turbulence theory, and a full solution is not available.

16.3.2 Eddy Viscosity

We might anticipate that the Reynolds stresses will depend on the intensity of the mean flow, and hence on the mean velocity gradients. Boussinesq suggested in 1877 that the Reynolds stresses be written in analogy to the viscous stresses as

$$\tau_{xy}^{(t)} = \eta_{\text{eddy}}\left(\frac{\partial \overline{v_x}}{\partial y} + \frac{\partial \overline{v_y}}{\partial x}\right) \tag{16.24}$$

and similarly for the other components. The notion of an *eddy viscosity* is often used in describing turbulent flows. Clearly, η_{eddy} cannot be a constant, however, for otherwise we would simply recover the usual Navier–Stokes equations with η replaced by $(\eta + \eta_{\text{eddy}})$, and we would lose the character of the turbulence. Furthermore, η_{eddy} must depend explicitly on position, for in a confined flow the fluctuating velocity $\mathbf{v}'$ must go to zero at a wall, and hence all components $\tau_{ij}^{(t)}$ must vanish. Some of the gradients of the mean flow in Eq. (16.24) will be finite at the wall, however, so η_{eddy} must go to zero there. Thus, the eddy viscosity is simply an alternative means of giving a name to our essential ignorance of the physics of the process. Moreover, it is a restrictive form of nomenclature in that it presumes that a single parameter or function which is independent of direction will suffice to describe the turbulence.

16.3.3 Mixing Length

We shall abandon any attempt at generality in our treatment of turbulence and concentrate on one-dimensional shear flows, where the problem of directionality does not become important. The usual means of analyzing the "shear" Reynolds stress $\tau_{xy}^{(t)}$ is based on an approach of Prandtl, published in 1933. He reasoned that the motion of fluid eddies in a turbulent flow field is analogous to the random motion of molecules in a gas. Using results from the kinetic theory of gases he was then able to reason that the eddy viscosity in a confined shear flow should have the form

$$\eta_{\text{eddy}} = \rho\ell^2 \left| \frac{d\bar{v}_x}{dy} \right| \tag{16.25}$$

where x is the direction of mean flow and y the direction normal to the wall. ℓ is known as the *mixing length* and is analogous to the mean free path in a gas; the physical interpretation of ℓ is that it is the distance over which a turbulent eddy retains its identity. The absolute value of $d\bar{v}_x/dy$ is required in Eq. (16.25) to ensure that $\tau_{xy}^{(t)}$ changes algebraic sign when the flow field changes direction.

The analogy between turbulence and molecular motion in a gas is, in fact, a poor one, but Eq. (16.25) is nevertheless quite useful. The equation may be inferred from several other starting points, including a dimensional argument which uses the fact that the turbulence is characterized by an eddy length scale ℓ and a fluctuating velocity v'. The several approaches are described in the references cited in the Bibliographical Notes. The problem still remains of ensuring that η_{eddy} go to zero at the wall. In the context of Eq. (16.24) this requires that the mixing length ℓ be a function of distance from the wall. The simplest such function, which we may consider to be a first-order estimate, is to take ℓ as proportional to the distance from the wall. This is the assumption made by Prandtl.

16.4 TURBULENT PIPE FLOW

16.4.1 Mean Flow Equation

We can now attempt to obtain the detailed mean velocity field for fully developed turbulent flow in a pipe. Although Eqs. (16.23) were derived using rectangular Cartesian coordinates, it is evident that the corresponding equations in cylindrical coordinates are obtained by replacing τ_{ij} with $\overline{\tau_{ij}} + \tau_{ij}^{(t)}$ in each term of Table 7-5.

We assume that there will be angular symmetry, and that there will be no variation in mean quantities in the z (flow) direction. In that case it follows (compare Sec. 8.4) that $\overline{v}_z$ is a function only of r, $\overline{v}_r$ and $\overline{v}_\theta$ are zero, $\partial\overline{\mathscr{P}}/\partial z$ is a constant, and the time-averaged z component of the momentum equation is

$$0 = -\frac{d\overline{\mathscr{P}}}{dz} + \frac{1}{r}\frac{d}{dr}r(\overline{\tau}_{rz} + \tau_{rz}^{(t)}) \qquad (16.26)$$

This can be integrated once to obtain

$$\overline{\tau}_{rz} + \tau_{rz}^{(t)} = \frac{r}{2}\frac{d\overline{\mathscr{P}}}{dz} \qquad (16.27)$$

If we perform a macroscopic momentum balance over a length of pipe Δz, we find, as in the footnote following Eq. (3.24), that the pressure gradient is balanced by the wall shear stress, τ_w (recall that $d\overline{\mathscr{P}}/dz < 0$):

$$\pi R^2\left(-\frac{d\overline{\mathscr{P}}}{dz}\right)\Delta z = 2\pi R\,\Delta z\tau_w \qquad (16.28)$$

We can then equivalently write Eq. (16.27) as

$$\overline{\tau}_{rz} + \tau_{rz}^{(t)} = -\frac{r}{R}\tau_w \qquad (16.29)$$

The mean viscous stress, $\overline{\tau}_{rz}$, is related to the mean velocity through the constitutive equation for a Newtonian fluid, which simplifies to

$$\overline{\tau}_{rz} = \eta\frac{d\overline{v}_z}{dr} \qquad (16.30)$$

The Reynolds stress, $\tau_{rz}^{(t)}$, is expressed in terms of the Prandtl mixing length, Eq. (16.25), as

$$\tau_{rz}^{(t)} = \rho\ell^2\left|\frac{d\overline{v}_z}{dr}\right|\frac{d\overline{v}_z}{dr} \qquad (16.31)$$

The mixing length, ℓ, is required to be a function of distance from the wall, so it is convenient to locate the origin of the coordinate system at the wall, $r = R$. We therefore define distance from the wall, y, as

$$y = R - r \qquad dy = -dr \qquad (16.32)$$

Equations (16.29) through (16.32) then combine to

$$\rho\ell^2\left(\frac{d\overline{v}_z}{dy}\right)^2 + \eta\frac{d\overline{v}_z}{dy} - \left(1 - \frac{y}{R}\right)\tau_w = 0 \qquad (16.33)$$

We have removed the absolute value sign by assuming that the flow is always in the positive z direction; in that case, $d\overline{v}_z/dy$ is always positive. Equation (16.33) is a first-order, nonlinear, ordinary differential equation for $\overline{v}_z$. The boundary condition is no-slip, $\overline{v}_z = 0$ at $y = 0$. Note that a solution cannot be obtained until the functional dependence of ℓ on y is defined.

16.4.2 Turbulence Variables

Turbulence analyses and data are traditionally reported in terms of a special set of dependent and independent variables. These turbulence variables arise naturally when one examines the important physical processes and scales *a posteriori*, but they may appear a bit strange on first exposure. We shall use the traditional variables here in order to retain consistency with the turbulence literature, fully recognizing that introduction of new variables at this point might provide some initial difficulties.

It is customary to replace the wall shear stress with the *friction velocity*, u_*, defined as

$$u_* = \sqrt{\frac{\tau_w}{\rho}} \tag{16.34}$$

If we substitute the pressure gradient for τ_w, using Eq. (16.28), and introduce the Fanning friction factor, Eq. (3.6), the friction velocity is equivalently defined as

$$u_* = \langle \overline{v_z} \rangle \sqrt{\frac{f}{2}} \tag{16.35}$$

Here, f is the friction factor and $\langle \overline{v_z} \rangle$ is the average velocity, defined as the flow rate divided by the pipe cross-sectional area.

The time-averaged axial velocity is made dimensionless with respect to the friction velocity:

$$u_+ = \frac{\overline{v_z}}{u_*} \tag{16.36}$$

Distance from the wall is made dimensionless with respect to the grouping of variables $\eta/\rho u_*$:

$$y_+ = y \left(\frac{\rho u_*}{\eta} \right) \tag{16.37}$$

The dimensionless pipe radius, R_+, is then

$$R_+ = R \left(\frac{\rho u_*}{\eta} \right) = \text{Re} \sqrt{\frac{f}{8}} \tag{16.38}$$

Here, the Reynolds number, Re, is based on pipe diameter, Eq. (3.5). Similarly, the dimensionless mixing length, ℓ_+, is

$$\ell_+ = \ell \left(\frac{\rho u_*}{\eta} \right) \tag{16.39}$$

Example 16.1

Estimate the range of u_+ and y_+.

We will take Re $= 10^5$ for the calculation. From Fig. 3-1 or Eq. (3.10), we find that f is approximately 4.5×10^{-3}, and $(f/2)^{1/2}$ is a bit less than 0.05.

Thus, from Eq. (16.35),

$$\text{Re} = 10^5: \quad u_+ \approx \frac{1}{0.05}\frac{\overline{v_z}}{\langle v_z \rangle} = 20\frac{\overline{v_z}}{\langle v_z \rangle}$$

The maximum velocity in turbulent flow will be about 20% greater than the mean; compare Fig. 6-7. Thus, u_+ will reach a value of about 24 at the centerline for $\text{Re} = 10^5$. At $\text{Re} = 10^7$ the friction factor drops to 2×10^{-3}, and in that case u_+ will reach a value of about 38 at the pipe centerline.

The maximum value of y_+ in a pipe is given by R_+. From Eq. (16.38), for $\text{Re} = 10^3$ and $f = 4.5 \times 10^{-3}$, $R_+ = 7.5 \times 10^3$. For $\text{Re} = 10^7$, $f = 2 \times 10^{-3}$, and $R_+ = 1.6 \times 10^5$.

16.4.3 Velocity Profile

Equation (16.33) for the time-averaged velocity can be rewritten in terms of the turbulence variables as

$$\ell_+^2 \left(\frac{du_+}{dy_+}\right)^2 + \frac{du_+}{dy_+} - \left(1 - \frac{y_+}{R_+}\right) = 0 \tag{16.40}$$

The equation is a quadratic for du_+/dy_+ and does not involve u_+; it can be solved explicitly for du_+/dy_+ to yield

$$\frac{du_+}{dy_+} = \frac{-1 + \sqrt{1 + 4\ell_+^2(1 - y_+/R_+)}}{2\ell_+^2} \tag{16.41}$$

This can be integrated from $y_+ = 0$, where the no-slip condition requires $u_+ = 0$, to yield

$$u_+ = \int_0^{y_+} \frac{-1 + \sqrt{1 + 4\ell_+^2(1 - y_+/R_+)}}{2\ell_+^2} \, dy_+ \tag{16.42}$$

An alternative form that will be helpful is to integrate from the centerline, $y_+ = R_+$, to yield

$$u_+ = u_{+,\text{max}} + \int_{y_+}^{R_+} \frac{1 - \sqrt{1 + 4\ell_+^2(1 - y_+/R_+)}}{2\ell_+^2} \, dy_+ \tag{16.43}$$

$u_{+,\text{max}}$ is the maximum dimensionless velocity, which occurs at the centerline. Note that in Eqs. (16.42) and (16.43) we have used the same symbol, y_+, for the limit of integration and the dummy variable. This should cause no difficulty, since the integration limit does not appear in the integrand.

Equations (16.42) or (16.43) cannot be integrated until we have specified the functional dependence of the mixing length on the distance from the wall. Available experimental data, which we shall discuss subsequently, suggest that ℓ_+ is a *unique function* of y_+, and further that ℓ_+ is of the order of y_+. Prandtl hypothesized that ℓ_+ is proportional to y_+:

$$\text{Prandtl:} \quad \ell_+ = \kappa y_+ \tag{16.44}$$

This relationship works quite well for y_+ greater than about 30, with $\kappa = 0.4$, but it greatly overestimates the value of ℓ_+ closer to the wall, where the solid surface hinders the mixing mechanisms. Van Driest, in 1956, found that the available data except near the pipe centerline were fit by a form

$$\text{Van Driest:} \quad \ell_+ = \kappa y_+[1 - \exp(-y_+/A)] \tag{16.45}$$

with $\kappa = 0.4$ and $A = 36$. This approaches Prandtl's equation for y_+ greater than A. The integration can be carried out numerically for a given function ℓ_+, but it is more useful to consider analytical relationships that are derived from asymptotic equations in the following sections.

16.4.4 Universality

The only way in which the process length scale enters the integral for the velocity profile, Eq. (16.42), is through the term $1 - y_+/R_+$. As long as y_+ is small compared to R_+ we may neglect y_+/R_+ relative to unity. From Example 16.1, the ratio y_+/R_+ will be negligible for y_+ up to about 10^3 for $\text{Re} = 10^5$ and up to about 10^4 for $\text{Re} = 10^7$. When y_+ is small compared to R_+, we can write Eq. (16.42) as

$$y_+ \ll R_+ = \text{Re}\sqrt{f/8}:$$

$$u_+ \approx \int_0^{y_+} \frac{-1 + \sqrt{1 + 4\ell_+^2}}{2\ell_+^2}\, dy_+ = 2\int_0^{y_+} \frac{dy_+}{1 + \sqrt{1 + 4\ell_+^2}} \tag{16.46}$$

Equation (16.46) involves only distance from the wall, and thus it can be expected to apply to *any* steady turbulent flow near a wall, not just flow near the wall of a circular pipe. We are thus led to the notion of a *universal velocity profile*, where measurements from a variety of flows should coincide when plotted in the form of u_+ versus y_+.

16.4.5 Viscous Sublayer

If we are willing to start with the hypothesis that the mixing length is of the order of the distance from the wall, then we can obtain the velocity profile directly adjacent to the wall without any further approximation. Close to the wall we have $y_+ \ll R_+$, so the starting point is Eq. (16.46). For $\ell_+ \ll 1$ [which corresponds to $y_+ \ll 10$ if we use the empirical Van Driest relation for y_+, Eq. (16.45), with $\kappa = 0.4$ and $A = 36$], the term $(1 + 4\ell_+^2)^{1/2}$ is close to unity, and we can write Eq. (16.46) as

$$\ell_+ \ll 1: \quad u_+ = \int_0^{y_+} (1 + \text{order of } \ell_+^2)\, dy_+ \approx y_+ \tag{16.47}$$

The same result can be obtained directly from Eq. (16.40). The Reynolds stresses must go to zero at the wall, so the viscous terms must dominate there.

Thus, the turbulent stress term, $\rho \ell_+^2 (du_+/dy_+)^2$, must be negligible relative to the viscous stress term, du_+/dy_+. In that case, together with the fact that $y_+/R_+ \ll 1$, we can write Eq. (16.40) as

$$\text{adjacent to wall:} \quad \frac{du_+}{dy_+} = 1 \tag{16.48}$$

which, of course, has the solution $u_+ = y_+$.

Because of the domination of the viscous stresses, the region directly adjacent to the wall is known as the *viscous sublayer*. The linear velocity profile is reminiscent of laminar plane Couette flow, and this region is sometimes called the laminar sublayer, but the name is misleading and should be avoided.

16.4.6 Turbulent Core

We can now obtain a solution for the time-averaged velocity profile away from the wall, $\ell_+ \gg 1$ and $y_+ \gg 10$. The most convenient starting point here is Eq. (16.43). For $\ell_+ \gg 1$, as long as $y_+ < R_+$, the square root term is approximately equal to $2\ell_+(1 - y_+/R_+)^{1/2}$, which is, in turn, large compared to unity. Thus, we can write

$$\ell_+ \gg 1: \quad 1 - \sqrt{1 + 4\ell_+^2(1 - y_+/R_+)} \approx -2\ell_+(1 - y_+/R_+)^{1/2} \tag{16.49}$$

Equation (16.43) can then be written

$$\ell_+ \gg 1: \quad u_+ \approx u_{+,\text{max}} - \int_{y_+}^{R_+} \frac{(1 - y_+/R_+)^{1/2}}{\ell_+} dy_+ \tag{16.50}$$

Equation (16.50) also follows directly from Eq. (16.40) if we assume that away from the laminar sublayer the turbulent stresses dominate the viscous stresses. In that case the term du_+/dy_+ can be neglected in Eq. (16.40), and Eq. (16.50) then follows.

We shall have to specify a form for the mixing length in order to carry out the integration in Eq. (16.50). Following the hypothesis of Prandtl, we shall assume that the mixing length is proportional to distance from the wall and use Eq. (16.44), $\ell_+ = \kappa y_+$; the Van Driest relation, Eq. (16.45), is equivalent for $y_+ \gg A$. We therefore have

$$\ell_+ \gg 1, \ell_+ = \kappa y_+: \quad u_+ = u_{+,\text{max}} - \int_{y_+}^{R_+} \frac{(1 - y_+/R_+)^{1/2}}{\kappa y_+} dy_+ \tag{16.51}$$

The integral can be evaluated analytically to yield

$$\ell_+ \gg 1: \quad u_+ = u_{+,\text{max}} + \frac{2}{\kappa}\sqrt{1 - \frac{y_+}{R_+}} + \frac{1}{\kappa}\ln\frac{1 - \sqrt{1 - y_+/R_+}}{1 + \sqrt{1 - y_+/R_+}} \tag{16.52}$$

Equation (16.52) contains two constants, κ and $u_{+,\text{max}}$; the former should be independent of flow conditions if the constitutive hypothesis regarding ℓ_+ is correct, while the latter will depend on flow conditions. The equation cannot apply in the region immediately adjacent to the wall, because the logarithmic term becomes unbounded as $y_+ \to 0$.

The comparison with experimental data is made easiest by considering the region away from the pipe centerline, say $y_+/R_+ < 0.5$. In that region we can approximate the square root as

$$\sqrt{1 - y_+/R_+} = 1 - \frac{1}{2}\frac{y_+}{R_+} + \ldots \tag{16.53}$$

and write Eq. (16.52) as

$$\ell_+ \gg 1: \quad u_+ = \frac{1}{\kappa}\ln y_+ + \left[\frac{2}{\kappa} - \frac{1}{\kappa}\ln 4R_+ + u_{+,\,\mathrm{max}}\right] \tag{16.54}$$
$$+ \text{ order of } (y_+/R_+)$$

Equation (16.54) suggests that experimental data should be plotted as u_+ versus the *logarithm* of y_+. Figure 16-2 shows data of four investigators over a three-decade range of Reynolds numbers, plotted as u_+ versus y_+ on a base 10 logarithmic scale (ln $y_+ = 2.303 \log_{10} y_+$)*. Data for y_+ less than about 20 are very difficult to obtain because of the proximity to the wall. It is evident that u_+ is nearly linear in log y_+ over three decades of y_+. The data are fit by the straight line

$$y_+ \gg 10: \quad u_+ = 5.75 \log_{10} y_+ + 5.5 = 2.5 \ln y_+ + 5.5 \tag{16.55}$$

It then follows from comparison of Eqs. (16.54) and (16.55) that κ and $u_{+,\,\mathrm{max}}$ are given by

$$\kappa = 0.4 \tag{16.56}$$

$$u_{+,\,\mathrm{max}} = 2.5 \ln \mathrm{Re}\sqrt{f} + 1.37 \tag{16.57}$$

The constant $\kappa = 0.4$ validates the assumption that, outside the viscous sublayer, ℓ is of the order of the distance from the wall. Note that Eq. (16.55) does not involve the macroscopic scale of the process, and thus it should apply universally to all turbulent shear flows near a wall. Equation (16.57) cannot be expected to provide a very good value of the centerline velocity, because we have excluded the centerline region in evaluating the parameters. Note that if Eq. (16.55) is extended to the centerline, it does not give a zero gradient. Nevertheless, the equation fits experimental data over nearly the entire pipe cross section, although there is an apparent tendency for the data to lie above the logarithmic profile at the largest values of y_+ for each set.

---

*The dashed horizontal lines on Nikuradse's data indicate a discrepancy between his tabulated and graphed results. The difference is important only at small values of y_+.

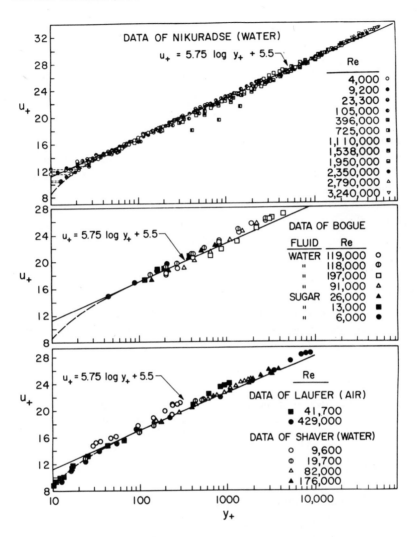

Figure 16-2. Turbulent mean velocity profile data, plotted as u_+ versus y_+, compared with Eq. (16.55).

Example 16.2

Estimate the centerline velocity using Eq. (16.57).

We shall take $Re = 10^5$, for which $f = 0.0045$. It then follows from Eq. (16.57) that $u_{+,\,\text{max}} = 23.4$. From Example 16.1, $u_+ = 20\overline{v_z}/\langle\overline{v_z}\rangle$ at $Re = 10^5$, so $\overline{v}_{z,\,\text{max}} = 1.17\langle\overline{v_z}\rangle$. This agrees reasonably well with the value $\overline{v}_{z,\,\text{max}} = 1.23\langle v_z\rangle$ for the data in Fig. 6-7.

16.4.7 Sublayer Thickness

We now have two asymptotic equations to describe the velocity profile, Eq. (16.47) for the viscous sublayer and Eq. (16.55) for the turbulent core:

$$\text{sublayer:} \quad u_+ = y_+ \tag{16.47}$$

$$\text{core:} \quad u_+ = 2.5 \ln y_+ + 5.5 \tag{16.55}$$

There is clearly a transition region between these two asymptotes, as can be seen in Fig. 16-2, so the thickness of the sublayer cannot be defined unambiguously, but it is useful to think in terms of the two regions. We therefore define the thickness of the viscous sublayer as the point at which the two asymptotes intersect. We will call the sublayer thickness δ_{vs}; we then obtain δ_{vs+} by equating the right-hand sides of Eqs. (16.47) and (16.55):

$$\delta_{vs+} = 2.5 \ln \delta_{vs+} + 5.5 \tag{16.58a}$$

This equation has a solution

$$\delta_{vs+} = 11.6 \tag{16.58b}$$

Some authors define three regions, including an intermediate region between the sublayer and the core. In that case the viscous sublayer thickness is about $\delta_{vs+} = 5$. The difference is unimportant.

Example 16.3

Estimate the fraction of the pipe radius occupied by the viscous sublayer.

We have $\delta_{vs}/R = \delta_{vs+}/R_+$. At $Re = 10^5$ we found in Example 16.1 that $R_+ = 7.5 \times 10^3$. With $\delta_{vs+} = 11.6$, we therefore find that $\delta_{vs}/R = 11.6/7.5 \times 10^3 = 1.5 \times 10^{-3}$. The fraction is even an order of magnitude less at $Re = 10^7$, where $R_+ = 1.6 \times 10^5$.

16.4.8 Friction Factor-Reynolds Number Relation

Now that we have an expression for the time-averaged velocity in turbulent pipe flow we can compute the relation between the friction factor and the Reynolds number. We first express the average velocity in Eq. (16.35) in terms of an integral over the cross section and write

$$u_* = \sqrt{\frac{f}{2}} \langle v_z \rangle = \sqrt{\frac{f}{2}} \frac{1}{\pi R^2} \int_0^R 2\pi r \overline{v_z} \, dr \tag{16.59}$$

We now write $u_+ = \overline{v_z}/u_*$, Eq. (16.36), and rearrange Eq. (16.59) to

$$\sqrt{\frac{2}{f}} = \frac{2}{R^2} \int_0^R r u_+ \, dr = \frac{2}{R^2} \int_0^R (R - y) u_+ \, dy \tag{16.60}$$

The second equality in Eq. (16.60) is obtained by the substitution $y = R - r$, $dy = -dr$.

To carry out the integration indicated in Eq. (16.60), we should break the integral up into three parts and use $u_+ = y_+$ from $y = 0$ to $y = \delta_{vs}$, the logarithmic velocity profile from $y = \delta_{vs}$ to a point about midway to the centerline, and a function characteristic of the centerline region velocity over the remainder of the radius. In fact, the viscous sublayer is so small a fraction of the pipe radius (compare Example 16.3) that there is negligible error involved in simply using the logarithmic profile right up to the pipe wall and ignoring the correction in the sublayer; the logarithmic singularity at $y = 0$ causes no problems in the integration. Similarly, the error in using the logarithmic velocity profile right up to the centerline is not large, and the error is further minimized by the radial weighting of the integrand in Eq. (16.20); the largest error is at the centerline, where $R - y$ goes to zero. Thus, for computational simplicity we replace u_+ by Eq. (16.55) and rewrite Eq. (16.60) as

$$\sqrt{\frac{2}{f}} = \frac{2}{R^2} \int_0^R (R - y)(2.5 \ln y_+ + 5.5) \, dy$$

$$= 2 \int_0^1 (1 - \xi)\left(2.5 \ln \xi + 5.5 + 2.5 \ln \text{Re} \sqrt{\frac{f}{8}}\right) d\xi \qquad (16.61)$$

The second equality follows from a change of variables from y to $\xi = y/R$, together with Eq. (16.38), $R_+ = \text{Re}\sqrt{f/8}$.

The quadrature in Eq. (16.61) is straightforward, as long as it is recalled that the limit as ϵ goes to zero of $\epsilon \ln \epsilon$ is zero. The final result is

$$\frac{1}{\sqrt{f}} = 1.8 \ln \text{Re} \sqrt{f} - 0.60 = 4.1 \log_{10} \text{Re} \sqrt{f} - 0.60 \quad (16.62)$$

The experimental data are fit better by slightly adjusting the coefficients to

$$\frac{1}{\sqrt{f}} = 1.7 \ln \text{Re} \sqrt{f} - 0.40 = 4.0 \log_{10} \text{Re} \sqrt{f} - 0.40 \quad (16.63)$$

Equation (16.63) is the *von Kármán–Nikuradse equation*, which was first introduced as Eq. (3.12). This equation is the line drawn through the turbulent flow data in Fig. 3-1 and fits all available data for smooth pipes.

16.4.9 Rough Pipes

In a rough pipe the turbulent fluctuations will be damped by the surface roughness. If the viscous sublayer extends into the flow beyond the surface roughness, the flow should be like that in a smooth pipe. If $\delta_+ \approx 12$ is closer to the wall than the deepest penetration of the surface roughness, however, then the turbulent eddies will be damped to a greater distance from the wall than in a smooth pipe. In that case the logarithmic profile in the turbulent core given by Eq. (16.54) can extend only to the vicinity of the edge of the surface roughness; that is, we expect the logarithmic profile to intersect the straight line $u_+ = y_+$ when $y = k$, where k is the roughness parameter introduced in Sec. 3.5.2. This leads to an intercept that is different from the 5.5 in Eq. (16.55) and to a friction factor which depends only on k/D and is independent of Re, as found by Nikuradse at high Reynolds numbers; compare Figs. 3-6 and 3-7. The details of the development are left as a problem.

16.4.10 Blasius Equation

Turbulent velocity profiles are often correlated empirically by power equations of the form

$$\frac{\overline{v_z}}{v_{z,\max}} = \left(\frac{y}{R}\right)^m \tag{16.64a}$$

or, equivalently,

$$\frac{\overline{v_z}}{\langle v_z \rangle} = \frac{(m+1)(m+2)}{2}\left(\frac{y}{R}\right)^m \tag{16.64b}$$

m is usually in the range $\frac{1}{10} < m < \frac{1}{6}$, with a value of $m = \frac{1}{7}$ often used to approximate behavior over several decades of Reynolds number. A typical plot is shown in Fig. 6-7 at a Reynolds number of about 10^5.

Equation (16.64) can be written in terms of usual turbulence variables as

$$u_+\sqrt{\frac{f}{2}} = \frac{(m+1)(m+2)}{2}\left(\frac{y_+}{R_+}\right)^m = \frac{(m+1)(m+2)y_+^m}{2\,\mathrm{Re}^m(\sqrt{f/8})^m} \tag{16.65}$$

This equation cannot be valid all the way to the wall, since, like the logarithmic profile, it gives an infinite wall stress. If we match it to the viscous sublayer $u_+ = y_+$ at $y_+ = \delta_{vs+}$, we obtain

$$\delta_{vs+}\sqrt{\frac{f}{2}} = \frac{(m+1)(m+2)\delta_{vs+}^m}{2\,\mathrm{Re}^m(\sqrt{f/8})^m} \tag{16.66}$$

or

$$f = \left\{\left[\frac{(m+1)(m+2)\sqrt{2}\,8^{m/2}}{2}\right]^{2/(m+1)}\delta_{vs+}^{(2m-2)/(m+1)}\right\}\mathrm{Re}^{-2m/(m+1)} \tag{16.67}$$

For $m = \frac{1}{7}$ we have

$$m = \tfrac{1}{7}: \quad f = 3.5\delta_+^{-3/2}\,\mathrm{Re}^{-1/4} \tag{16.68}$$

This is the form of the empirical Blasius equation, Eq. (3.10),

$$f = 0.079\mathrm{Re}^{-1/4} \tag{16.69}$$

In that case we must have

$$\delta_{vs+} = 12.6 \tag{16.70}$$

which is essentially the same as the sublayer thickness found with the logarithmic profile, Eq. (16.58).

It is interesting to note that the relation

$$m = 2\sqrt{f} \tag{16.71}$$

is a good approximation to available data on both smooth and rough pipes. This gives only the exponent, however, and not the value of δ_{vs+}, which will be of order R_+k/R in rough pipes.

16.5 TURBULENT BOUNDARY LAYER OVER A FLAT PLATE

Turbulent boundary layers are analyzed by means of the integral momentum method. It is readily established that the Reynolds stress terms integrate to zero and Eq. (15.44) applies to the time-averaged boundary layer equations over a flat plate at zero angle of incidence as long as we use $\overline{v_x}$ in place of v_x. We can therefore write

$$\tau_w = \rho \frac{d}{dx} \int_0^{\delta_{BL}} \overline{v_x}(U - \overline{v_x}) \, dy \tag{16.72}$$

We shall use δ_{BL} to denote the boundary layer thickness in order to avoid confusion with the sublayer thickness.

It is inconvenient to perform the integrations in Eq. (16.72) using the logarithmic velocity profile. The one-seventh power equation is an adequate approximation and allows the integrations to be carried out easily. The maximum velocity is the free stream velocity, U, at $y = \delta_{BL}$, so Eq. (16.64a) with $m = \frac{1}{7}$ is written

$$\frac{\overline{v_x}}{U} = \left(\frac{y}{\delta_{BL}}\right)^{1/7} \tag{16.73}$$

If we neglect the viscous sublayer and assume that Eq. (16.73) applies right up to the wall, Eq. (16.72) is written

$$\tau_w = \rho U^2 \frac{d}{dx} \int_0^{\delta_{BL}} \left(\frac{y}{\delta_{BL}}\right)^{1/7}\left[1 - \left(\frac{y}{\delta_{BL}}\right)^{1/7}\right] dy = \frac{7}{72}\rho U^2 \frac{d\delta_{BL}}{dx} \tag{16.74}$$

We need an independent expression for τ_w in terms of δ_{BL}. Equation (16.73) cannot be differentiated at the wall to provide the stress because the equation does not apply right at the wall. However, the one-seventh-power equation can be written

$$u_+ = \text{constant } (y_+)^{1/7} \tag{16.75a}$$

and the requirement that $u_+ = y_+$ at $y_+ = \delta_{vs+} = 12.6$ fixes the constant at $(12.6)^{6/7}$, or 8.8:

$$u_+ = 8.8 y_+^{1/7} \tag{16.75b}$$

At the edge of the boundary layer, where $u_+ = U/u_*$ and $y_+ = \rho u_* \delta_{BL}/\eta$, we then have

$$\frac{U}{u_*} = 8.8 \left(\frac{\rho u_*}{\eta} \delta_{BL}\right)^{1/7} \tag{16.76}$$

and setting $u_* = (\tau_w/\rho)^{1/2}$, we obtain an equation for τ_w:

$$\tau_w = \rho u_*^2 = 0.022 \rho U^2 \left(\frac{\eta}{\rho U \delta_{BL}}\right)^{1/4} \tag{16.77}$$

Equations (16.74) and (16.77) combine to give an equation for δ_{BL}:

$$\delta_{BL}^{1/4} \frac{d\delta_{BL}}{dx} = 0.23 \left(\frac{\eta}{\rho U}\right)^{1/4} \tag{16.78}$$

The equation can be integrated directly to yield

$$\delta_{BL}(x) = \left[\delta_{BL}^{5/4}(x_{tr}) + 0.29 \left(\frac{\eta}{\rho U}\right)^{1/4} (x - x_{tr})\right]^{4/5} \approx 0.37 \left(\frac{\eta}{\rho U}\right)^{1/5} x^{4/5} \tag{16.79}$$

where x_{tr} is the position of transition from a laminar to a turbulent boundary layer, and $\delta_{BL}(x_{tr})$ is given by Eq. (15.32) with Re_x equal to about 2×10^5. The four-fifths-power dependence of boundary layer growth is a very different pattern from that of a laminar boundary layer and results in markedly different heat and mass transfer characteristics.

16.6 ENTRY LENGTH FOR TURBULENT FLOW

It is instructive to repeat the entry length calculation in Sec. 15.7 for turbulent flow. We set $\delta_{BL} = D/2$, the pipe midpoint, and $x = L_e$. Equation (16.79) then becomes

$$\frac{D}{2} \approx 0.37 \left(\frac{\eta}{\rho U}\right)^{1/5} L_e^{4/5} \tag{16.80}$$

or

$$\frac{L_e}{D} \approx 1.4 \left(\frac{\rho U D}{\eta}\right)^{1/4} = 1.4 \, Re^{1/4} \tag{16.81}$$

This result is quite insensitive to Re. For $Re = 10^4$ the estimate is $L_e \approx 14D$, while for $Re = 10^8$ the estimate is $L_e \approx 140D$. The experimental rule of thumb for turbulent entry lengths, Eq. (3.14b), is $L_e \approx 40D$, which is consistent with this calculation.

16.7 CONCLUDING REMARKS

Time averaging, with the mixing length theory for the Reynolds stresses, provides considerable insight into the structure of turbulent shear flow. The concepts of the viscous sublayer and the logarithmic core region provide useful estimates of the scales over which phenomena occur and the functional forms to be expected.

Time averaging must lose much of the information about the physical processes that is contained in the full fluctuating equations, and the results obtained in this chapter give only the broadest picture of turbulent flow. Much effort has been expended in measurements and analyses of the structure and spatial variation of the random processes that characterize turbulence. Modern turbulence theory is closely related in method to the statistical mechanics of dilute gases that is studied in advanced courses in physical chemistry.

BIBLIOGRAPHICAL NOTES

There are good introductions to turbulence in

BRODKEY, R. S., *The Phenomena of Fluid Motions*, Addison-Wesley Publishing Company, Inc., Reading, Mass., 1967.

SCHLICHTING, H., *Boundary Layer Theory*, 6th ed., McGraw-Hill Book Company, New York, 1968.

The basic text is

HINZE, J. O., *Turbulence*, 2nd ed., McGraw-Hill Book Company, New York, 1975

and a good recent text is

TENNEKES, H., and LUMLEY, J. L., *A First Course in Turbulence*, MIT Press, Cambridge, Mass., 1972.

PROBLEMS

16.1. Starting with Eqs. (16.23), and without making any specific assumptions about the form of the Reynolds stresses, derive the velocity profile for pressure-driven turbulent flow between parallel planes. Express your solution in the form of a deviation from the parabolic profile:

$$\overline{v_x} = \overline{v}_{x,\max}\left[1 - \left(\frac{2y}{H}\right)^2\right] + \phi(y, (y, \overline{v'_x v'_y}(y)))$$

Show that $\phi = 0$ at $y = 0$, $H/2$. Can you also show that $\phi > 0$ for $0 < y < H/2$?

16.2. Von Kármán suggested that the mixing length should have the form

$$\ell = \kappa \left| \frac{d\overline{v}_x/dy}{d^2\overline{v}_x/dy^2} \right|$$

Show that this form leads to the logarithmic universal velocity profile for $y/R \ll 1$. (Hint: It is easier to solve this problem by working backwards from the solution.)

16.3. In a pipe with uniform roughness k, the velocity profile in the turbulent core is found to have the form

$$u_+ = \frac{1}{\kappa} \ln (y/k) + C$$

where C is a constant and $\kappa = 0.4$.

a) Justify the fact that κ has the same value in both smooth and rough pipes.
b) Obtain the friction factor-Reynolds number relation, and evaluate C by comparison with the Colebrook formula, Eq. (3.37).
c) This result can be expected to apply only when $k_+ = k\rho u_*/\eta > 12$. Explain, and compare to the estimate obtainable from the Colebrook formula.

16.4. Multiply the x-component of the Navier–Stokes equations by v_y' and time-average the resulting equation. Show that this technique, applied to each component equation with each component of the velocity fluctuation, results in a set of equations for the Reynolds stresses. Describe the problem of solution of these equations, and (if you have studied statistical mechanics in physics or physical chemistry) relate the problem to the "closure problem" in classical statistical mechanics.

16.5. Derive Eq. (16.72) for the integral momentum method for turbulent boundary layers.

Perturbation and 17
Numerical Solution

17.1 INTRODUCTION

It is sometimes necessary to obtain solutions to the Navier–Stokes equations for conditions that are outside the ranges of applicability of the approximations in Chapters 12 through 15, or for geometries for which analytical solutions cannot be obtained even if the approximations do apply. Two approaches are commonly used, perturbation methods and numerical solution. Both procedures are time consuming and limited in scope.

We shall provide brief introductions to both approaches in this chapter. The purpose is simply to illustrate these means of extending the analyst's ability to obtain solutions to problems in fluid mechanics. A full treatment is beyond the scope of this introductory text. Section 17.3 on numerical methods does not depend on Sec. 17.2 on perturbation methods and can be read separately.

17.2 PERTURBATION SOLUTION

17.2.1 Corrections to Creeping Flow

We will illustrate perturbation solutions to the Navier–Stokes equations by considering steady flows for which the Reynolds number is small but nonzero. We will suppose that a creeping flow solution exists and that it applies over the entire flow field.

The starting point is the dimensionless steady-state Navier–Stokes equation, Eq, (12.1), which we write in the form

$$\text{Re } \tilde{\mathbf{v}} \cdot \tilde{\nabla}\tilde{\mathbf{v}} + \tilde{\nabla}\tilde{\mathcal{P}} - \tilde{\nabla}^2\tilde{\mathbf{v}} = 0 \tag{17.1}$$

We assume that the solution can be expressed as a power series in Reynolds number,

$$\tilde{\mathbf{v}} = \tilde{\mathbf{v}}^{(0)} + \text{Re } \tilde{\mathbf{v}}^{(1)} + \text{Re}^2 \, \tilde{\mathbf{v}}^{(2)} + \dots \tag{17.2a}$$

$$\tilde{\mathcal{P}} = \tilde{\mathcal{P}}^{(0)} + \text{Re } \tilde{\mathcal{P}}^{(1)} + \text{Re}^2 \, \tilde{\mathcal{P}}^{(2)} + \dots \tag{17.2b}$$

The dimensionless functions $\tilde{\mathbf{v}}^{(0)}, \tilde{\mathbf{v}}^{(1)}, \tilde{\mathbf{v}}^{(2)}, \tilde{\mathcal{P}}^{(0)}, \tilde{\mathcal{P}}^{(1)}, \tilde{\mathcal{P}}^{(2)}, \dots$ are of order unity. Substituting into Eq. (17.1) and grouping terms by the power of Re then gives

$$\text{Re}^0 \,[\tilde{\nabla}\tilde{\mathcal{P}}^{(0)} - \tilde{\nabla}^2\tilde{\mathbf{v}}^{(0)}] + \text{Re}^1 \,[\tilde{\mathbf{v}}^{(0)} \cdot \tilde{\nabla}\tilde{\mathbf{v}}^{(0)} + \tilde{\nabla}\tilde{\mathcal{P}}^{(1)} - \tilde{\nabla}^2\tilde{\mathbf{v}}^{(1)}]$$
$$+ \text{Re}^2 \,[\tilde{\mathbf{v}}^{(0)} \cdot \tilde{\nabla}\tilde{\mathbf{v}}^{(1)} + \tilde{\mathbf{v}}^{(1)} \cdot \tilde{\nabla}\tilde{\mathbf{v}}^{(0)} + \tilde{\nabla}\tilde{\mathcal{P}}^{(2)} - \tilde{\nabla}^2\tilde{\mathbf{v}}^{(2)}] + \dots = 0 \tag{17.3}$$

Equation (17.3) is a power series summing to zero that must be valid for arbitrary (small) Re. Thus, the coefficient of each power of Re must separately equal zero:

$$\text{Re}^0: \quad -\tilde{\nabla}\tilde{\mathcal{P}}^{(0)} + \tilde{\nabla}^2\tilde{\mathbf{v}}^{(0)} = 0 \tag{17.4a}$$

$$\text{Re}^1: \quad -\tilde{\nabla}\tilde{\mathcal{P}}^{(1)} + \tilde{\nabla}^2\tilde{\mathbf{v}}^{(1)} = \tilde{\mathbf{v}}^{(0)} \cdot \tilde{\nabla}\tilde{\mathbf{v}}^{(0)} \tag{17.4b}$$

$$\text{Re}^2: \quad -\tilde{\nabla}\tilde{\mathcal{P}}^{(2)} + \tilde{\nabla}^2\tilde{\mathbf{v}}^{(2)} = \tilde{\mathbf{v}}^{(0)} \cdot \tilde{\nabla}\tilde{\mathbf{v}}^{(1)} + \tilde{\mathbf{v}}^{(1)} \cdot \tilde{\nabla}\tilde{\mathbf{v}}^{(0)} \tag{17.4c}$$

Equation (17.4a) is the equation for creeping flow, Eq. (12.3). Equation (17.4b) is a *linear* equation for the first-order correction term, with a forcing term depending on $\tilde{\mathbf{v}}^{(0)}$. Each successive equation is linear and depends only on the preceding solutions.

The normalization for each order of solution is the same, so we can rewrite Eqs. (17.2) and (17.4) in dimensional form, which is sometimes more convenient for actual solution:

$$\mathbf{v} = \mathbf{v}^{(0)} + \text{Re } \mathbf{v}^{(1)} + \text{Re}^2\mathbf{v}^{(2)} + \dots \tag{17.5a}$$

$$\mathcal{P} = \mathcal{P}^{(0)} + \text{Re } \mathcal{P}^{(1)} + \text{Re}^2 \, \mathcal{P}^{(2)} + \dots \tag{17.5b}$$

$$\text{Re}^0: \quad -\nabla\mathcal{P}^{(0)} + \eta\nabla^2\mathbf{v}^{(0)} = 0 \tag{17.6a}$$

$$\text{Re}^1: \quad -\nabla\mathcal{P}^{(1)} + \eta\nabla^2\mathbf{v}^{(1)} = \rho\mathbf{v}^{(0)} \cdot \nabla\mathbf{v}^{(0)} \tag{17.6b}$$

$$\text{Re}^2: \quad -\nabla\mathcal{P}^{(2)} + \eta\nabla^2\mathbf{v}^{(2)} = \rho[\mathbf{v}^{(0)} \cdot \nabla\mathbf{v}^{(1)} + \mathbf{v}^{(1)} \cdot \nabla\mathbf{v}^{(0)}] \tag{17.6c}$$

It readily follows that the continuity equation becomes

$$\nabla \cdot \mathbf{v}^{(i)} = 0 \quad i = 0, 1, 2, \dots \tag{17.7}$$

The boundary conditions must also be expanded in Re and applied to the appropriate order terms.

17.2.2 Flow Between Rotating Disks

A creeping flow solution to flow between rotating disks was obtained in Sec. 12.2. The configuration is shown in Fig. 12-1. The zero-order solution is given by Eqs. (12.11) and (12.12):

$$v_\theta^{(0)} = r\Omega\left(1 - \frac{z}{H}\right) \tag{17.8a}$$

$$v_r^{(0)} = v_z^{(0)} = 0 \qquad \mathcal{P}^{(0)} = \text{constant} \tag{17.8b}$$

The first-order equation, Eq. (17.6b), is obtained by putting the zero-order solution into the left-hand side of the Navier–Stokes equations and writing the right-hand side for the first-order solution. We assume that there is no θ variation. From Table 7-7, we then obtain

$$r: \quad -\rho r\Omega^2\left(1 - \frac{z}{H}\right)^2 = -\frac{\partial \mathcal{P}^{(1)}}{\partial r} + \eta\left[\frac{\partial}{\partial r}\left(\frac{1}{r}\frac{\partial}{\partial r}rv_r^{(1)}\right) + \frac{\partial^2 v_r^{(1)}}{\partial z^2}\right] \tag{17.9a}$$

$$\theta: \quad 0 = \eta\left[\frac{\partial}{\partial r}\left(\frac{1}{r}\frac{\partial}{\partial r}rv_\theta^{(1)}\right) + \frac{\partial^2 v_\theta^{(1)}}{\partial z^2}\right] \tag{17.9b}$$

$$z: \quad 0 = -\frac{\partial \mathcal{P}^{(1)}}{\partial z} + \eta\left[\frac{1}{r}\frac{\partial}{\partial r}\left(r\frac{\partial v_z^{(1)}}{\partial r}\right) + \frac{\partial^2 v_z^{(1)}}{\partial z^2}\right] \tag{17.9c}$$

The first-order term for the continuity equation is

$$\frac{1}{r}\frac{\partial}{\partial r}(rv_r^{(1)}) + \frac{\partial v_z^{(1)}}{\partial z} = 0 \tag{17.10}$$

$v_\theta^{(0)}$ satisfies the boundary conditions for no-slip at the upper and lower plates, so $v_\theta^{(1)}$, $v_\theta^{(2)}$, ... must be zero at both plates. Similarly, $v_r^{(1)}$, $v_r^{(2)}$, $v_z^{(1)}$, $v_z^{(2)}$, ... must be zero at both surfaces.

It readily follows from Eq. (17.9b) and the zero boundary conditions for $v_\theta^{(1)}$ that $v_\theta^{(1)} = 0$. The remainder of the solution is obtained by assuming that $v_z^{(1)}$ is a function only of z. The continuity equation can then be integrated to give $v_r^{(1)}$ in terms of $v_z^{(1)}$:

$$v_r^{(1)} = -\frac{r}{2}\frac{dv_z^{(1)}}{dz} \tag{17.11}$$

Equations (17.9a) and (c) then become, respectively,

$$\rho r\Omega^2\left(1 - \frac{z}{H}\right)^2 = \frac{\partial \mathcal{P}^{(1)}}{\partial r} + \frac{r}{2}\eta\frac{d^3 v_z^{(1)}}{dz^3} \tag{17.12a}$$

$$\frac{\partial \mathcal{P}^{(1)}}{\partial z} = \eta\frac{d^2 v_z^{(1)}}{dz^2} \tag{17.12b}$$

We see from Eq. (17.12b) that $\partial\mathcal{P}^{(1)}/\partial z$ is a function only of z, so $\mathcal{P}^{(1)}$ must be a sum of r- and z-dependent terms. Thus, $\partial\mathcal{P}^{(1)}/\partial r$ must depend only on r.

Each of the other terms in Eq. (17.12a) is proportional to r, so $\partial \mathcal{P}^{(1)}/\partial r$ must be of the form

$$\frac{\partial \mathcal{P}^{(1)}}{\partial r} = C_1 r \tag{17.13}$$

Equation (17.12a) then simplifies to

$$\frac{\eta}{2} \frac{d^3 v_z^{(1)}}{dz^3} = \rho \Omega^2 \left(1 - \frac{z}{H}\right)^2 - C_1 \tag{17.14}$$

Equation (17.14) can be integrated three times. The no-slip condition requires that $v_z^{(1)}$ and $dv_z^{(1)}/dz$ be zero at $z = 0$ and $z = H$; the derivative condition follows from Eq. (17.11). We thus obtain

$$v_z^{(1)} = \frac{\rho \Omega^2 H^3}{60 \eta} \left[-4\left(1 - \frac{z}{H}\right)^2 + 6\left(1 - \frac{z}{H}\right)^3 - 2\left(1 - \frac{z}{H}\right)^5 \right] \tag{17.15a}$$

$v_r^{(1)}$ is then obtained from Eq. (17.11):

$$v_r^{(1)} = -\frac{r \rho \Omega^2 H^2}{60 \eta} \left[4\left(1 - \frac{z}{H}\right) - 9\left(1 - \frac{z}{H}\right)^2 + 5\left(1 - \frac{z}{H}\right)^4 \right] \tag{17.15b}$$

The velocity field can thus be written

$$v_\theta = r\Omega\left(1 - \frac{z}{H}\right) + \text{order of } \left(\frac{\rho H^2 \Omega}{\eta}\right)^2 \tag{17.16a}$$

$$v_r = -\left(\frac{\rho H^2 \Omega}{\eta}\right) \frac{r\Omega}{60} \left[4\left(1 - \frac{z}{H}\right)^2 - 9\left(1 - \frac{z}{H}\right)^2 + 5\left(1 - \frac{z}{H}\right)^4 \right]$$
$$+ \text{order of } \left(\frac{\rho H^2 \Omega}{\eta}\right)^2 \tag{17.16b}$$

$$v_z = \left(\frac{\rho H^2 \Omega}{\eta}\right) \frac{H\Omega}{60} \left[-4\left(1 - \frac{z}{H}\right)^2 + 6\left(1 - \frac{z}{H}\right)^3 - 2\left(1 - \frac{z}{H}\right)^5 \right]$$
$$+ \text{order of } \left(\frac{\rho H^2 \Omega}{\eta}\right)^2 \tag{17.16c}$$

We shall not carry the expansion out any further. The next term in v_θ is second order in $(\rho H^2 \Omega/\eta)$, while the next term in v_r and v_z is third order. The v_r and v_z terms describe a radial secondary flow superimposed on the primary circular streamlines, with inflow along the upper (stationary) plate and outflow along the lower (rotating) plate. The secondary flow could play a role in parallel plate viscometry. This point is discussed by Savins and Metzner, who present data that are well described by Eqs. (17.16).

17.2.3 Comments on Perturbation Methods

The procedure outlined here is tedious, and solutions are rarely carried beyond first- or second-order corrections. While it reduces the original non-linear problem to a sequence of linear problems, even these linear problems can rarely be solved in complex geometries.

Only *regular* problems can be solved by the straightforward expansion method. Regularity requires that the highest derivative be retained at each order and that the lower-order solutions be uniformly valid over the entire flow regime. We noted in Sec. 12.3.4 that the creeping flow solution for flow past a sphere is not self-consistent at large distances from the sphere surface. A first-order correction to this problem cannot be obtained from Eq. (17.7b) by substituting the Stokes flow solution for $\mathbf{v}^{(0)}$, and special techniques must be used. This apparent paradox regarding flow past a sphere was one of the reasons for the development of the modern theory of matched asymptotic expansions. The material in Sec. 11.8 on singular perturbations is relevant to this topic, which requires different perturbation expansions in different regions of the flow field.

17.3 NUMERICAL METHODS

17.3.1 Finite Differences and Finite Elements

Numerical methods are approximate procedures by which differential equations are represented by a set of algebraic equations that define the process variables at discrete points over the flow field. The solution to the algebraic equations then provides an approximate solution to the differential equation. Most numerical methods for solution of the Navier–Stokes equations use either *finite differences* or *finite elements*.

The underlying idea behind finite-difference methods is deceptively simple. Suppose that the y coordinate extends from zero to H, and we break it into a grid with N points. The grid points are thus spaced a distance H/N from one another. We can estimate, say, the derivative $\partial v_x/\partial y$ at the nth grid point by

$$\frac{\partial v_x}{\partial y} \approx \frac{v_x(\text{at } (n+1)\text{st grid point}) - v_x(\text{at } n\text{th grid point})}{H/N} \quad (17.17)$$

We carry out a similar process for all derivatives. In this way all derivatives are expressed in terms of values of the function at neighboring grid points. By doing such finite differencing at each grid point we obtain a set of algebraic equations that must be solved simultaneously.

Because of the complexity of the Navier–Stokes equations the finite differencing must, in fact, be done in sophisticated ways that would not be obvious to someone inexperienced in numerical computation. The difficulties involve numerical instability and convergence. Instability is the tendency of numerical approximations to "blow up" even though the physical flow is stable. Convergence is the requirement that the numerical solution approach the true solution to the Navier–Stokes equations as the grid spacing gets closer and closer. Note that a large grid spacing means a poor approximation to the derivative, while a very close grid spacing can lead to computer round-off problems.

Very few three-dimensional finite-difference solutions of the full Navier–Stokes equations have been carried out. If each coordinate direction requires 10 grid points for an adequate approximation, for example, then each equation must be written in finite-difference form at each of *one thousand* points in the three-dimensional grid. This means solution of 4000 coupled nonlinear algebraic equations for a complete solution. Clearly, problems of computer time, storage, and accuracy are important considerations.

Finite-element methods represent a different approach in principle to numerical solution, although the magnitude of the ultimate computational problem will be about the same as for finite differences. Here, the spatial region is broken up into a number of smaller regions, usually triangular in shape. The solution is approximated in each triangle by a function of specified form but with unknown coefficients, usually using the method of weighted residuals; this method is briefly introduced in Sec. 9.6 and is a generalization of the integral momentum method described in Secs. 9.5 and 15.6. The integration over each volume, together with the requirement that the velocities and stresses be continuous on adjacent elements, gives rise to the algebraic equations for the coefficients. Velocities are usually approximated in each element by quadratic functions and stresses by linear functions.

Rather general finite-element programs exist, and they enable the use of small triangular elements in regions where rapid changes are taking place and large elements in regions where there is little change. The programming logic for a comparable uneven grid spacing in a finite-difference solution is far more difficult, although it is possible. Much has been made in the literature of the differences between finite-element and finite-difference methods, but in fact the differences are more apparent than real. Basic finite-element and finite-difference schemes lead to the same algebraic equations for solution of Laplace's equation [Eq. (14.8)] in a rectangular region, for example. The finite-element method probably has an advantage in terms of versatility in the application of one computer code to a variety of problems.

In the next several sections we will show some numerical solutions to the Navier–Stokes equations in order to give a flavor of this type of solution. No details are included.

17.3.2 Flow Between Rotating Disks

A finite-difference solution of flow between rotating disks is shown in Figs. 17-1 and 17-2 for disks with radii equal to twice and 10 times the spacing, respectively, at $\rho H^2 \Omega / \eta = 1$. The geometry is shown in Fig. 12-1. The computed θ component of velocity never differs from the solution given by Eq.

z = H

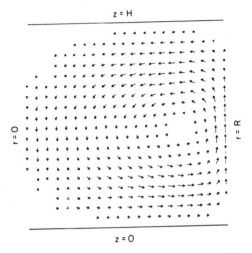

z = 0

Figure 17-1. Inertial secondary flow between rotating disks, $R/H = 2$. Reproduced from McCoy and Denn, *Rheol. Acta*, **10**, 408 (1971), copyright by Dr. Dietrich Steinkopff Verlag, by permission.

z = H

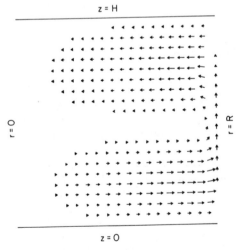

z = 0

Figure 17-2. Inertial secondary flow between rotating disks, $R/H = 10$. Reproduced from McCoy and Denn, *Rheol. Acta*, **10**, 408 (1971), copyright by Dr. Dietrich Steinkopff Verlag, by permission.

(12.11) or (17.8a) by more than 1%. The inertial secondary flow is shown in the figures as the projection of the velocity vector on an *rz* plane. The sizes of the vectors are scaled to represent the relative magnitude of the velocity.

The secondary flow consists of a single eddy, with outflow along the lower (rotating) plate and inflow along the upper (stationary) plate. For disks with radii greater than five gap spacings, the secondary flow is well represented by Eqs. (17.16) until just before the outer edge, where the secondary flow must turn around. The perturbation solution does not account for the finite radius of the disks and contains a source and sink of fluid at infinite radius.

The solution shown here is incomplete in that the problem was solved in a rectangular cross-sectional plane, with the free surface constrained to be flat. The free surface will, in fact, adjust to a curved shape in order to allow continuity of the normal stress at the interface. An estimate of the surface shape can be obtained from the radial stress variation, and the process could be repeated iteratively until the assumed and computed surface shapes agree. This type of iterative calculation is difficult with finite-difference methods.

17.3.3 Flow Through a Contraction

Flow through a plane contraction is shown schematically in Fig. 10-1. A finite-difference solution to the *creeping flow* equations in this geometry is shown in Figs. 17-3 to 17-6. The flow is between two regions of fully developed Poiseuille flow. The numerical technique used here is one that exploits the linearity of the creeping flow equations in order to express the solution entirely in terms of integrals over the boundary of the region, so the spatial grid is only over the solid boundary and the fully developed cross sections. The upstream and downstream boundaries were taken at one large and small channel spacing, respectively.

The streamlines and values of the stream function are shown in Figs. 17-3 and 17-4 for a 5:1 contraction with entry half-angles of 90° and 45°, respectively. There is a weak recirculating corner eddy for the 90° contraction whose approximate center is located along the dotted line in Fig. 17-3; the numerical

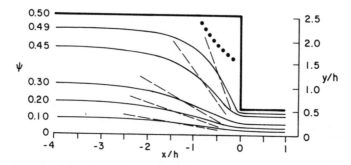

Figure 17-3. Streamlines for flow through a 5:1 plane contraction, $\alpha = 90°$. Solid lines are the numerical solution, dashed lines are Hamel flow, Sec. 10.4. After Black and Denn, *J. Non-Newtonian Fluid Mech.*, **1**, 83 (1976), copyright by Elsevier Scientific Publishing Co., by permission.

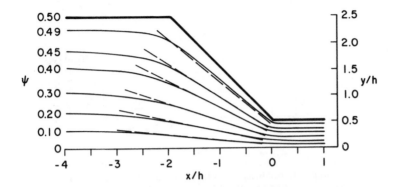

Figure 17-4. Streamlines for flow through a 5:1 plane contraction, $\alpha = 45°$. Solid lines are the numerical solution, dashed lines are Hamel flow, Sec. 10.4 Reproduced from Black and Denn, *J. Non-Newtonian Fluid Mech.*, **1**, 83 (1976), copyright by Elsevier Scientific Publishing Co., by permission.

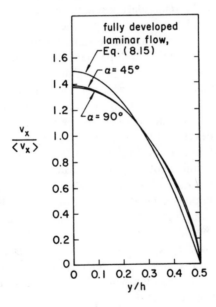

Figure 17-5. Velocity profile at entrance to downstream channel in a 5:1 plane contraction. Reproduced from Black, Denn, and Hsiao, in *Theoretical Rheology*, Applied Science Publishers Ltd., Barking, Essex, England, 1975, by permission.

precision is inadequate in that region to get a better definition of the eddy. There is no recirculation for the 45° contraction.

The straight dashed lines in the figures are the corresponding streamlines obtained from the Hamel flow, Sec. 10.4. The Hamel flow approximation is

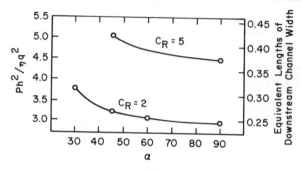

Figure 17-6. Entry losses in 5:1 and 2:1 plane contractions as a function of entry angle. Reproduced from Black, Denn, and Hsiao, in *Theoretical Rheology*, Applied Science Publishers Ltd., Barking, Essex, England 1975 by permission.

adequate over much of the converging section for the 45° angle, but it is quite poor for the 90° contraction. This is because the Hamel flow solution cannot account for the recirculating region in the corner, which moves the streamlines closer to the center of the channel.

The axial velocity profile at the entrance to the downstream channel is shown in Fig. 17-5. The fully developed laminar profile is the parabola given by Eq. (8.15). Evidently, the flow is nearly fully developed before it enters the downstream section, even for the large contraction angles.

The entry losses are shown in Fig. 17-6 for contraction ratios (C_R) of 2:1 and 5:1. The entry loss is defined here as the additional power beyond that required for fully developed laminar flow in the upstream channel followed by fully developed laminar flow in the downstream channel. The result is also given in terms of equivalent lengths of downstream channel. The insensitivity of the entry losses to angle of convergence is observed in practice. The similarity in shape to Fig. 10-7 should be noted. The relevant equation here is Eq. (10.62), and in order to put the two figures on the same basis, the scale on Fig. 17-6 should be divided by $C_R^2/4(C_R^2 - 1)$. In that case the curves for the two contraction ratios are close to one another and 30 to 40% below the curve for P^* in Fig. 10-7. The velocity profile development evidently leads to lower entry losses then would be expected if Hamel flow occurred over the entire contraction region.

17.3.4 Free Jet

The calculation of the diameter of a free jet is described in Sec. 6.7, and the process is shown schematically in Fig. 6-10. That calculation is valid at high Reynolds numbers, where the inertial terms dominate, and the conclusion is reached that the jet diameter is less than the tube diameter. At very small Reynolds numbers, where inertial effects are negligible, the jet is observed to *swell* by about 13% after leaving the tube; the jet diameter decreases as a function of Re, reaching the tube diameter only at Re ≈ 20 and approaching the asymptotic value of $0.82D$ at Re ~ 80.

Figures 17-7 and 17-8 show the results of a finite-element calculation of a free Newtonian jet in the absence of gravity and surface tension effects; the calculations shown here are for the limit Re = 0, although finite Reynolds numbers have also been computed. The flow is taken to be fully developed one diameter upstream of the exit.

The finite element grid is shown in Fig. 17-7. Note the close spacing near the tube exit, where the major rearrangement is expected to occur, and the large elements used away from the exit. The calculation is iterative because of the existence of a free surface, and the program adjusts the location of the free surface after each calculation. The initial value of D_j/D was taken to be 1.000, followed by values of 1.116, 1.126, 1.128, and 1.128 on successive interations.

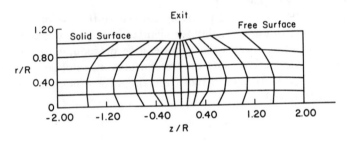

Figure 17-7. Finite element grid for jet expansion, final iteration. Reproduced from Nickell, Tanner, and Caswell, *J. Fluid Mech.*, **65**, 189 (1974), copyright by Cambridge University Press, by permission.

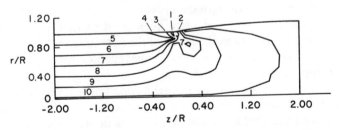

Figure 17-8. Contours of constant values of dimensionless shear stress, $\tau_{rz}R/\eta\langle v_z\rangle$, for free jet problem. Contours 1 through 10 have values as follows: (1) −6.0; (2) −5.3; (3) −4.7; (4) −4.0; (5) −3.4; (6) −2.7; (7) −2.0; (8) −1.4; (9) −0.7; (10) 0. Reproduced from Nickell, Tanner, and Caswell, *J. Fluid Mech.*, **65**, 189 (1974), copyright by Cambridge University Press, by permission.

Lines of constant value of the shear stress are shown in Fig. 17-8. Note the deviation from the linear shear stress distribution at about one-half diameter upstream from the exit, and the stress concentration near the tip of the tube. There is a stress discontinuity at the tip, where an abrupt change is required from the no-slip condition inside to the no-shear stress condition on the free surface. The pressure at the exit plane is not uniform; the value at the surface is positive and the centerline value is negative.

17.3.5 Comments on Numerical Methods

The three examples of numerical solutions give some indication of the types of problems that can be solved. Most numerical techniques have difficulties at high Reynolds numbers. Few three-dimensional or transient problems have been solved, because of computer time and memory requirements. Turbulent flows can be handled only in a time-averaged sense, using an eddy viscosity in place of η.

New computer codes will probably emphasize the use of finite-element methods because of the versatility for changes in geometry. Experience with available programs indicates that the iterative procedures for dealing with the nonlinear inertial terms and the free surfaces may not converge well for flows that are qualitatively different from those for which the program was developed and tested.

17.4 CONCLUDING REMARKS

The methods described in this chapter are generally beyond the scope of an introduction to process fluid mechanics, but it is important to be aware of them. Perturbation solutions usually require a significant effort for what may be small returns, but they do provide insight into the influence of the perturbation parameter. It is important to keep in mind that a perturbation solution is good only as long as the terms of the next order are unimportant, and these higher-order terms are usually unavailable. Thus, perturbation solutions should be interpreted with care.

The role of numerical solutions in engineering applications is changing. It is not long since any numerical solution of the Navier–Stokes equations comprised a complete graduate thesis. With the rapid developments in computer technology and the growing availability of general-purpose codes, we may be approaching the day when numerical solutions of the Navier–Stokes equations will be obtainable with sufficient ease to be widely applicable. We have not arrived there yet, however,

BIBLIOGRAPHICAL NOTES

For a good introduction to perturbation methods in applied mathematics, see

BELLMAN, R., *Perturbation Techniques in Mathematics, Physics, and Engineering*, Holt, Rinehart and Winston, Inc., New York, 1964.

The solution for rotating disks is from

MELLOR, G. L., CHAPPLE, P. J., and STOKES, V. K., *J. Fluid Mech.*, **31**, 95 (1968).

For a comparison with radial flow data, see

SAVINS, J. G., and METZNER, A. B., *Rheol. Acta*, **9**, 365 (1970).

Textbooks on finite-difference and finite-element methods for partial differential equations are, respectively,

FORSYTH, G. E., and WASOW, W. R., *Finite Difference Methods for Partial Differential Equations*, John Wiley & Sons, Inc., New York, 1960.

ZIENKIEWICZ, O. C., *The Finite Element Method*, 3rd ed., McGraw-Hill, London, 1977.

The rotating disk example is from

McCOY, D. H., and DENN, M. M., *Rheol. Acta*, **10**, 408 (1971).

The perturbation solution reported in that paper contains some misprints. The numerical scheme used was developed by Chorin,

CHORIN, A. J., *J. Comp. Phys.*, **2**, 12 (1967).

The converging flow example is from

BLACK, J. R., and DENN, M. M., *J. Non-Newtonian Fluid Mech.*, **1**, 83 (1976).

BLACK, J. R., DENN, M. M., and HSIAO, G. C., in *Theoretical Rheology*, J. F. Hutton, J. R. A. Pearson, and K. Walters, eds., Applied Science Publishers Ltd., Barking, Essex, England, 1975.

The numerical scheme is described in the second of these papers.

The jet example is from

NICKELL, R. E., TANNER, R. I., and CASWELL, B., *J. Fluid Mech.*, **65**, 189 (1974).

TANNER, R. I., *Appl. Polymer Symp.*, **20**, 201 (1973).

TANNER, R. I., NICKELL, R. E., and BILGER, R. W., *Computer Methods Appl. Mech. Eng.*, **6**, 155 (1975).

The numerical method is described in the first and third of these papers.

The papers cited here contain many references to numerical solutions of fluid mechanics problems. Others can be found in recent issues of the journals published by AIChE, ASME, and AIAA, in *Journal of Fluid Mechanics* and *Physics of Fluids*, and specialized journals such as *International Journal of Numerical Methods in Engineering, Journal of Computational Physics*, and *Computer Methods in Applied Mechanics and Engineering*.

Two-Phase Gas-Liquid Flow* 18

18.1 INTRODUCTION

Many processing applications require cocurrent flow of a gas and a liquid. These two-phase flows are far more complex than the flow of either pure phase, and two-phase flow is an area of current research interest. We shall touch briefly on the subject in this chapter to give some insight into the present understanding of gas–liquid configurations in a pipeline and of pressure drop; the former information is essential for estimates of interphase mass transfer, while the latter is required for even the most elementary designs. The discussion will be limited to flow of Newtonian fluids in horizontal pipes; vertical flow is similar, but there are important differences in detail.

18.2 FLOW PATTERNS

A number of different flow patterns can be distinguished for cocurrent flow of a gas and a liquid of low or moderate viscosity (up to about 0.1 Pa·s). The flow pattern can range from the extremes of dispersed gas bubbles in a continuous liquid phase at low relative gas rates to a mist, or dispersed liquid droplets in a continuous gas phase, at high relative rates. The flow regimes that are observed with increasing relative gas rate are sketched in Fig. 18-1.

*This chapter was coauthored by T. W. F. Russell.

342

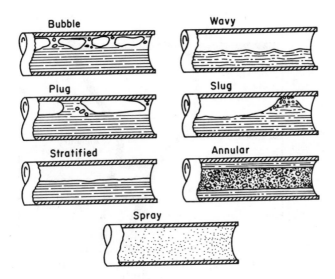

Figure 18-1. Flow patterns in horizontal gas-liquid flow in a pipe. After Alves, *Chem. Eng. Progr.*, **50**, 449 (1954), copyright by the American Institute of Chemical Engineers, by permission.

Bubble flow occurs at high liquid to gas rates. The bubbles move with a velocity that is close to the liquid velocity, and their size and distribution is determined by the level of liquid turbulence. When the bubbles become large, the flow is referred to as *plug flow;* the distinction between bubble and plug flow is not well defined.

In *stratified flow* there are continuous gas (upper) and liquid (lower) phases, with a smooth interface. At sufficiently high gas velocities the interface becomes *wavy*; when the wave crests are sufficiently high to reach the top of the pipe, the flow is in the *slug* regime. The liquid slugs move through the pipe at a speed greater than the mean liquid velocity, and the impact of liquid slugs on bends and other fittings can cause dangerous vibrations.

Annular flow consists of a liquid film on the pipe wall and a high-velocity gas stream containing entrained liquid droplets in the center. At very high gas velocities nearly all the liquid is entrained as small droplets, and the regime is called variously *dispersed*, *spray*, or *mist*.

Data from a large number of visual observations of flow regimes are summarized in Fig. 18-2. The boundaries are not well defined, since the distinction between regimes is a matter of judgment and may vary from observer to observer. Subscripts L and G denote liquid and gas phases, respectively. The subscript ∞ denotes a *superficial velocity*, calculated by assuming that each phase occupies the entire cross section (i.e., $v_{L,\infty} = 4Q_L/\pi D^2$). Most of the data used to construct Fig. 18-2 are for air–water systems in pipes of diameters of 100 mm and less. The influence of physical properties is not

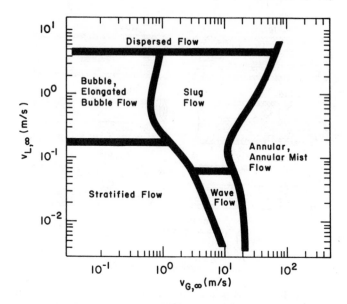

Figure 18-2. Boundaries of flow regimes for cocurrent gas–liquid flow in horizontal pipes. Reproduced from Mandhane, Gregory, and Aziz, *Int. J. Multiphase Flow*, **1**, 537 (1974), copyright by Pergamon Press, Ltd., by permission.

well defined, although it is included in some flow maps. Extrapolation to large pipes and to systems with very different physical properties must be done with caution.

18.3 LAMINAR–LAMINAR STRATIFIED FLOW

18.3.1 Flow Between Parallel Plates

It is most convenient to begin the analytical treatment of two-phase flows with stratified flow between parallel plates when both phases are laminar. Such flow is not very important in practice, but it provides a good basis for introducing some important concepts. The flow configuration is shown in Fig. 18-3. The single-phase analog of this problem was considered in Sec. 8.2.

We assume that the pressure drop is sufficiently small to take the gas density as a constant. The Navier–Stokes equations for an incompressible fluid thus apply to each phase. In laminar flow the streamlines will be parallel

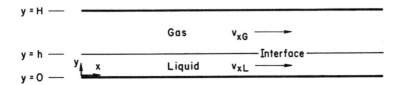

Figure 18-3. Schematic of stratified flow between plane walls.

to the walls, so v_x is the only nonzero velocity component in each phase and is a function only of y. The x components of the Navier–Stokes equations in the gas and liquid phases are then, respectively,

$$\eta_G \frac{d^2 v_{xG}}{dy^2} = \frac{d\mathcal{P}}{dx} = \text{constant} \tag{18.1a}$$

$$\eta_L \frac{d^2 v_{xL}}{dx} = \frac{d\mathcal{P}}{dx} = \text{constant} \tag{18.1b}$$

Here we have made use of the fact that the y component requires that $\mathcal{P}$ be independent of y. Furthermore, since σ_{yy} must be continuous at the interface $y = h$, and $\sigma_{yy} = -p$ in this flow, it then follows that $\mathcal{P}$ is continuous across the plane interface and *the equivalent pressure is therefore the same in the two phases at each value of x.*

Equations (18.1) can each be integrated twice to give

$$v_{xG} = \frac{1}{2\eta_G} \frac{d\mathcal{P}}{dx} y^2 + C_1 y + C_2 \tag{18.2a}$$

$$v_{xL} = \frac{1}{2\eta_L} \frac{d\mathcal{P}}{dx} y^2 + C_3 y + C_4 \tag{18.2b}$$

Two boundary conditions are thus required for each phase in order to evaluate the constants C_1 to C_4. Each phase must satisfy the no-slip condition:

$$v_{xG} = 0 \text{ at } y = H \tag{18.3a}$$

$$v_{xL} = 0 \text{ at } y = 0 \tag{18.3b}$$

Since the two phases must flow together without slipping at the interface, we must have

$$v_{xG} = v_{xL} \text{ at } y = h \tag{18.3c}$$

Finally, the shear stress τ_{yx} in the liquid phase and in the gas phase must be equal at the interface:

$$\eta_G \frac{dv_{xG}}{dy} = \eta_L \frac{dv_{xL}}{dy} \text{ at } y = h \tag{18.3d}$$

Equations (18.3a) through (18.3d) combine with Eqs. (18.2) to give, respectively,

$$0 = \frac{1}{2\eta_G}\frac{d\mathcal{P}}{dx}H^2 + C_1 H + C_2 \tag{18.4a}$$

$$0 = C_4 \tag{18.4b}$$

$$\frac{1}{2\eta_G}\frac{d\mathcal{P}}{dx}h^2 + C_1 h + C_2 = \frac{1}{2\eta_L}\frac{d\mathcal{P}}{dx}h^2 + C_3 h + C_4 \tag{18.4c}$$

$$\frac{d\mathcal{P}}{dx}h + \eta_G C_1 = \frac{d\mathcal{P}}{dx}h + \eta_L C_3 \tag{18.4d}$$

Equations (18.4) can be solved for the constants C_1 through C_4, although we shall not record the results here. It is more convenient simply to express the results for the average phase velocities, $\langle v_{xG}\rangle$ and $\langle v_{xL}\rangle$, as follows:

$$\langle v_{xG}\rangle = \frac{1}{H-h}\int_h^H v_{xG}\,dy = \frac{H^3}{12(H-h)\eta_G}\left(-\frac{d\mathcal{P}}{dx}\right)\left(\frac{\alpha_1\eta_G + \alpha_2\eta_L}{\alpha_3\eta_G + \alpha_4\eta_L}\right) \tag{18.5a}$$

$$\langle v_{xL}\rangle = \frac{1}{h}\int_0^h v_{xL}\,dy = \frac{H^3}{12h\eta_L}\left(-\frac{d\mathcal{P}}{dx}\right)\left(\frac{\alpha_5\eta_G + \alpha_6\eta_L}{\alpha_3\eta_G + \alpha_4\eta_L}\right) \tag{18.5b}$$

The constants α_1 through α_6 are defined as follows:

$$\alpha_1 = \left(\frac{h}{H}\right)\left(4 - \frac{h}{H}\right)\left(1 - \frac{h}{H}\right)^2 \tag{18.6a}$$

$$\alpha_2 = \left(1 - \frac{h}{H}\right)^4 \tag{18.6b}$$

$$\alpha_3 = \frac{h}{H} \tag{18.6c}$$

$$\alpha_4 = 1 - \frac{h}{H} \tag{18.6d}$$

$$\alpha_5 = \left(\frac{h}{H}\right)^4 \tag{18.6e}$$

$$\alpha_6 = \left(\frac{h}{H}\right)^2\left(3 + \frac{h}{H}\right)\left(1 - \frac{h}{H}\right) \tag{18.6f}$$

The relative volumes of liquid and gas that are in the conduit are different from the feed ratio. The *holdup** of each phase in the conduit determines the time available for interphase processes such as mass transfer, so it is important to have this information available in design calculations. For this case the *in situ* liquid holdup, h/H, is simply related to the feed volumetric ratio, $h\langle v_{xL}\rangle/(H-h)\langle v_{xG}\rangle$, by eliminating the pressure gradient between Eqs.

Holdup is usually taken to mean the fraction of the pipe occupied by each phase. Thus, liquid holdup is h/H and gas holdup is $(H-h)/H$.

(18.5a) and (18.5b) to obtain

$$\frac{Q_L}{Q_G} = \left(\frac{\eta_G}{\eta_L}\right)\left[\frac{\alpha_5(\eta_G/\eta_L) + \alpha_6}{\alpha_1(\eta_G/\eta_L) + \alpha_2}\right] \tag{18.7}$$

$\alpha_1, \alpha_2, \alpha_5,$ and α_6 are functions of h/H, so Eq. (18.7) can be solved for h/H as a function of Q_L/Q_G and the viscosity ratio, η_G/η_L.

Example 18.1

Air and water flow cocurrently through a 50-mm plane channel at close to atmospheric pressure and 20°C. The water flows at 0.24 kg/s per meter of channel width and air at 0.017 kg/s per meter of width. Compute the holdup h/H of liquid in the channel.

We may take $\rho_L = 10^3$, $\rho_G = 1.2$, $\eta_L = 10^{-3}$, and $\eta_G = 1.7 \times 10^{-5}$. Equation (18.7) can be written

$$\frac{w_L}{w_G} = \frac{\rho_L\eta_G}{\rho_G\eta_L}\left[\frac{\alpha_5(\eta_G/\eta_L) + \alpha_6}{\alpha_1(\eta_G/\eta_L) + \alpha_2}\right]$$

or, substituting the given values,

$$\frac{0.24}{0.017} = \frac{(10^3)(1.7 \times 10^{-5})}{(1.2)(10^{-3})}\frac{1.7 \times 10^{-2}\alpha_5 + \alpha_6}{1.7 \times 10^{-2}\alpha_1 + \alpha_2}$$

where the α_i are given by Eqs. (18.6). This fourth-order equation for h/H can be solved to give $h/H \sim \frac{1}{3}$. Note that the volumetric *feed* ratio Q_L/Q_G is three orders of magnitude smaller than the *in situ* volumetric ratio! This is because the less viscous gas moves much faster than the liquid under the same pressure gradient and therefore requires less cross-sectional area.

Example 18.2

Verify that the flow described in Example 18.1 is stratified.

In the absence of a flow regime map for parallel plates we will use Fig. 18-2, which was constructed entirely from pipe flow data. The liquid and gas superficial velocities are obtained as follows:

$$v_{L,\infty} = \left(0.24\,\frac{\text{kg}}{\text{s}\cdot\text{m}}\right)\left(\frac{1}{10^3\,\text{kg/m}^3}\right)\left(\frac{1}{50 \times 10^{-3}\,\text{m}}\right) = 4.8 \times 10^{-3}\,\text{m/s}$$

$$v_{G,\infty} = \left(0.017\,\frac{\text{kg}}{\text{s}\cdot\text{m}}\right)\left(\frac{1}{1.2\,\text{kg/m}^3}\right)\left(\frac{1}{50 \times 10^{-3}\,\text{m}}\right) = 0.28\,\text{m/s}$$

Figure 18-2 shows that a flow with these superficial velocities is well within the stratified regime for flow in a pipe.

18.3.2 Interfacial Stress

The existence of the gas–liquid interface is what makes two-phase flow different from the single-phase flows studied earlier, and the nature of the interactions determines the flow behavior of each phase. We can obtain some useful insight into this interfacial stress by examining the limiting behavior for $\eta_L \gg \eta_G$, which corresponds to most situations of interest. We first consider the gas phase and write Eq. (18.5a) for flow between parallel plates in the equivalent form

$$\langle v_{xG} \rangle = \frac{(H-h)^2}{12\eta_G}\left(-\frac{d\mathcal{P}}{dx}\right)\left[\frac{1 + \alpha_1\eta_G/\alpha_2\eta_L}{1 + \alpha_3\eta_G/\alpha_4\eta_L}\right] \qquad (18.8)$$

The term in brackets will be close to unity when $\alpha_1\eta_G/\alpha_2\eta_L$ and $\alpha_3\eta_G/\alpha_4\eta_L$ are both small compared to unity; the former is the more restrictive and requires that

$$\frac{h}{H} \ll 1 - \left(\frac{\eta_G}{\eta_L}\right)^{1/2} \qquad (18.9)$$

For an air–water system, for example, $(\eta_G/\eta_L)^{1/2} \sim 0.1$, so we would require that $h/H \ll 0.9$; this ratio is always less than 0.6 for stratified flow in pipes. When the term in brackets is close to unity, Eq. (18.8) is identical to Eq. (8.14) for flow in a plane channel of height $H - h$ with two rigid no-slip surfaces; that is, as long as the gas film is not too thin, *the gas flows as though the liquid had been replaced by a solid surface.*

The behavior of the liquid phase is unfortunately not as simple. Equation (18.5b) can be rearranged to

$$\langle v_{xL} \rangle = \frac{h^2}{3\eta_L}\left(-\frac{d\mathcal{P}}{dx}\right)\left\{\frac{1 + 3(H/h)}{4}\right\}\left[\frac{1 + \alpha_5\eta_G/\alpha_6\eta_L}{1 + \alpha_3\eta_G/\alpha_4\eta_L}\right] \qquad (18.10)$$

The term in brackets is negligible provided that

$$\frac{h}{H} \ll 1 - \frac{1}{4}\frac{\eta_G}{\eta_L} \qquad (18.11)$$

which is even less restrictive than Eq. (18.9).

Equation (18.10) can be interpreted in terms of a somewhat different problem, in which we require that the liquid phase experience a shear stress τ_I at the interface. In that case we evaluate the constants C_3 and C_4 in Eq. (18.2b) from the no-slip condition and the condition

$$\eta_L \frac{dv_{xL}}{dy} = \tau_I \text{ at } y = h \qquad (18.12)$$

The result is

$$v_{xL} = \frac{1}{2\eta_L}\left(-\frac{d\mathcal{P}}{dx}\right)(2hy - y^2) + \frac{\tau_I}{\eta_L}y \qquad (18.13)$$

It is useful to have an expression for the shear stress in the liquid at the wall, τ_{wL}:

$$\tau_{wL} = \eta_L \frac{dv_{xL}}{dy}\bigg|_{y=0} = h\left(-\frac{d\mathcal{P}}{dx}\right) + \tau_I \qquad (18.14)$$

The average velocity for the problem with a specified stress at the interface is

$$\langle v_{xL} \rangle = \frac{1}{h} \int_0^h v_{xL}\, dy = \frac{h^2}{3\eta_L}\left(-\frac{d\mathcal{P}}{dx}\right) + \frac{\tau_I h}{2\eta_L} \tag{18.15}$$

Comparison with Eq. (18.10) indicates that for H/h close to unity, but still satisfying the inequality in Eq. (18.11), the interfacial stress is nearly zero. Thus, for a thin gas film *the liquid flows as though the upper surface were nearly unstressed.* This last point can be made quantitative by use of Eq. (18.14) to replace $(-d\mathcal{P}/dx)$ in Eqs. (18.10) and (18.15); we then find that τ_I and τ_{wL} are related to H/h as follows:

$$\tau_I = \tau_{wL}\left[\frac{(H/h) - 1}{(H/h) + 1}\right] \tag{18.16}$$

Thus, *when the interface lies in the upper portion of the channel cross section, the interfacial stress can be neglected relative to the wall shear stress and the pressure drop can be attributed to wall drag.*

18.3.3 Hydraulic Diameter and Reynolds Number

It is necessary to be able to establish conditions under which the two phases are laminar. The transition to turbulence in rectilinear flows usually occurs when the Reynolds number is about 2100, where Re is based on the hydraulic diameter. The hydraulic diameter is defined by Eq. (3.40):

$$D_H = \frac{4 \times \text{cross-sectional area}}{\text{wetted perimeter}} \tag{18.17}$$

The definition of the wetted perimeter for flow with a fluid–fluid interface is not obvious, but the results of the preceding section provide a reasonable basis for selection.

We continue to focus on flow in a channel (Fig. 18-3). The gas flows as though it were in a channel of depth $H - h$ with two no-slip surfaces. Thus, both the upper and lower boundaries must comprise the wetted perimeter. The wetted perimeter is then twice the channel width, while the area is the width times the height. Thus,

$$D_{HG} = \frac{4(H - h) \times \text{width}}{2 \times \text{width}} = 2(H - h) \tag{18.18}$$

The Reynolds number for the gas phase is

$$\mathrm{Re}_G = \frac{D_{HG}\langle v_{xG}\rangle \rho_G}{\eta_G} = \frac{2(H - h)\langle v_{xG}\rangle \rho_G}{\eta_G} \tag{18.19}$$

According to Eq. (18.16) the interfacial shear stress is small relative to the wall shear stress in the liquid for $H/h < 2$, while for $H/h \rightarrow \infty$ the two stresses become comparable. An effective wetted perimeter would then be to take the solid surface and the fraction to the liquid–gas surface defined by Eq. (18.16):

$$\text{liquid wetted perimeter} = \left[1 + \frac{(H/h) - 1}{(H/h) + 1}\right] \times \text{width} \qquad (18.20)$$

The Reynolds number for the liquid phase is then

$$Re_L = \frac{2h(H + h)\langle v_{xL}\rangle \rho_L}{H\eta_L} \qquad (18.21)$$

Note that $(H - h)\langle v_{xG}\rangle \rho_G$ and $h\langle v_{xL}\rangle \rho_L$ are the mass flow rates of gas and liquid per unit width, respectively. These quantities are independent of the *in situ* volumetric holdup. Thus, *the gas-phase Reynolds number can be computed by assuming that the gas phase occupies the entire channel cross section, and the liquid-phase Reynolds number can be computed to within a factor of 2 by assuming that the liquid occupies the entire channel cross section.* It is probably for this reason that the superficial velocities work as correlating variables, since for the narrow range of available physical properties the Reynolds numbers for both phases are determined by $v_{L,\infty}$ and $v_{G,\infty}$.

Example 18.3

Compute the gas- and liquid-phase Reynolds numbers for Examples 18.1 and 18.2.

For the gas phase, $(H - h)\langle v_{xG}\rangle \rho_G = 0.017$ kg/s/m of width, and $\eta_G = 1.7 \times 10^{-5}$. Thus,

$$Re_G = \frac{(2)(1.7 \times 10^{-2})}{1.7 \times 10^{-5}} = 2000$$

The gas phase is probably laminar, but close to the transition. For the liquid, $h\langle v_{xL}\rangle \rho_L = 0.24$ kg/s/m and $\eta_L = 10^{-3}$. Thus,

$$Re_L = \frac{2(1 + h/H)(2.4 \times 10^{-1})}{10^{-3}} = 480\left(1 + \frac{h}{H}\right) = 640$$

Here we have used the result of Example 18.1 that $h/H = \frac{1}{3}$. Without this information we could have bounded Re_L between 480 and 960, which is entirely in the laminar range.

18.3.4 Pipe Flow

The analysis has been limited thus far to flow between parallel plates, because analytical solutions are available for this geometry, and these analytical solutions help in developing physical insight. The geometrical parameters for flow in a pipe are shown in Fig. 18-4. The axial velocity now varies in both r and θ directions, so the left-hand sides of Eqs. (18.1) must now contain both r and θ derivatives. The approach to solution of this problem for two laminar phases is identical to that for parallel plates, but a

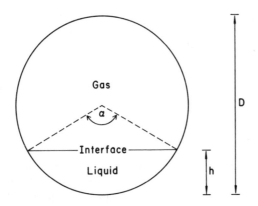

Figure 18-4. Schematic of stratified flow in a pipe.

closed-form analytical solution can be obtained for the more complex pipe flow geometry only when $h = D/2$. For all other cases the solution must be obtained by numerical methods of the type described in Chapter 17. Yu and Sparrow have published such a solution, where the analogs of Eqs. (18.5) and (18.7) for the flow rates and holdup are given in graphical form.

The absence of an analytical solution for laminar–laminar stratified pipe flow requires that we rely on the analogy to the plane channel for computation of the phase Reynolds numbers. It is assumed for pipe flow that the wetted perimeter for the gas phase includes both the wall and the gas–liquid interface; in that case the gas-phase Reynolds number becomes

$$\text{pipe:} \quad \text{Re}_G = \frac{4w_G}{\pi D \eta_G [1 - \alpha/2\pi + (\sin \alpha/2)/\pi]} \tag{18.22}$$

The term in brackets is close to unity for $h/D < \frac{1}{2}$, so the gas-phase Reynolds number may be estimated by assuming that the gas occupies the entire pipe cross section and using the superficial gas velocity.

A result comparable to Eq. (18.16) does not exist for pipe flow, and it is *assumed* that the stress at the gas–liquid interface is negligible relative to the wall shear stress. This assumption should be less serious in a pipe then in a channel, since the gas–liquid surface area in a pipe will be smaller than the liquid–solid surface. With the assumption that the wetted perimeter is simply the arc length, $D\alpha/2$, the liquid-phase Reynolds number is

$$\text{pipe:} \quad \text{Re}_L = \frac{4w_L}{\pi D \eta_L (\alpha/2\pi)} \tag{18.23}$$

The liquid-phase Reynolds number may differ significantly from that computed assuming that the liquid fills the entire pipe if the liquid film is very thin ($\alpha \ll 2\pi$), but the error will be less than a factor of 4 for $h/D > 0.1$.

It is possible, using Yu and Sparrow's numerical solution for pipe flow, to compute h for various flow conditions, and then to use Eqs. (18.22) and

(18.23) to calculate the Reynolds numbers for points within the stratified region on Fig. 18-2. There is a finite region in which one or both phases may be turbulent in stratified flow.

18.4 LOCKHART–MARTINELLI METHOD

18.4.1 Stratified Flow

Stratified flow in which one or both phases is turbulent can be treated in a manner that follows from the development in Sec. 18.3. We will deal here with the case in which both phases are turbulent; the laminar liquid–turbulent gas is left as a problem. (Turbulent liquid–laminar gas is handled in an identical way, but it is of no practical importance.) Turbulent–turbulent stratified flow probably does not occur often, but the analysis of such a flow configuration forms the basis for an approach by Lockhart and Martinelli that applies to nonstratified flows.

When both phases are laminar, the pressure drop-flow rate relationship in each phase can be calculated from the Navier–Stokes equations, as in the preceding section. When the phases are turbulent, it is necessary to use the turbulent friction factor–Reynolds number correlation to obtain the same information. We will limit ourselves here to cases for which the conduit is smooth and $Re \leq 10^5$, so that we may use the Blasius equation, Eq. (3.10), written in terms of the hydraulic diameter:

$$f = \left(-\frac{\Delta \mathcal{P}}{L}\right)\frac{D_H}{2\rho \langle v_x \rangle^2} = 0.079 \, Re^{-1/4} = 0.079\left(\frac{\eta}{D_H \langle v_x \rangle \rho}\right)^{1/4} \quad (18.24)$$

It is again convenient first to consider flow in a plane channel, Fig. 18-3. As in the preceding section, we assume that the wetted perimeter includes both the solid surface and the liquid–gas interface, so $D_{HG} = 2(H - h)$. The friction factor equation for the gas phase is then

$$f_G = \left(-\frac{\Delta \mathcal{P}}{L}\right)\frac{H - h}{\rho_G \langle v_{xG} \rangle^2} = 0.079\left[\frac{\eta_G}{2(H - h)\langle v_{xG} \rangle \rho_G}\right]^{1/4} \quad (18.25)$$

The wetted perimeter for the liquid phase includes the solid surface and a portion of the gas–liquid interface, since the stress at the interface is less than the stress at the solid surface. We therefore take $D_{HL} = 4h/\beta$, where $\beta = 1$ for a no-shear interface with the gas and $\beta = 2$ if the stress at the gas interface is equal to the wall stress; we generally expect β to be closer to unity. The friction factor equation for the liquid phase is then

$$f_L = \left(-\frac{\Delta \mathcal{P}}{L}\right)\frac{2h}{\beta \rho_L \langle v_{xL} \rangle^2} = 0.079\left[\frac{\beta \eta_L}{4h \langle v_{xL} \rangle \rho_L}\right]^{1/4} \quad (18.26)$$

If h is known, the pressure drop can be computed from either Eq. (18.25) or (18.26). An equation for h is obtained by eliminating $(-\Delta\mathcal{P}/L)$ between Eqs. (18.25) and (18.26). The algebra is straightforward and leads to the explicit result

$$\frac{h}{H} = \left\{1 + \left(\frac{2}{\beta}\right)^{5/12}\left(\frac{\eta_G}{\eta_L}\right)^{1/12}\left(\frac{\rho_L}{\rho_G}\right)^{1/3}\left(\frac{w_G}{w_L}\right)^{7/12}\right\}^{-1} \tag{18.27}$$

The result is relatively insensitive to β, since the factor $(2/\beta)^{5/12}$ ranges from 1.33 for $\beta = 1$ to 1.0 for $\beta = 2$; if we assume that $\beta = 1.5$ is the "most probable" value, the coefficient equals 1.13.

An equivalent result follows for pipe flow, using the hydraulic diameters in Eqs. (18.22) and (18.23). A simple expression for h cannot be obtained because of the added complexity of the trigonometric relationships in the pipe, however, so the details of the calculation are more involved.

Example 18.4

Air and water flow cocurrently through a 50-mm plane channel at close to atmospheric pressure and 20°C. The water flows at 1.44 kg/s per meter of channel width and the air at 0.034 kg/s per meter of width. Compute the relative holdup h/H of liquid in the channel.

This is essentially the same problem as Example 18.1, but the water flow has been increased by a factor of 6 and the air flow has been doubled. It then follows from the calculation in Example 18.3 that $Re_G = 4000$ and $Re_L \geq 2900$, so both phases may be assumed to be turbulent. Substituting the numerical values for densities and viscosities, together with $w_G/w_L = 0.034/1.44$, into Eq. (18.27) gives $h/H = 0.57$ for $\beta = 2$ and $h/H = 0.50$ for $\beta = 1$, with $h/H = 0.54$ for the "most probable" value of $\beta = 1.5$. We thus find that $h \approx 27$ mm. The superficial velocities are $v_{L,\infty} = 2.9 \times 10^{-2}$ m/s and $v_{G,\infty} = 0.56$, which is within the stratified region in Fig. 18-2.

18.4.2 Nonstratified Flow

Satisfactory analytical procedures are not available for flows outside the stratified regime, and empirical procedures must be used to predict holdup and pressure drop. The most commonly used method is the correlation by Lockhart and Martinelli. The choice of correlating variables is best motivated by a reformulation of the turbulent stratified flow equations.

We shall be dealing with three different pressure gradients, so it is helpful to use subscripts. The pressure gradient for the two-phase flow is denoted $(-d\mathcal{P}/dx)_{TPF}$. The pressure gradient that would exist if the liquid phase occupied the entire pipe cross section and the gas were absent is denoted $(-d\mathcal{P}/dx)_{L,FP}$. The pressure gradient that would exist if the gas phase occupied the entire pipe and the liquid phase were absent is denoted $(d\mathcal{P}/dx)_{G,FP}$.

If we limit ourselves to smooth pipe and $Re \leq 10^5$, we may compute

$(-d\mathcal{P}/dx)_{L,FP}$ and $(-d\mathcal{P}/dx)_{G,FP}$ from the Blasius equation, Eq. (3.10) or (18.24), as

$$\left(-\frac{d\mathcal{P}}{dx}\right)_{L,FP} = (0.079)\frac{32w_L^2}{\pi^2\rho_L D^5}\left(\frac{\pi D\eta_L}{4w_L}\right)^{1/4} \tag{18.28a}$$

$$\left(-\frac{d\mathcal{P}}{dx}\right)_{G,FP} = (0.079)\frac{32w_G^2}{\pi^2\rho_G D^5}\left(\frac{\pi D\eta_G}{4w_G}\right)^{1/4} \tag{18.28b}$$

The ratio of these two fictitious pressure drops is defined in the two-phase literature as X_{tt}^2:

$$X_{tt}^2 = \frac{(-d\mathcal{P}/dx)_{L,FP}}{(-d\mathcal{P}/dx)_{G,FP}} = \left(\frac{\rho_G}{\rho_L}\right)\left(\frac{\eta_L}{\eta_G}\right)^{1/4}\left(\frac{w_L}{w_G}\right)^{7/4} \tag{18.29}$$

According to Eq. (18.27), assuming that we can employ an analogy between pipe and parallel plate flow, *the holdup in stratified turbulent–turbulent flow is a unique function of X_{tt}.*

Equation (18.24) is written for either phase in the two-phase flow, so the pressure gradient that appears in that equation is $(-d\mathcal{P}/dx)_{TPF}$. If we write the equation for the gas phase, and combine it with Eq. (18.28a) and the definition of D_{HG}, we obtain a ratio of pressure gradients defined in the two-phase literature as ϕ_{Gtt}^2:

$$\phi_{Gtt}^2 = \frac{(-d\mathcal{P}/dx)_{TPF}}{(-d\mathcal{P}/dx)_{G,FP}} = \left(\frac{\text{gas wetted perimeter}}{\text{pipe perimeter}}\right)^{5/4}\left(\frac{\text{pipe area}}{\text{gas flow area}}\right)^3 \tag{18.30}$$

The right-hand side of Eq. (18.30) depends only on the holdup; thus, *in stratified turbulent–turbulent flow the ratio of the two-phase pressure gradient to the pressure gradient that would be experienced by the gas phase alone in the full pipe is a unique function of X_{tt}.*

In the Lockhart–Martinelli correlation *it is assumed that the holdup and ϕ_{Gtt} are unique functions of X_{tt} for all flow regimes*[*]. The resulting correlation for ϕ_{Gtt} and the *in situ* liquid volume holdup ratio, R_L, is shown in Fig. 18-5. The correlation predicts pressure drops to within about $\pm 50\%$ for pipes up to 100 mm in diameter. Predictions tend to be high for the pressure drop in stratified, wavy, and slug flow, and low in annular flow.

Gas–liquid pipeline contactors for chemical reaction are usually designed to operate in the bubble regime for effective interphase mass transfer. Since bubbles are normally approximately 4 to 5 mm in diameter, the effective interfacial area can be estimated from the holdup. The Lockhart–Martinelli correlation is adequate for holdup in the bubble regime, but a correlation by Hughmark and Prestberg is preferable.

---

[*]It is customary to define ϕ_{Gtl}^2 and ϕ_{Gll}^2, but these are not of interest to us, since the approaches in Secs. 18.3 and 18.4 can be used to deal with laminar–laminar and laminar–turbulent flows.

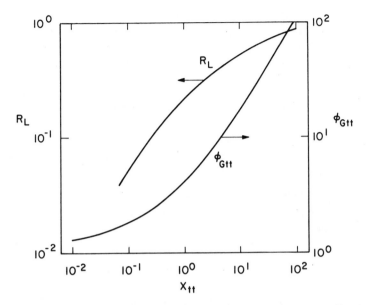

Figure 18-5. Correlation for horizontal gas-liquid flow. After Lockhart and Martinelli, *Chem. Eng. Progr.*, **45**, 39 (1949), copyright by the American Fightitute of Chemical Engineers, by permission.

Example 18.5

Air and water flow cocurrently in a pipe with a mass flow ratio $w_L/w_G = 42.4$. Estimate the liquid-phase holdup and the pressure gradient.

This is a slight modification of Example 18.4. We can compute X_{tt} from the Blasius equation, using the relation in Eq. (18.29) and the air–water data in Example 18.1:

$$X_{tt} = \left(\frac{\rho_G}{\rho_L}\right)^{1/2}\left(\frac{\eta_L}{\eta_G}\right)^{1/8}\left(\frac{w_L}{w_G}\right)^{7/8} = \left(\frac{1.2}{10^3}\right)^{1/2}\left(\frac{10^{-3}}{1.7 \times 10^{-5}}\right)^{1/8}(42.4)^{7/8} = 1.5$$

From Fig. 18-5 we obtain $R_L \approx 0.27$ and $\phi_{Gtt} \approx 5.2$. The pressure gradient is then calculated from Eq. (18.28b):

$$\left(-\frac{d\mathcal{P}}{dx}\right)_{TPF} = \phi_{Gtt}^2\left(-\frac{d\mathcal{P}}{dx}\right)_{G,FP} = (5.2)^2(0.079)\frac{32w_G^2}{\pi^2\rho_G D^5}\left(\frac{\pi D\eta_G}{4w_G}\right)^{1/4}$$

Note that the pressure drop for the two-phase system is *twenty-seven* times (ϕ_{Gtt}^2) the pressure drop for the gas phase flowing alone.

18.5 CONCLUDING REMARKS

There are many other correlations for two phase flow in addition to the Lockhart–Martinelli method, but the improvement in predictive ability is not substantial. It has been our intention in this chapter to present only an introduction to the general topic of two-phase flow. The more specialized literature cited in the Bibliographical Notes should be consulted before undertaking process design calculations for two-phase systems, to ensure that the best correlations for particular flow configurations are used.

BIBLIOGRAPHICAL NOTES

The most complete textbook is

GOVIER, G. W., and AZIZ, K., *The Flow of Complex Mixtures in Pipes*, Van Nostrand Reinhold, New York, 1972.

There is an older discussion in Chapter 16 of

BRODKEY, R. S., *The Phenomena of Fluid Motions*, Addison-Wesley Publishing Co., Inc., Reading, Mass., 1967.

The flow map in Fig. 18-2 is from

MANDHANE, J. M., GREGORY, G. H., and AZIZ, K., *Int. J. Multiphase Flow*, **1**, 537 (1974).

References to other flow regime correlations and extensive data banks can be found in that paper.

For stratified flow, see

RUSSELL, T. W. F., ETCHELLS, A. W., JENSEN, R. H., and ARRUDA, P. J., *AIChEJ.*, **20**, 664 (1974).

YU, H. S., and SPARROW, E. M., *AIChE J.*, **13**, 10 (1967).

The original paper by Lockhart and Martinelli should be consulted in order to see the range of data from which the correlation in Fig. 18-5 was constructed:

LOCKHART, R. W., and MARTINELLI, R. C., *Chem. Eng. Progr.*, **45**, 39 (1949).

Most two-phase flow correlations require iterative calculation. A computer program, TWOFAZFLOW, that performs the calculations described in this chapter and others using correlations by Hughmark and Dukler is contained in

MORRIS, F. R., "A Computer Program for the Calculation of Two Phase Flow Design Parameters," M.Ch.E. thesis, University of Delaware, Newark, Del., 1975.

PROBLEMS

18.1. Show that the holdup in stratified turbulent-turbulent flow in a round pipe is a unique function of X_{tt} and describe a procedure for calculating it.

18.2. Air and water are in horizontal cocurrent flow in a pressurized 127 mm diameter pipeline oxygenator that is part of a wastewater treatment process. The viscosity and density of the air are 1.8×10^{-5} Pa·s and 6.6 kg/m^3, respectively. Flow rates are $w_L = 1.58 \times 10^5$ kg/hr, $w_G = 45.4$ kg/hr. Determine the flow regime, the liquid phase holdup, and the overall pressure drop.

18.3. A steam stripping operation takes place in a horizontal 41 mm diameter pipeline. 450 kg/hr of a liquid with $\rho_L = 10^3$ kg/m³, $\eta_L = 10^{-3}$ Pa·s are to be processed, and the slug flow regime must be avoided because of unacceptable vibrations. Estimate the minimum steam flow rate. Steam properties may be taken as $\rho_G = 1.9$, $\eta_G = 1.2 \times 10^{-5}$.

18.4. The process described in Problem 18.3 is operated with 450 kg/hr of steam and liquid. Determine the flow regime, the liquid phase holdup, and the overall pressure drop.

18.5. Repeat the analysis in Sec. 18.4.1 for a laminar liquid and a turbulent gas.

Viscoelasticity **19**

19.1 INTRODUCTION

We have assumed throughout this text that time-dependent effects of the type shown in Fig. 2-13 are absent, and that the stress in the fluid responds instantaneously to changes in the deformation rate. This assumption is an excellent one for gases and for low-molecular-weight liquids, but it may be quite poor for solutions and melts of high-molecular-weight polymers.

High-molecular-weight materials sometimes show stress behavior in the liquid state that is similar to the behavior of elastic solids. For this reason these polymeric liquids are often referred to as *viscoelastic fluids*. The study of the flow behavior of viscoelastic fluids (and, in fact, all non-Newtonian fluids) is a part of the branch of mechanics known as *rheology*. The complex rheological behavior of such fluids can play an important role in their processing.

The purpose of this chapter is to provide a brief introduction to the principal physical concepts and some examples. The interested reader can then consult the more specialized texts and periodical literature for more information about the flow behavior of this class of materials.

19.2 COMPLEX VISCOSITY AND MODULUS

Viscoelastic behavior is frequently studied experimentally in a flow with induced oscillations. Consider a fluid contained between two plates, as shown in Fig. 19-1. The relative position of the bottom plate moves back and forth

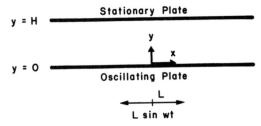

Figure 19-1. Schematic of oscillatory shear.

sinusoidally with a frequency ω and a small amplitude L; this displacement, normalized with respect to the spacing, H, can be considered a *strain*, so we have

$$\text{strain} = \frac{L}{H} \sin \omega t \tag{19.1}$$

The velocity of the lower plate is simply the rate of change of displacement:

$$\text{at } y = 0: \quad v_x = L\omega \cos \omega t \tag{19.2}$$

In order to analyze this motion we must solve the momentum equation. We assume that the only nonzero velocity component is v_x, and that v_x is independent of x and z. The momentum equation is then found from Table 7-2 to simplify to

$$\rho \frac{\partial v_x}{\partial t} = \frac{\partial \tau_{yx}}{\partial y} \tag{19.3}$$

The discussion on ordering in Sec. 11.5 can now be applied here, particularly that leading to Eqs. (11.17). For a Newtonian fluid we may neglect the inertial term provided that

$$\omega \ll \frac{\rho H^2}{\eta} \tag{19.4}$$

Clearly, this is not a very restrictive condition. For a polymer melt the viscosity is very high, and even for dilute low-viscosity solutions we can make H extremely small. [Equation (19.4) is not rigorously correct for non-Newtonian fluids, but it suffices for our purposes here.] If the inertial term can be neglected, we have $\partial \tau_{yx}/\partial y = 0$, or τ_{yx} is *independent of position at any time.*

The behavior shown in Fig. 2-13 requires that τ_{yx} depend in some way on time as well as on deformation rate. The simplest form we can hypothesize that is consistent with the initial transient and ultimate steady state in Fig. 2-13 is

$$\lambda \frac{d\tau_{yx}}{dt} + \tau_{yx} = \eta \frac{\partial v_x}{\partial y} \tag{19.5}$$

When the time constant λ is zero, we recover the Newtonian fluid. A fluid described by Eq. (19.5) is called a *Maxwell fluid.*

For the experiment described in Fig. 19-1 we have established that, as long as ω satisfies the inequality in Eq. (19.4), the shear stress τ_{yx} is independent of y. If the left-hand side of Eq. (19.5) is independent of y, so must be the right-hand side. Thus, $\partial v_x/\partial y$ is independent of y, so v_x must be linear. In order to satisfy Eq. (19.2) at $y = 0$ and the no-slip condition at $y = H$, it then follows that

$$v_x = L\omega\left(1 - \frac{y}{H}\right)\cos \omega t \tag{19.6}$$

$$\frac{\partial v_x}{\partial y} = -\frac{L\omega}{H}\cos \omega t \tag{19.7}$$

Equation (19.5) is now a rather elementary ordinary differential equation for the shear stress:

$$\lambda \frac{d\tau_{yx}}{dt} + \tau_{yx} = -\frac{\eta L\omega}{H}\cos \omega t \tag{19.8}$$

The solution for t greater than about 3λ, when any initial transients have died out, is

$$\tau_{yx} = -\frac{\eta L\omega}{H(1 + \lambda^2\omega^2)}\cos \omega t - \frac{\lambda\eta L\omega^2}{H(1 + \lambda^2\omega^2)}\sin \omega t \tag{19.9}$$

This is often written in an alternative form,

$$\tau_{yx} = -\eta'\left(\frac{L\omega}{H}\cos \omega t\right) - G'\left(\frac{L}{H}\sin \omega t\right) \tag{19.10}$$

Here, η' and G' are called the real parts of the complex viscosity and modulus, respectively, and are defined

$$\eta' = \frac{\eta}{1 + \lambda^2\omega^2} \tag{19.11a}$$

$$G' = \frac{\eta}{\lambda}\frac{\lambda^2\omega^2}{1 + \lambda^2\omega^2} \tag{19.11b}$$

Note that the shear stress consists of two terms. The first is proportional to the shear rate, as in a Newtonian fluid, and is in fact the only term remaining in the limit $\lambda\omega \rightarrow 0$. The second term is proportional to the *strain*, which is the type of behavior expected of a solid. At very high frequencies, $\lambda\omega \rightarrow \infty$, only the second, solidlike term remains, and the stress is related to the strain through a modulus η/λ.

Data for η' and G' on a silicone polymer are shown in Fig. 19-2, where the solid lines represent Eqs. (19.11). This general qualitative behavior is shown by all polymeric liquids as long as the strain amplitude is small, but Eq. (19.5) is usually too simple for a quantitative representation and the approach must be generalized. The form of Eq. (19.10) is a consequence of a small-amplitude (linear) experiment, and is independent of the assumption of a Maxwell fluid. (We shall not prove this.) Thus, η' and G' are defined by Eq. (19.10), but the experimental functions are not always well represented

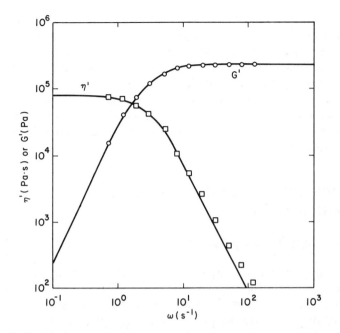

Figure 19-2. Oscillatory shear measurements on a silicone polymer. The solid lines are Eqs. (19.11) with $\eta = 8 \times 10^4$ Pa·s and $\lambda = 0.31$ s. Data of F. N. Cogswell.

by Eqs. (19.11); equivalently, Eq. (19.5) represents the behavior of most real liquids only qualitatively.

19.3 DEBORAH NUMBER

Equations (19.10) and (19.11) can be summarized as follows:

$$\lambda\omega \ll 1: \quad \text{stress} \sim \text{viscosity} \times \text{shear rate} \qquad (19.12a)$$

$$\lambda\omega \gg 1: \quad \text{stress} \sim \text{modulus} \times \text{strain} \qquad (19.12b)$$

ω is the reciprocal of the time scale of the process, so we can conclude that the nature of the material response depends on the ratio of fluid response (or relaxation) time to process time. This ratio is often called the *Deborah number*.* Thus, we conclude that *a viscoelastic liquid will behave like a purely viscous liquid when the Deborah number is small and like an elastic solid when the Deborah number is large.* All the flow problems described in the preceding chapters apply only to small Debroah number situations.

*From the Song of Deborah, Judges V, 5: "The mountains quaked (sometimes translated 'flowed') at the presence of the Lord," suggesting that even solid mountains flow like a liquid on an appropriately long process time scale.

The relaxation time for commercially important molten polymers at processing temperatures ranges from less than 10^{-2} s for polyesters such as polyethylene terephthalate to one to 10 s for polyolefins such as polyethylene. The process time is roughly the length scale in the flow direction divided by the mean velocity. The Deborah number for flow in an extruder will usually be small, while the Deborah number for fiber drawing may sometimes be large. Each process must be analyzed individually, and simplifications may be possible in various limits, just as we found for Newtonian fluids in Chapters 11 through 15.

19.4 OBSERVABLE VISCOELASTIC PHENOMENA

19.4.1 Turbulent Drag Reduction

Turbulent flow contains rapid fluctuations. Thus, even though the characteristic response time for polymer solutions is quite small (10^{-6} to 10^{-3} s), the Deborah number may sometimes be large. In that case there is at least one additional dimensionless group, involving λ, that must be considered in any dimensional analysis; a suitable independent group is $\lambda \langle \bar{v} \rangle / D$, which is sometimes called the *Weissenberg number*, We. The friction factor cannot be expected to be a unique function of Reynolds number in that case, but must be expected to depend on Re and We in a possibly complex way.

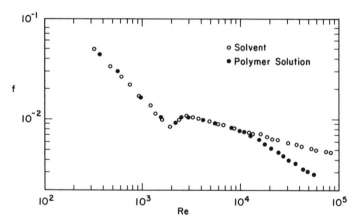

Figure 19-3. Friction factor-Reynolds number data for distilled water and a solution of 296 weight parts per million of polyethylene oxide in distilled water in an 8.46-mm-diameter pipe. Note the significantly reduced frictional drag in the polymer solution for Re greater than 10^4. Reproduced from Virk, *AIChEJ.*, **21**, 625 (1975), copyright by the American Institute of Chemical Engineers, by permission.

Figure 19-3 shows friction factor-Reynolds number data for a very dilute aqueous polymer solution. The data deviate from the single line characteristic of turbulent flow of a Newtonian fluid, and in fact the friction factor (and hence the pressure drop) can be greatly reduced; drag reduction of more than 90% has been observed. There is often a strong pipe diameter effect, however, and lower levels of drag reduction are obtained in large-diameter pipes. This drag reduction was first reported in the open literature by Toms in 1948, and is sometimes known as the *Toms effect*, but it had been observed earlier by Mysels and others and described in World War II documents that were not declassified until later.

The onset of an influence of fluid elasticity should occur when the "elastic" term in Eq. (19.9) becomes comparable to the viscous term, say when $\lambda\omega$ is of order unity. In that case the "elastic" stress is of order η/λ [Eq. (19.11b)]; the viscous stress in turbulent flow is most easily represented by the wall shear stress, τ_w. We thus expect the deviation from Newtonian turbulent behavior to occur when the dimensionless ratio $\lambda\tau_w/\eta$ (or, more usually, $\lambda\rho u_*^2/\eta$) is of order unity, and this is indeed the case. A scale-up procedure by Savins is based on this dimensional analysis consideration and allows pressure drops in large-diameter pipes to be estimated from convenient laboratory measurements on small systems.

19.4.2 Die Swell

When a Newtonian liquid is extruded from a long, round tube into air at very small Reynolds number, the emerging jet expands to a diameter that is about 10 to 15% larger than the tube diameter; compare Sec. 17.3.4. Polymer melts and concentrated solutions, on the other hand, may expand to two or more times the tube diameter. Expansion in a horizontal jet is shown in Fig. 19-4, where the "droop" is caused by gravity.

This expansion of the extruded jet, usually called *die swell*, is clearly of processing significance; consider, for example, the problem of designing a die to produce an extrudate of a desired shape and size. If the Reynolds number is indeed negligible, then the only remaining independent dimensionless group is the Weissenberg number, We $= \lambda\langle v\rangle/D$. The die swell ratio does correlate with We and follows an approximate theory developed by Tanner:

$$\frac{D_j}{D} = 0.1 + \left(1 + \frac{\lambda\tau_w}{\eta}\right)^{1/6} \tag{19.13}$$

The group $\lambda\tau_w/\eta$ is proportional to We. A comparison of some data and the theory is shown in Fig. 19-5. The theoretical analysis follows from the discussion of principles in Sec. 19.5.

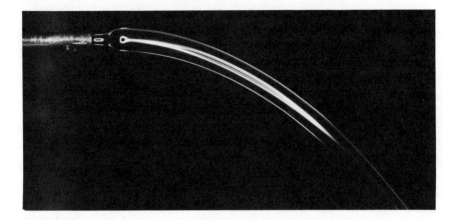

Figure 19-4. Expansion of a jet of 8% polyisobutylene in decalin. Photograph of Ginn, reproduced from Metzner, White and Denn, *Chem. Eng. Progress*, **67**, No. 12, 81 (Dec., 1966), copyright by the American Institute of Chemical Engineers, by permission.

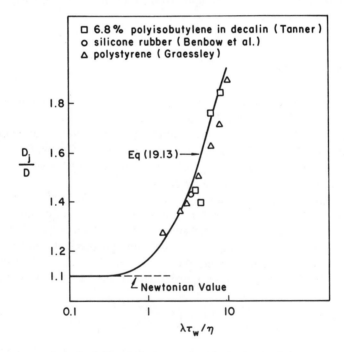

Figure 19-5. Die swell data for three viscoelastic liquids and the theoretical curve of Tanner. Reproduced from Tanner, *J. Polymer Sci.*, A-2, 8, 2067 (1970), copyright by John Wiley and Sons, Inc., by permission.

19.4.3 Melt Fracture

Extruded molten polymers and concentrated solutions exhibit a flow instability at a critical throughput rate. The instability manifests itself in a distorted extrudate. An example is shown in Fig. 19-6, where a low-density

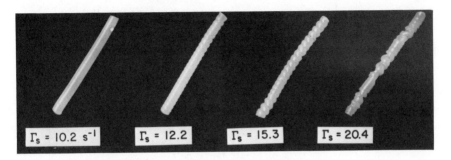

$\Gamma_s = 10.2 \text{ s}^{-1}$ $\Gamma_s = 12.2$ $\Gamma_s = 15.3$ $\Gamma_s = 20.4$

Figure 19-6. Extrudates of low-density polyethylene from a circular die at increasing throughput rates. Note the transition from a smooth extrudate to a helical distortion. Data of F. N. Cogswell.

polyethylene has been extruded through a round die. At sufficiently low throughput the surface is smooth, but with increasing shear rate a helical distortion can be seen on the surface, and gross irregularities appear at still higher throughputs. This instability is often called *melt fracture*, because the liquid sometimes seems to fracture with accompanying crackling noises. The term "elastic turbulence" is also used, but it should be avoided in view of the fact that turbulence cannot be a factor; Reynolds numbers are always much less than unity, typically of order 10^{-2}, and melt fracture has been observed in polytetrafluoroethylene at Re $= 10^{-15}$!

From dimensional-analysis considerations, in the absence of a Reynolds number effect, melt fracture might be expected to occur at a critical value of We. The instability is usually observed for We of order unity, although there is a range of reported critical values of about an order of magnitude. A quantitative theoretical explanation is still lacking. Melt fracture is a limiting factor in throughput for many polymer processing operations.

19.4.4 Normal Stresses

When a viscoelastic liquid is sheared between two plates, a stress is generated in the fluid that tends to push the plates apart. This *normal stress* can be measured as a function of shear rate by simply measuring the force necessary to keep the plates at a constant spacing, as in Fig. 19-7. Commercial instruments are available to make such measurements.

Normal stress measurements on a relatively viscous polymer solution are

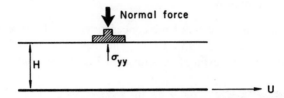

Figure 19-7. Schematic of normal stress in shear flow of a viscoelastic liquid.

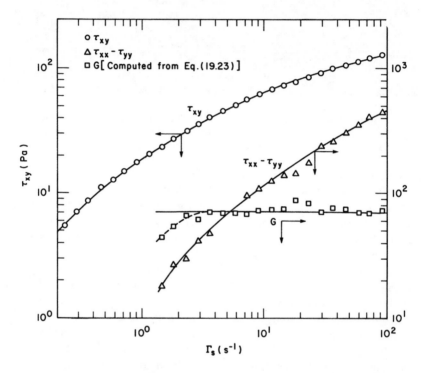

Figure 19-8. Shear and normal stresses functions for a 6% by weight aqueous solution of a sodium cellulose sulfate polymer. G is calculated from Eq. (1914). Data of J. C. Chang.

shown in Fig. 19-8, together with shear stress measurements for comparison. Most data follow the relationship

$$\text{normal stress} = \frac{2}{G}(\text{shear stress})^2 \tag{19.14}$$

where G is a constant modulus that is equal to the ratio η/λ. Equation (19.14) is, in fact, the usual means of measuring λ, since the shear and normal stresses can be measured simultaneously in a shearing experiment. There is some theoretical basis for Eq. (19.14), as discussed in the next section.

This normal stress phenomenon provides a qualitative explanation for die swell. At the end of the tube there is no wall to support the stresses normal to the direction of flow, so the fluid "pushes out" against the nonresistant atmosphere and swells until the normal stresses have relaxed. This stress relaxation requires a time of order λ.

19.5 MAXWELLIAN LIQUIDS

19.5.1 Constitutive Equation

The Newtonian fluid was introduced for one-dimensional steady shearing flows in Sec. 2.4.2 and generalized to arbitrary three-dimensional motions in Sec. 7.4.2. We have similarly introduced the Maxwell material through Eq. (19.5) in terms of a small amplitude one-dimensional oscillatory motion, and this constitutive equation must also be generalized. The procedures for the generalization use symmetry arguments like those for the Newtonian fluid, as well as certain invariance principles that are discussed in specialized texts. The requirement that the stress state be independent of the frame of reference of the observer gives rise to certain nonlinear terms that may be unexpected.

The generalization of the Maxwell fluid to arbitrary motions is not unique, in that several similar forms satisfy all mathematical invariance requirements and reduce to Eq. (19.5) for small amplitude one-dimensional oscillations. We shall present the form that has some basis in molecular theories of polymer solutions. For simplicity we will write the equation for two-dimensional motions; we will assume that $v_z = 0$ and that all z derivatives are zero. We then have

$$\sigma_{xx} = -\hat{p} + \tau_{xx} \tag{19.15a}$$

$$\sigma_{yy} = -\hat{p} + \tau_{yy} \tag{19.15b}$$

$$\tau_{xx} + \frac{\eta}{G}\left[\frac{\partial \tau_{xx}}{\partial t} + v_x \frac{\partial \tau_{xx}}{\partial x} + v_y \frac{\partial \tau_{xx}}{\partial y} - 2\frac{\partial v_x}{\partial x}\tau_{xx} - 2\frac{\partial v_x}{\partial y}\tau_{yx}\right] = 2\eta \frac{\partial v_x}{\partial x} \tag{19.16a}$$

$$\tau_{yy} + \frac{\eta}{G}\left[\frac{\partial \tau_{yy}}{\partial t} + v_x \frac{\partial \tau_{yy}}{\partial x} + v_y \frac{\partial \tau_{yy}}{\partial y} - 2\frac{\partial v_y}{\partial x}\tau_{xy} - 2\frac{\partial v_y}{\partial y}\tau_{yy}\right] = 2\eta \frac{\partial v_y}{\partial y} \tag{19.16b}$$

$$\tau_{xy} + \frac{\eta}{G}\left[\frac{\partial \tau_{xy}}{\partial t} + v_x \frac{\partial \tau_{xy}}{\partial x} + v_y \frac{\partial \tau_{xy}}{\partial y} - \frac{\partial v_x}{\partial y}\tau_{yy} - \frac{\partial v_y}{\partial x}\tau_{xx}\right] = \eta\left(\frac{\partial v_x}{\partial y} + \frac{\partial v_y}{\partial x}\right) \tag{19.16c}$$

$$\tau_{xy} = \tau_{yx} \tag{19.16d}$$

The fluid is taken to be incompressible, and we have written $\hat{p}$ in place of p in Eqs. (19.15) to emphasize that the isotropic pressure term need *not* satisfy Eq. (7.40); compare Sec. 7.4.3. η may be a function of deformation rate, as for example in the power law viscosity defined by Eq. (8.86). The modulus, G, is taken as a constant, which is only approximately in accordance with experimental data. The relaxation time, λ, equals η/G. Note that in the absence of nonlinear terms, as would be appropriate for small amplitude oscillations, Eq. (19.16c) reduces to Eq. (19.5).

A small number of flow problems has been solved for Eqs. (19.16) and similar, sometimes more general, constitutive equations. The die swell result in Fig. 19-5 is one such result, although some severe approximations have been made there.

19.5.2 Viscometric Flow

Consider steady shearing between two plates, as shown in Fig. 19-7. The bottom plate moves with a velocity U relative to the top plate. If the only nonzero component of velocity is v_x, and v_x is assumed to depend only on y, the momentum equation in Table 7-2 simplifies to*

$$0 = -\frac{\partial \hat{\mathcal{P}}}{\partial x} + \frac{\partial \tau_{xx}}{\partial x} + \frac{\partial \tau_{yx}}{\partial y} \tag{19.17a}$$

$$0 = -\frac{\partial \hat{\mathcal{P}}}{\partial y} + \frac{\partial \tau_{yy}}{\partial y} + \frac{\partial \tau_{xy}}{\partial x} \tag{19.17b}$$

We can assume that $\hat{\mathcal{P}}$ is independent of x, since there is no imposed pressure gradient and no change in area. We can further assume that τ_{xx} and τ_{xy} are independent of x, since the stresses depend only on the velocity gradient and v_x is independent of x. Thus, the momentum equation simplifies to

$$0 = \frac{\partial \tau_{yx}}{\partial y} \qquad \tau_{yx} = \text{constant} \tag{19.18a}$$

$$0 = \frac{\partial}{\partial y}(-\hat{\mathcal{P}} + \tau_{yy}) \qquad \sigma_{yy} = -\hat{\mathcal{P}} + \tau_{yy} = \text{constant} \tag{19.18b}$$

At some point the flow field must come to an end, and in the absence of surface tension effects we can assume that σ_{xx} is balanced across an interface by atmospheric pressure, which can be taken as zero. Since σ_{xx} is independent of x, we therefore obtain

$$\sigma_{xx} = 0 = -\hat{\mathcal{P}} + \tau_{xx} \tag{19.19a}$$

or

$$\hat{\mathcal{P}} = \tau_{xx} \tag{19.19b}$$

It then follows from Eq. (19.18b) that the total stress normal to the upper plate is

$$\text{normal stress} = -\sigma_{yy} = \tau_{xx} - \tau_{yy} \tag{19.20}$$

---

*We denote the equivalent pressure by $\hat{\mathcal{P}}$ to emphasize that the pressure portion, $\hat{p}$, does not necessarily follow Eq. (7.40).

The negative sign is required by the stress convention adopted in Sec. 7.3.2. Note that the development thus far is perfectly general and is independent of the choice of constitutive equation. For a Newtonian fluid, $\tau_{xx} = \tau_{yy} = 0$ in this flow, so the normal stress is zero.

When we substitute into Eqs. (19.16), using the fact that v_x is a function only of y, and x derivatives are zero, we obtain

$$\tau_{xx} + \frac{\eta}{G}\left(-2\frac{\partial v_x}{\partial y}\tau_{xy}\right) = 0 \tag{19.21a}$$

$$\tau_{yy} = 0 \tag{19.21b}$$

$$\tau_{xy} + \frac{\eta}{G}\left(-2\frac{\partial v_x}{\partial y}\tau_{yy}\right) = \eta\frac{\partial v_x}{\partial y} \tag{19.21c}$$

Combination of Eqs. (19.21b) and (19.21c) gives

$$\tau_{xy} = \eta\frac{\partial v_x}{\partial y} \tag{19.22}$$

This equation is expected, for it simply states that the shear stress equals viscosity times shear rate. Since the shear stress is a constant, and the viscosity depends only on shear rate, the right-hand side of Eq. (19.22) states that a function of $\partial v_x/\partial y$ is independent of y; this can only be true if $\partial v_x/\partial y$ is independent of y, so we have established that the velocity profile is linear regardless of the viscosity-shear rate relation.

Equations (19.20) through (19.22) combine to give

$$\text{normal stress} = -\sigma_{yy} = \tau_{xx} - \tau_{yy} = \frac{2}{G}\eta\frac{\partial v_x}{\partial y}\tau_{xy} = \frac{2}{G}(\tau_{xy})^2 \tag{19.23}$$

We thus obtain Eq. (19.14), which represents much experimental data. G is typically of order 10^5 Pa for polymer melts at processing conditions.

19.6 CAPILLARY VISCOMETER

The use of flow rate-pressure drop measurements in a long capillary to obtain the viscosity of a Newtonian fluid was first described in Sec. 8.4. The development was extended to fluids with a power-law viscosity in Sec. 8.7. We now show how the viscosity function for any fluid can be obtained. We assume that the pipe is sufficiently long that entrance and exit effects can be neglected, including viscoelastic effects in the developing region. The Deborah number is $\lambda\langle v\rangle/L$; this can be written $(\lambda\langle v\rangle/D)(D/L)$, and, since $\lambda\langle v\rangle/D$ will usually be of order unity and D/L of order 10^{-2}, the Deborah number will be sufficiently small to neglect fluid elasticity.

We assume that v_z is the only nonzero velocity component, and that v_z depends only on radial position, r. We know from the analysis in the preceding section and the existence of normal stresses that $\mathcal{P}$ may depend on r in a viscoelastic liquid, but it is reasonable to assume that $\partial\mathcal{P}/\partial z$ is independent of r, and no contradiction results from this assumption. The axial component of the momentum equation in Table 7-5 is then

$$0 = -\frac{\partial\mathcal{P}}{\partial z} + \frac{1}{r}\frac{d}{dr}(r\tau_{rz}) \tag{19.24}$$

This is integrated once to obtain

$$\tau_{rz} = \frac{r}{2}\frac{\partial\mathcal{P}}{\partial z} = \frac{r}{2}\left(\frac{\Delta\mathcal{P}}{L}\right) \tag{19.25}$$

The shear stress at the wall, τ_w, is

$$\tau_w = \frac{R}{2}\left(-\frac{\Delta\mathcal{P}}{L}\right) \tag{19.26}$$

Note that the wall stress is taken by convention to be a positive number ($\Delta\mathcal{P} < 0$). When we measure flow rate versus pressure drop we therefore obtain the wall shear stress for a given flow rate. We now need the corresponding shear rate at the wall in order to obtain the shear stress-shear rate relationship that provides the viscosity.

The viscosity is defined as the ratio of shear stress to shear rate, so Eq. (19.25) can be written

$$\tau_{rz} = \eta\frac{\partial v_z}{\partial r} = \frac{r}{2}\left(\frac{\Delta\mathcal{P}}{L}\right) \tag{19.27}$$

We now multiply both sides by πr^2, divide by viscosity, and integrate from zero to R:

$$\int_0^R \pi r^2\frac{\partial v_z}{\partial r}\,dr = \left(\frac{\Delta\mathcal{P}}{L}\right)\int_0^R \frac{\pi r^3}{2\eta}\,dr = \pi r^2 v_z\Big|_0^R - \int_0^R 2\pi r v_z\,dr = -Q \tag{19.28}$$

Equation (19.28) is combined with Eq. (19.26) to write

$$Q = \frac{3\tau_w}{R}\int_0^R \frac{\pi r^3}{2\eta}\,dr \tag{19.29}$$

The integration cannot be carried out if η is a function of shear rate, since $\partial v_z/\partial r$ varies in an unknown manner over the cross section.

The solution is now obtained through a clever trick. The viscosity can be considered to be a function of the stress. We can combine Eqs. (19.25) and (19.26) to express r in terms of τ_{rz} and τ_w, where the latter is a constant with respect to the integration:

$$r = -\frac{R}{\tau_w}\tau_{rz} \qquad dr = -\frac{R}{\tau_w}d\tau_{rz} \tag{19.30}$$

The integral in Eq. (19.29) can therefore be written in terms of integration with respect to τ_{rz}:

$$Q = \frac{2\tau_w}{R} \int_0^{-\tau_w} \frac{\pi R^3 \tau_{rz}^3}{2\eta\tau_w^3} \frac{R \, d\tau_{rz}}{\tau_w} = \frac{\pi R^3}{\tau_w^3} \int_0^{-\tau_w} \frac{\tau_{rz}^3 \, d\tau_{rz}}{\eta} \qquad (19.31)$$

We presumably have data for Q versus τ_w, so we can obtain a slope if desired. If we differentiate Q in Eq. (19.31) with respect to τ_w we obtain two terms, one from the τ_w^{-3} term and one from the τ_w in the upper limit; the latter occurs from application of the Leibnitz rule for differentiation of an integral. Thus,

$$\frac{dQ}{d\tau_w} = -\frac{3\pi R^3}{\tau_w^4} \int_0^{-\tau_w} \frac{\tau_{rz}^3 \, d\tau_{rz}}{\eta} + \frac{\pi R^3}{\eta_w} = -\frac{3Q}{\tau_w} + \frac{\pi R^3}{\eta_w} \qquad (19.32)$$

Here, η_w is the viscosity evaluated at the wall shear rate. We now have almost the required result, for the ratio τ_w/η_w gives the wall shear rate, $\Gamma_{s,w}$. Equation (19.32) is easily rearranged to give

$$\Gamma_{s,w} = \frac{8\langle v \rangle}{D} \left(\frac{3}{4} + \frac{1}{4} \frac{d \ln Q}{d \ln \tau_w} \right) \qquad (19.33)$$

Equation (19.33) is sometimes known as the *Rabinowitsch equation*. The procedure, then, is to plot Q versus $\Delta \mathcal{P}$ on logarithmic coordinates. The shear stress is given by Eq. (19.26) and the shear rate by Eq. (19.33). The data at higher shear rates in Fig. 2-8 were in fact obtained in this way. The pressure drops are usually not sufficient to obtain low-shear-rate data in a capillary viscometer.

The modulus and normal stresses cannot be measured conveniently in pipe flow, although some authors believe that they can be inferred from entry and exit losses. There is a theoretical basis for the measurement from exit losses, but the experiment is difficult and has never been carried out with the required precision in a range where the theory applies. There is no existing theoretical basis for the measurement from entry losses.

Experiments must often be carried out in capillaries that are too short to neglect the entry and exit losses. Techniques exist for removing these effects from the data. These are discussed in the literature on viscometry cited in the Bibliographical Notes; see also Problem 5.5.

19.7 CONCLUDING REMARKS

The object of this short chapter has been to provide a brief introduction to the flow phenomena of viscoelastic liquids and experimental methods for fluid characterization. The concept of the Deborah number is useful for flow classification, and it is important to recall the existence of a material parameter with the dimension of time when undertaking a dimensional analysis. The Reynolds number is often an unimportant parameter in polymer flow problems, and scale-up and process analysis can frequently be based on a group like $\lambda \langle v \rangle / D$.

The Maxwell liquid described in this chapter is often an adequate description of the stress-deformation behavior of real polymeric materials, but it cannot describe some phenomena quantitatively. There is, in fact, a distribution of response times characteristic of any polymeric liquid, and the portions of the distribution respond in different ways in different types of flows. The Maxwell liquid does describe most of the important qualitative observations about polymer fluid flow, however, and it has been used most often in the solution of flows of engineering significance. Fluid characterization and the solution of flow problems are both active areas of current research, and much remains to be learned.

BIBLIOGRAPHICAL NOTES

For good introductory discussions of viscoelasticity, see

METZNER, A. B., "Flow of Non-Newtonian Fluids," in *Handbook of Fluid Mechanics*, V. L. Streeter, ed., McGraw-Hill Book Company, New York, 1961.

MIDDLEMAN, S., *The Flow of High Polymers*, John Wiley & Sons, Inc., New York, 1968.

More advanced treatments may be found in

ASTARITA, G., and MARRUCCI, G., *Principles of Non-Newtonian Fluid Mechanics*, McGraw-Hill Book Company, New York, 1974.

BIRD, R. B., ARMSTRONG, R. C., and HASSAGER, O., *Dynamics of Polymeric Liquids*, John Wiley & Sons, New York, 1977.

LODGE, A. S., *Elastic Liquids*, Academic Press, Inc., New York, 1964.

SCHOWALTER, W. R., *Mechanics of Non-Newtonian Fluids*, Pergamon Press, Oxford, 1978.

Rheological measurement is the topic of

WALTERS, K., *Rheometry*, Chapman & Hall Ltd., London, 1975.

The Deborah number was introduced by Reiner,

REINER, M., *Physics Today*, January 1964, p. 62.

It is discussed and expanded on in

METZNER, A. B., WHITE, J. L., and DENN, M. M., *Chem. Eng. Prog.*, **62**, No. 12, 81 (December, 1966).

TANNER, R. I., *AIChE J.*, **22**, 910 (1976).

For a general review of the effect of fluid elasticity on flow patterns, see

ASTARITA, G., and DENN, M. M., "The Effect of Non-Newtonian Properties of Polymer Solutions on Flow Fields," in *Theoretical Rheology*, J. F. Hutton, J.R.A. Pearson, and K. Walters, eds, Applied Science Publishers Ltd., Barking, Essex, England, 1975.

Some of the topics in this chapter, and solutions of other flow problems for Maxwell materials, are covered in

MIDDLEMAN, S., *Fundamentals of Polymer Processing*, McGraw-Hill Book Company, New York, 1977. [See in particular Chapter 14 on elastic phenomena.]

Polymer processing instabilities, including melt fracture, are reviewed in

PETRIE, C. J. S., and DENN, M. M., *AIChE J.*, **22**, 209 (1976).

Drag reduction is reviewed in

HOYT, J. W., *J. Basic Eng. (Trans. ASME)*, **94D**, 258 (1972).
VIRK, P. S., *AIChE J.*, **21**, 625 (1975).

There is an interesting historical discussion of die swell in

METZNER, A. B., *Trans. Soc. Rheol.*, **13**, 467 (1969).

The die swell analysis and data are from

TANNER, R. I., *J. Polymer Sci.*, *A-2*, **8**, 2067 (1970).

Viscoelasticity is important in extensional flows such as fiber drawing. See

DENN, M. M., "Extensional Flows: Theory and Experiment," in *The Mechanics of Viscoelastic Fluids*, R. Rivlin, ed., AMD-22, American Society of Mechanical Engineers, New York, 1977.
DENN, M. M., *Ann. Rev. Fluid Mech.*, **12**, 365 (1980).

For numerical solutions of the flow of a Maxwell fluid through a contraction, see

CROCHET, M. J., and BEZY, M., *J. Non-Newtonian Fluid Mech*, **5**, 201 (1979)

PERERA, M. G. N., and WALTERS, K., *J. Non-Newtonian Fluid Mech.*, **2**, 191 (1977).

PERERA, M. G. N., and STRAUSS, K., *J. Non-Newtonian Fluid Mech*, **5**, 269 (1979)

The relevant periodical literature includes *Polymer Engineering and Science, Journal of Applied Polymer Science, Rheologica Acta, Transactions of the Society of Rheology, Journal of Non-Newtonian Fluid Mechanics*, and the *AIChE Journal*.

Author Index

Subject Index